Edward Frankland

Lecture Notes for Chemical Students

Embracing mineral and organic Chemistry

Edward Frankland

Lecture Notes for Chemical Students
Embracing mineral and organic Chemistry

ISBN/EAN: 9783337176891

Printed in Europe, USA, Canada, Australia, Japan

Cover: Foto ©ninafisch / pixelio.de

More available books at **www.hansebooks.com**

LECTURE NOTES

FOR

CHEMICAL STUDENTS:

EMBRACING

MINERAL AND ORGANIC CHEMISTRY.

BY

EDWARD FRANKLAND, F.R.S., For. Sec. C.S.,

CORRESPONDING MEMBER OF THE IMPERIAL INSTITUTE OF FRANCE,
PROFESSOR OF CHEMISTRY IN THE ROYAL INSTITUTION
OF GREAT BRITAIN, AND IN THE GOVERNMENT
SCHOOL OF MINES.

LONDON:
JOHN VAN VOORST, PATERNOSTER ROW.
MDCCCLXVI.

PRINTED BY TAYLOR AND FRANCIS,
RED LION COURT, FLEET STREET.

PREFACE.

THE contents of this little book are to a great extent a transcript of the notes of my lectures delivered at the Royal College of Chemistry during the Winter Session of 1865–66. These notes have been considerably amplified only in the earlier chapters, on nomenclature, notation, and the atomicity of elements—subjects which, in their modern developments, have undergone such profound changes as to render their somewhat more extended treatment necessary to the comprehension of the remainder of the book.

To render the work as concise as possible, all formal description of the properties of the bodies treated of has been, for the most part, entirely omitted. Such a description (which is moreover easily accessible elsewhere) would, even if brief, have swollen the book to more than double its present size. For the same reason I have been compelled to treat the metallic elements in a manner which will doubtless seem, to many, unworthy of their importance; but their number is so great, that any attempt to give more than the names and formulæ of their chief compounds would have extended the work far beyond the limits I had assigned to it. My aim has been to classify and systematize rather than to describe, and I have endeavoured to furnish the student with a kind of skeleton of

the science, which it is intended he should himself clothe
with the already known and daily increasing facts of expe-
rimental research.   To aid him in this, he has the choice
of numerous standard treatises, amongst which may be
mentioned Watts's 'Dictionary of Chemistry,' Gmelin's
'Handbook of Chemistry,' Graham's 'Elements of Chemistry,' Miller's 'Elements of Chemistry,' Odling's
'Manual of Chemistry,' Gerhardt's 'Traité de Chimie
Organique,' 'Traité de Chimie Générale' by Pelouze and
Frémy, Kolbe's 'Lehrbuch der organischen Chemie,' and
Kekulé's 'Lehrbuch der organischen Chemie.'

I have often noticed with regret the great amount of
labour which an earnest student expends in noting down
the reactions and the names and formulæ of substances
which are presented to his notice in the lecture-theatre.
He is thus greatly interrupted in following the arguments
and explanations of the speaker, and he often loses more
important generalizations in securing a record of details.
One of my chief objects in the preparation of this book
has been to relieve him from such distractions.   For this
purpose very full lists of names and formulæ are given,
and a comparatively large amount of space is devoted
to equations expressing the reactions occurring in the formation and decomposition of the substances treated of.

Such being the chief objects of the book, it would
obviously have been impossible to give in all, or even in
many cases the reasons which have induced me to adopt
such views of the constitution of both mineral and organic
compounds as are either novel or not generally recognized.
Thus, I am aware that the atomicity which is assigned
to many of the elements may be called in question; but
it is hoped that, in thus giving for the first time a thorough
and consistent scheme of the combining-powers of atoms,

the advantages of the simplicity of symbolic expression, thereby secured, will more than outweigh the evil resulting from the few errors which future research may reveal.

Whilst the rapid progress of organic chemistry has on the one hand enormously increased the number of organic compounds, it has, on the other, revealed new relations between the different groups of these compounds, and opened up many new paths, both from one group to another and from member to member of the same group. The relative importance of individual compounds has thus gradually diminished in comparison with that of the family to which they belong; and the time has now arrived for recognizing this condition of things, by so classifying organic bodies as to make the description of the individual members subsidiary to that of the family to which they are attached. The student can thus more easily gain a general view of the otherwise almost hopelessly vast array of organic substances.

To illustrate important constitutional formulæ, I have extensively adopted the graphic notation of Crum Brown, which appears to me to possess several important advantages over that first proposed by Kekulé. Graphic notation affords most valuable aid to the teacher in rendering intelligible the constitution of chemical compounds, especially when it is supplemented by what may be called the glyptic formulæ of Hofmann. The system of symbolic notation, which I have explained in Chapter III., is so framed as to express the same ideas, of the chemical functions of atoms, as the graphic and glyptic formulæ, with which, therefore, it harmonizes completely; whilst it enables the student gradually to dispense with the last two forms of constitutional notation.

I am aware that graphic and glyptic formulæ may be

objected to, on the ground that students, even when spe-
cially warned against such an interpretation, will be liable
to regard them as representations of the actual physical
position of the atoms of compounds. In practice I have
not found this evil to arise; and even if it did occasionally
occur, I should deprecate it less than ignorance of all
notion of atomic constitution.

In conclusion, I have much pleasure in thanking my
assistant, Mr. Herbert McLeod, for his valuable help,
both in compilation and in the revision of the proofs.
Mr. McLeod has devoted much attention to the constitutional formulæ of minerals; and most of the symbolic
and graphic expressions for these compounds are from his
pen. To my assistant, Mr. W. Valentin, I am also much
indebted for aid in the laborious work of revising proofs.
In a book which is so full of formulæ, it would be too
much to expect entire freedom from errors; but every
care has been taken to reduce their number as much as
possible.

E. F.

Royal College of Chemistry, London,
    September 15, 1866.

TABLE OF CONTENTS.

CHAPTER IX.

DYAD ELEMENTS.

CHAPTER X.

TRIAD ELEMENTS.

CHAPTER XI.

TETRAD ELEMENTS.

CHAPTER XII.

PENTAD ELEMENTS.

CHAPTER XIII.

HEXAD ELEMENTS.

CHAPTER XXXI.

THE ALCOHOLS (*continued*).

CHAPTER XXXII.

THE ALCOHOLS (*continued*).

CHAPTER XXXIII.

THE ALCOHOLS (*continued*).

CHAPTER XXXIV.

THE ALCOHOLS (*continued*).

CHAPTER XXXIX.

THE ACIDS. *Acrylic or Oleic Series.*

CHAPTER XL.

THE ACIDS. *Lactic Series.*

CHAPTER XLI.

THE ACIDS. *Pyruvic Series.*

CHAPTER XLVII.

THE ANHYDRIDES.

CHAPTER XLVIII.

THE KETONES.

CHAPTER XLIX.

ETHEREAL SALTS.

CHAPTER L.

ORGANIC COMPOUNDS CONTAINING TRIAD AND PENTAD NITROGEN AND THEIR ANALOGUES.

CHAPTER LI.

ORGANIC COMPOUNDS OF TRIAD NITROGEN (*continued*).

CHAPTER LII.

COMPOUNDS OF PENTAD NITROGEN AND ITS ANALOGUES.

CHAPTER LIII.

ORGANOMETALLIC BODIES.

ERRATA.

Page 12, line 3 from bottom, and in a few other places, the atomicity marks ought to be placed to the left of atom coefficient. Thus *loc. cit. for* " Ca₃'''' " *read* " Ca''₃."

— 45, top line, *for* " OXIDES AND ACIDS OF CHLORINE " *read* " OXIDES AND OXACIDS OF CHLORINE."

— 46, line 16 from bottom, *for* " 79·4 criths " *read* " 39·7 criths."

— 244, — 17 from top, for " *Ethyl series* " read " *Methyl series.*"

— 292, — 5 from bottom, *for* " Œnanthic " *read* " Œnanthylic."

LECTURE NOTES

FOR

CHEMICAL STUDENTS.

~~~~~~~~~~~~~~~~~~~~~~~~

CHAPTER I.

INTRODUCTORY.

DEFINITION.—Chemistry is the science which treats of the atomic composition of bodies, and of those changes in matter which result from an alteration in the relative position of atoms.

SIMPLE AND COMPOUND MATTER.—All kinds of matter are divided into two great classes,—*simple substances*, and *compound substances*. A simple substance is one out of which it is impossible to obtain, by any known process, two or more essentially different kinds of matter. A compound substance, on the other hand, is one which can be resolved into two or more simple substances. The simple substances at present known are sixty-two in number, and are termed *elements*. By the combination of these elements with each other, all the infinitely varied forms of terrestrial matter are produced.

MODES OF CHEMICAL ACTION.—Matter undergoes chemical change in five different ways, viz. :—

1st. By the direct combination of elements or compounds with each other.

2nd. By the displacement of one element or group of elements in a body by another element or group of elements.

3rd. By a mutual exchange of elements or groups of elements in two or more bodies.

B

4th. By the re-arrangement of the elements or groups of elements already contained in a body.

5th. By the resolution of a compound into its elements, or into two or more less complex compounds.

ATOMIC WEIGHT.—Chemists assign to every element a number called its *atomic weight*. This number is made to represent, as far as possible,—

1st. The smallest proportion by weight in which the element enters into or is expelled from a chemical compound,—the smallest weight of hydrogen so entering or leaving a chemical compound being taken as unity.

2nd. The weight of the element in the solid condition which, at any given temperature, contains the same amount of heat as seven parts by weight of solid lithium at the same temperature.

3rd. The weight of the element which, in the form of gas or vapour, occupies, under like conditions of temperature and pressure, the same volume as one part by weight of hydrogen.

The atomic weight of a compound is the sum of the atomic weights of its elements.

The atomic weights of the elements are given in the Table at page 6.

ATOMS AND MOLECULES.—The proportional amount of any element represented by its atomic weight, as above described, is commonly called an *atom* of that element.

When an element is isolated, or separated from every other kind of matter, its atoms still exist, except in a few cases, in combination with each other. In many instances the atoms of isolated elements are associated in pairs when thus combined. Such an isolated atom or group of atoms constitutes an *elementary molecule*.

It follows from what has been said that the bulk of a molecule, or the *molecular volume* of an element in the gaseous or vaporous condition, must be the same as the molecular volume of hydrogen at the same temperature and pressure, and that the *molecular weight* of an element is in a large number of cases twice its own atomic weight.

The following is a list of those elements whose molecular volumes have been determined.

Molecules containing of the element				
One atom. *Monatomic Molecules.*	Two atoms. *Diatomic Molecules.*	Three atoms. *Triatomic Molecules.*	Four atoms. *Tetratomic Molecules.*	Six atoms. *Hexatomic Molecules.*
Mercury. Cadmium. Zinc.	Hydrogen. Oxygen. Chlorine. Bromine. Iodine. Fluorine. Nitrogen. Sulphur. Selenium.	Oxygen (as Ozone).	Phosphorus. Arsenic.	Sulphur.

It will be perceived from the above Table that an element may have two distinct molecular weights. This is known to be the case with oxygen and sulphur.

The molecular weight of a compound is, with very few exceptions, identical with its atomic weight. The molecular volume or the space occupied by the combining proportion of a compound is, with very few exceptions, equal to that occupied by two combining proportions, or one molecule, of hydrogen. Hence the law—*equal volumes of all gases and vapours contain, at the same temperature and pressure, an equal number of molecules.*

With very few exceptions, therefore, the molecules of all compounds, no matter how great may be the aggregate volume of their constituents, occupy, when compared at the same temperature and pressure, one uniform volume, which is exactly the same as that filled by one molecule of hydrogen. Thus:

vol.	vol.		vols.
1 of Hydrogen	+1 of Chlorine	form	2 of Hydrochloric acid.
1 of Hydrogen	+1 of Bromine vapour	„	2 of Hydrobromic acid.
2 of Hydrogen	+1 of Sulphur vapour.............	„	2 of Sulphuretted Hydrogen.
2 of Hydrogen	+1 of Oxygen	„	2 of Steam.
3 of Hydrogen	+1 of Nitrogen....................	„	2 of Ammonia.
4 of Hydrogen	+x of Carbon vapour	„	2 of Marsh-gas.
6 of Hydrogen	+1 of Oxygen +2x of Carbon vapour	„	2 of Alcohol vapour.
12 of Hydrogen	+1 of Oxygen +5x of Carbon vapour	„	2 of Amylic alcohol vapour.

B 2

CHEMICAL AFFINITY.—The force or power which holds together the elements of a compound is termed *chemical affinity*.

Elements which readily combine with each other, and develope much heat on combination, are said to have a powerful affinity for each other. The elements which thus exhibit towards each other a great affinity are possessed of widely different properties; and when their compounds are decomposed by an electric current, the constituents are evolved at the opposite poles. Those elements which, under such circumstances, make their appearance at the *positive* pole are termed *electro-negative* or *negative* elements, whilst those disengaged at the negative pole are called *electro-positive* or *positive* elements. For reasons which will appear hereafter, the negative are sometimes called *chlorous*, and the positive *basylous* elements. It must be remembered, however, that the difference between these two classes is one of degree only; they insensibly merge into each other, since the members of both classes exhibit a graduated intensity of the positive or negative quality. Thus potassium is more positive than sodium, and oxygen more negative than sulphur, whilst mercury is negative to sodium but positive to iodine.

The following eight elements are negative or chlorous towards the remaining fifty-four elements, which are more or less positive or basylous:—

Fluorine.	Oxygen.
Chlorine.	Sulphur.
Bromine.	Selenium.
Iodine.	Tellurium.

Although two positive or two negative elements can combine together chemically, yet their union is rarely attended with such striking phenomena as are manifested when the combination takes place between a positive and a negative element.

CHAPTER II.

CHEMICAL NOMENCLATURE.

THE study of every science necessitates an acquaintance with the system of names and peculiar modes of expression which have been found most convenient to denote the materials and to describe the phenomena which form its objects. Such names and modes of expression form the groundwork of the language of every science, upon the right employment of which depend the precision and accuracy of scientific definition.

The nomenclature of a science ought to be distinguished for its clearness and simplicity; but it is by no means easy to secure these conditions in a science like chemistry, where the rapid progress of discovery necessitates the continual addition of new and the frequent alteration of old names. The chemical name of a substance should not only identify and individualize that substance, but it should also express the composition and constitution of the body, if a compound, to which it is applied. The first of these conditions is readily attained; but the second is much more difficult to secure, inasmuch as our ideas of the constitution of chemical compounds—the mode in which they are built up as it were—require frequent modification. On this account all attempts to frame a perfectly consistent system of chemical nomenclature have hitherto been only partially successful.

It has been already mentioned that the number of elements at present known is sixty-two. These have received the names given in the following Table, in which the twenty-one most important elements are distinguished by the largest type, those next in importance by medium type, whilst the names of elements which are either of rare occurrence, or of which our knowledge is yet very imperfect, are printed in the smallest type.

Name.	Symbol.	Atomic weight.	Name.	Symbol.	Atomic weight.
ALUMINIUM	Al	27·5	Molybdenum ...	Mo	92
ANTIMONY	Sb	122	NICKEL	Ni	58·8
ARSENIC	As	75	Niobium.........	Nb	97·6
BARIUM	Ba	137	**NITROGEN** ...	N	14
BISMUTH	Bi	208	Osmium	Os	199
BORON	B	11	**OXYGEN**	O	16
BROMINE	Br	80	PALLADIUM ...	Pd	106·5
Cadmium	Cd	112	**PHOSPHORUS**	P	31
Cæsium	Cs	133	PLATINUM	Pt	197·4
CALCIUM	Ca	40	**POTASSIUM** ...	K	39
CARBON	C	12	RHODIUM	Rh	104
Cerium	Ce	92	Rubidium	Rb	85·5
CHLORINE ...	Cl	35·5	Ruthenium......	Ru	104
CHROMIUM......	Cr	52·5	Selenium	Se	79
COBALT	Co	58·8	**SILICON**	Si	28·5
COPPER	Cu	63·5	**SILVER**.........	Ag	108
Didymium	D	96	**SODIUM**	Na	23
FLUORINE ...	F	19	STRONTIUM ...	Sr	87·5
Glucinum	G	14	**SULPHUR**......	S	32
GOLD	Au	196·7	Tantalum	Ta	137·5
HYDROGEN ...	H	1	Tellurium	Te	128
Indium	In	74	Thallium	Tl	204
IODINE	I	127	Thorium.........	Th	231·5
IRIDIUM	Ir	198	TIN	Sn	118
IRON	Fe	56	TITANIUM	Ti	50
Lanthanium ...	L	92	TUNGSTEN	W	184
LEAD	Pb	207	URANIUM	U	120
Lithium	Li	7	Vanadium	V	137
MAGNESIUM ...	Mg	24	Yttrium	Y	68
MANGANESE	Mn	55	**ZINC**	Zn	65
MERCURY ...	Hg	200	Zirconium	Zr	90

These elementary substances have been long divided into two great classes—*metals* and *non-metals*, the latter being also sometimes termed *metalloids*. The metals are by far the most numerous, the non-metals numbering only the following thirteen elements:—Boron, Bromine, Carbon, Chlorine, Fluorine, Hydrogen, Iodine, Nitrogen, Oxygen, Phosphorus, Selenium, Silicon, Sulphur.

The names of the elements can scarcely be said to have been given according to any rule; many of them are derived from the most prominent property of the bodies themselves, whilst others have a mythological origin. An attempt has been made to distinguish the metals by the termination *um*, as potassium, sodium, &c.; but the common metals, such as gold, copper, iron, &c., still retain their original names; and one substance, selenium, which at the time of its discovery was regarded as a metal, has had no change made in its name, although further research has divested it of all metallic attributes. An important group of electro-negative non-metals—fluorine, chlorine, bromine, and iodine—have received the termination *ine*; three are distinguished by the terminal syllable *on*, viz. carbon, silicon, and boron; and three others have *gen* for their final syllable, viz. oxygen, hydrogen, and nitrogen, these last names being derived from Greek words denoting the property possessed by these elements of generating respectively acid, water, and nitre.

When two elementary bodies unite together, they form a chemical compound of the first order, to which the name *binary compound* has been applied. The names of these compounds are formed from those of their constituents, the name of the positive constituent with the terminal *ic*, or some abbreviation thereof, preceding that of the negative constituent, which is made to terminate in *ide*, thus:—

Potassium and Sulphur form Potassic sulphide.
Sodium   „   Oxygen   „   Sodic oxide.
Silver   „   Chlorine   „   Argentic chloride.
Zinc   „   Iodine   „   Zincic iodide.
Calcium   „   Chlorine   „   Calcic chloride.

But the same elements frequently form with each other two compounds, in which case the one which contains the smaller proportion of the negative element is distinguished by changing the terminal syllable of the name of its positive constituent into *ous*, the terminal *ic* being retained for the compound containing the larger proportion of the negative element. Thus,

  One atom of tin and two atoms of chlorine form Stannous chloride.
  One atom of tin and four atoms of chlorine form Stannic chloride.

Sometimes, however, the same elements form with each other more than two compounds. In these cases the prefixes *hypo* and *per* are employed as marks of distinction; but their use is very rarely required.

If a binary compound contains oxygen, and forms an acid when made to unite with water, or a salt when added to a base, it is termed an *anhydride* or *anhydrous acid*. Thus,

One atom of carbon and two atoms of oxygen form carbonic anhydride.
Two atoms of nitrogen and five atoms of oxygen form nitric anhydride.
Two atoms of nitrogen and three atoms of oxygen form nitrous anhydride.
One atom of sulphur and three atoms of oxygen form sulphuric anhydride.
One atom of sulphur and two atoms of oxygen form sulphurous anhydride.

In the following cases, the systematic names have not displaced the trivial and irregular names used for the same substances :—

Systematic name.	Trivial or irregular name.
Hydric oxide	Water.
Hydric sulphide	Sulphuretted hydrogen.
Hydric selenide	Seleniuretted hydrogen.
Hydric telluride	Telluretted hydrogen.
Hydric chloride	Hydrochloric acid.
Hydric bromide	Hydrobromic acid.
Hydric iodide	Hydriodic acid.
Hydric fluoride	Hydrofluoric acid.
Hydric carbide	{ Marsh-gas or light carburetted hydrogen.
Hydric nitride	Ammonia.
Hydric phosphide	Phosphuretted hydrogen.
Hydric arsenide............	Arsenuretted hydrogen.
Hydric antimonide	Antimonuretted hydrogen.

The term *acid* was originally applied only to substances possessing a sour taste like vinegar; but analogy has necessitated the application of the same name to a large number of compounds which have not this property. In the modern acceptation of the name, an acid may be defined as a compound containing one or more atoms of hydrogen, which become displaced by a metal when the latter is presented to the compound in the form of a hydrate. The hydrogen capable of being so displaced may be conveniently termed *displaceable hydrogen*. An acid containing one such atom of hydrogen is said to be *monobasic*. two such atoms *dibasic*, &c. Acids of a greater basicity than unity are frequently termed *polybasic acids*.

Thus nitric acid gives, with sodic hydrate, sodic nitrate:

$$NO_3H \ + \ ONaH \ = \ NO_3Na \ + \ OH_2.$$
Nitric acid.   Sodic hydrate.   Sodic nitrate.   Water.

Sulphuric acid gives, with potassic hydrate, potassic sulphate:

$$SO_4H_2 \ + \ 2OKH \ = \ SO_4K_2 \ + \ 2OH_2:$$
Sulphuric acid.   Potassic hydrate.   Potassic sulphate.   Water.

and hydrochloric acid gives, with potassic hydrate, potassic chloride:

$$HCl \ + \ OKH \ = \ KCl \ + \ OH_2.$$
Hydrochloric acid.   Potassic hydrate.   Potassic chloride.   Water.

When an acid contains oxygen, its name is generally formed by adding the terminal *ic* either to the name of the element with which the oxygen is united, or to an abbreviation of that name; thus sulphur forms, with oxygen, sulphuric acid; nitrogen, nitric acid; and phosphorus, phosphoric acid. But it frequently happens that the same element forms two acids with oxygen; and when this occurs, the acid containing the larger amount of oxygen receives the terminal syllable *ic*, whilst that containing less oxygen is made to end in *ous*. Thus we have sulphurous acid, nitrous acid, and phosphorous acid, each containing a smaller proportion of oxygen than that necessary to form respectively sulphuric, nitric, and phosphoric acids.

In some instances, however, the same element forms more than two acids with oxygen, in which case the two Greek words *hypo*, under, and *hyper*, over, are prefixed to the name of the acid. Thus an acid of sulphur containing less oxygen than sulphurous acid is termed hyposulphurous acid; and another acid of the same element containing, in proportion to sulphur, more oxygen than sulphurous acid and less than sulphuric, might be named either hypersulphurous acid, or hyposulphuric acid; but the latter term has been universally adopted. The prefix *per* is frequently substituted for *hyper*; thus in the case of chlorine, which forms the following four acids with oxygen, viz. hypochlorous acid, chlorous acid, chloric acid, and hyperchloric acid, the latter is generally named perchloric acid; but *per* can only be used as a prefix to the acid containing the largest proportion of oxygen.

Some acids do not contain oxygen amongst their constituents, but consist of sulphur or hydrogen united with other elements. This peculiarity of composition is expressed in their nomenclature by the prefixes *sulpho* or *sulph*, and *hydro* or *hydr*: thus sulpharsenic acid and sulphostannic acid denote acids composed respectively of sulphur, hydrogen, and arsenic, and sulphur, hydrogen, and tin; whilst the names hydrochloric acid and hydriodic acid are given to acids composed, the first of hydrogen and chlorine, and the second of hydrogen and iodine. The terminals *ous* and *ic* are also applied to these acids in exactly the same manner as to the oxygen acids: thus we have sulpharsenious and sulpharsenic acid, the latter containing a larger proportion of sulphur than the former; but the application of the second of these terminals has not hitherto been found necessary in the case of hydrogen acids, since no element has yet been observed to form more than one acid with hydrogen.

The term *anhydride* or *anhydrous acid* is applied to the residue obtained by the abstraction of water from one or two molecules of an oxygen acid.   Thus,

$$SO_4H_2 \quad - \quad OH_2 \quad = \quad SO_3:$$
Sulphuric acid.   Water.   Sulphuric anhydride.

$$2NO_3H \quad - \quad OH_2 \quad = \quad N_2O_5.$$
Nitric acid.   Water.   Nitric anhydride.

The term *base* is applied to three classes of compounds, all of which are converted into salts by the action of acids. These are—

1st. Certain compounds of metals with oxygen, such as sodic oxide (Na_2O), zincic oxide (ZnO), &c.

2nd. Certain compounds of metals with the compound radical hydroxyl (HO), such as sodic hydrate $\left(OHNa \text{ or } Na(HO)\right)$, zincic hydrate $\left(Zn(HO)_2\right)$, &c.

3rd. Certain compounds of nitrogen, phosphorus, arsenic, and antimony, such as ammonia (NH_3).

There are also some organic compounds to which the name base is sometimes given, but which are not included in the above classes; it is, however, unnecessary further to allude to them here.

The bases of the first class are named in accordance with the rules already given for compounds of two elements. The following bases, however, still retain their irregular names:—

Systematic names.	Irregular names.
Baric oxide	Baryta.
Strontic oxide	Strontia.
Calcic oxide	Lime.
Magnesic oxide	Magnesia.
Aluminic oxide...........	Alumina.
Glucinic oxide	Glucina.
Zirconic oxide	Zirconia.

The names of the bases belonging to the second class are formed by changing the terminal syllable of the metal into *ic* or *ous*, and the word hydroxyl into *hydrate*. Thus cæsium and hydroxyl form cæsic hydrate $\left(Cs(HO)\right)$; barium and hydroxyl, baric hydrate, $\left(Ba\,(HO)_2\right)$; and iron and hydroxyl, ferric hydrate $\left(Fe_2\,(HO)_6\right)$.

A few of these bases have trivial or irregular names, which are almost invariably used instead of the systematic names:—

Systematic names.	Irregular names.
Potassic hydrate	Potash.
Sodic hydrate..............	Soda.
Lithic hydrate	Lithia.

The bases of the third class are distinguished by the terminal syllable *ine*, except nitrine, (NH_3), which retains its trivial name ammonia. These bases belong almost exclusively to the department of organic chemistry, and their nomenclature could not be advantageously discussed here.

It has been already mentioned that by the mutual action of an acid and a base upon each other, a *salt* is produced.   If the salt be free from oxygen and sulphur, like common salt, ($NaCl$), it is termed a *haloid salt*; if it contain oxygen it is termed an *oxysalt*; and if this oxygen be replaced by sulphur, it is distinguished as a *sulphosalt*.

The haloid salts are named according to the rules for binary compounds above given, thus:

Name.	Formula.
Sodic chloride	$NaCl$.
Calcic iodide	CaI_2.
Ferrous bromide	$FeBr_2$.
Ferric bromide	Fe_2Br_6.

Oxysalts are divided into *normal, acid,* and *basic.*

A normal salt is one in which the displaceable hydrogen of the acid (see page 9) *is all exchanged for an equivalent amount of a metal or of a positive compound radical.*

The following examples will serve to illustrate this definition of a normal, or as it is sometimes incorrectly called, a *neutral* salt, the displaceable atoms of hydrogen in the acid, and the metal by which they have been displaced in the salt, being printed in italics :—

Acid.		Normal salt.	
Nitric acid	NO_3H	Sodic nitrate	NO_3Na.
		Calcic nitrate	$(NO_3)_2Ca''$.
Sulphuric acid	SO_4H_2	Potassic sulphate	SO_4K_2.
		Calcic sulphate	SO_4Ca''.
Phosphoric acid	PO_4H_3	Potassic phosphate	PO_4K_3.
		Calcic phosphate	$(PO_4)_2Ca_3''$.
Hypophosphorous acid	PO_2H_2H	Sodic hypophosphite	PO_2H_2Na.
Phosphorous acid	PO_3HH_2	Potassic phosphite	PO_3HK_2.

Acid.		Normal salt.	
Metaphosphoric acid	... PO_3H	Lithic metaphosphate .	PO_3Li.
Pyrophosphoric acid	... $P_2O_7H_4$ ...	Calcic pyrophosphate ..	$P_2O_7Ca_2$".
Nordhausen sulphuric acid	$\left\{\begin{array}{l} SO_3H \\ O \\ SO_3H \end{array}\right.$	Sodic bisulphate	$\left\{\begin{array}{l} SO_3Na \\ O \\ SO_3Na \end{array}\right.$.
Unknown acid	$\left\{\begin{array}{l} CrO_3H \\ O \\ CrO_3H \end{array}\right.$	Potassic bichromate .	$\left\{\begin{array}{l} CrO_3K \\ O \\ CrO_3K \end{array}\right.$.

An acid salt is one in which the displaceable hydrogen of the acid is only partially exchanged for a metal or positive compound radical.

The following examples illustrate the constitution and nomenclature of these salts :—

Acid.		Acid salt.	
Sulphuric acid	... SO_4H_2	Hydric sodic sulphate	SO_4HNa.
Carbonic acid	... CO_3H_2?	Hydric potassic carbonate..	CO_3HK.
Phosphoric acid...	$PO_4H_3 \left\{\begin{array}{l} \\ \\ \\ \end{array}\right.$	Hydric disodic phosphate..	PO_4HNa_2.
		Dihydric sodic phosphate...	PO_4H_2Na...
		Microcosmic salt	$PO_4H(NH_4)Na$.

(Hydric ammonic sodic phosphate.)

Acid salts are produced almost exclusively from polybasic acids.

When the number of bonds* of the metal or compound positive radical contained in a salt exceeds the number of atoms of displaceable hydrogen in the acid, the compound is usually termed a basic salt—as, for instance,

Acid.		Basic salt.	
Carbonic acid	$CO_3H_2 \left\{\begin{array}{l} \\ \\ \end{array}\right.$	Malachite	$CO_5H_2Cu_2$".
		Blue cupric carbonate	$C_2O_7H_2Cu_2$".
Sulphuric acid	$SO_4H_2 \left\{\begin{array}{l} \\ \\ \end{array}\right.$	Tribasic cupric sulphate	$SO_xH_4Cu_3$".
		Turpeth mineral	SO_6Hg_3".

These and most, if not all, other basic salts do not differ essentially in their constitution from the normal and acid salts. This will be seen from the arrangement of their atoms given under the different metals entering into their composition.

* For an explanation of this term see Chap. III. p. 18.

The nomenclature of organic bodies is founded upon the same principles as that of inorganic compounds; but its discussion could not be conveniently introduced here.

CHAPTER III.

SYMBOLIC NOTATION.—Every element is represented by a symbol, which is frequently the initial letter of the name of the element; but as in some cases the names of two or more elements begin with the same letter, it is necessary to distinguish them by the use of a second letter in small type, which is either the second letter of the word, or some other letter prominently heard in its pronunciation; thus carbon, cadmium, cobalt, and cerium all begin with the same letter; but they are distinguished by the symbols C, Cd, Co, and Ce. In the use of the single letters, the non-metallic elements have the preference; thus oxygen, hydrogen, nitrogen, sulphur, phosphorus, boron, carbon, iodine, and fluorine are expressed by the single letters O, H, N, S, P, B, C, I, and F; whilst the metals osmium, mercury, nickel, strontium, platinum, bismuth, cobalt, iridium, and iron are symbolized by two letters each; thus Os, Hg (hydrargyrum), Ni, Pt, Bi, Co, Ir, and Fe (ferrum). In the selection of the single letter for other cases, preference is given to the most important element; thus sulphur, selenium, and silicon are all non-metallic elements, beginning with the same letter, but sulphur being the most important, the single letter S is assigned to it; whilst selenium and silicon are denoted respectively by Se and Si.

The symbols of compounds are formed by the simple juxtaposition of the symbols of their constituent elements. Such a group of two or more symbols is termed a *chemical formula*.

Thus:

Argentic chloride .... AgCl.

Zincic oxide .... ZnO.

The symbols not only represent the elements for which they are used, but they also denote a certain definite proportion by weight of each element; the formula HCl, for instance, does not merely denote a compound of hydrogen and chlorine, but it signifies a molecule of that compound containing one atom (1 part by weight) of hydrogen, and one atom ($35\cdot5$ parts by weight) of chlorine. When, therefore, the molecule of a compound contains more than one atom or combining proportion of any element, it is necessary to express such fact in the formula: this is done by the use of a coefficient placed after the symbol of the element:

Zincic chloride .... $ZnCl_2$.

Ferric chloride .... Fe_2Cl_6.

Stannous chloride........................... $SnCl_2$.

Stannic chloride .... $SnCl_4$.

When it is necessary to denote two or more molecules of any compound, a large figure is placed before the formula of the compound; such a figure then affects every symbol in that formula: thus $3SO_4H_2$ means three molecules of the compound SO_4H_2.

The changes which occur during chemical action are expressed by equations, in which the symbols of the elements or compounds, as they exist before the change, are placed on the left, and those which result from the reaction on the right. Thus, taking an example from each of the five kinds of chemical action above mentioned, we have

(1) $$Zn \ + \ Cl_2 \ = \ ZnCl_2.$$
Zinc. Chlorine. Zincic chloride.

(2) $$2HCl \ + \ Na_2 \ = \ 2NaCl \ + \ H_2.$$
Hydrochloric acid. Sodium. Sodic chloride. Hydrogen.

(3) $$SO_4Cu \ + \ (NO_3)_2Ba \ = \ SO_4Ba \ + \ (NO_3)_2Cu.$$
Cupric sulphate. Baric nitrate. Baric sulphate. Cupric nitrate.

$$(4) \quad (CN)O(NH_4) = N_2H_4(CO).$$
Ammonic cyanate.   Urea.

$$(5) \quad 2OH_2 = O_2 + 2H_2.$$
Water.   Oxygen.   Hydrogen.

The sign $+$, as seen from the foregoing examples, is placed between the formulæ of the molecules of the different substances which are brought into contact before the reaction, and of those which result from the change. This sign must never be used to connect together the constituents of one and the same chemical compound.

The sign $-$ is only very rarely used in chemical notation, but when employed it has the ordinary signification of abstraction; thus,

$$SO_4H_2 - H_2O = SO_3.$$
Sulphuric acid.   Water.   Sulphuric anhydride.

Use of the bracket.—The bracket has been employed in various senses in chemical formulæ; but in the following pages it is used in notation for one purpose only, viz. for expressing chemical combination between two or more elements which are placed perpendicularly with regard to each other and next to the bracket in a formula. Thus in the following cases,

I.   II.   III.

$$\left\{ \begin{array}{l} CH_3 \\ CH_3 \end{array} \right. \quad \cdot \quad \left\{ \begin{array}{l} CH_3 \\ O \\ CH_3 \end{array} \right. \quad \left. \begin{array}{l} COO \\ Ba \\ COO \end{array} \right\}$$

the formula No. I. signifies that two atoms of carbon are directly united with each other, No. II. that two atoms of carbon are linked, as it were, together by an atom of oxygen, the latter being united to both carbon atoms, whilst in like manner No. III. expresses the fact that one atom of oxygen in the formula of the upper line is linked to another atom of oxygen in the formula of the lower line, by an atom of barium.

Use of thick letters.—As a rule, the formulæ in this book are so written as to denote that the element represented by the first symbol of a formula is directly united with all the active bonds of the other elements or compound radicals following upon the same line: thus the formula SO_2Ho_2 (sulphuric acid) signifies that the hexad atom of sulphur is combined with the four bonds of the two atoms of oxygen, and also with the two bonds of the two atoms of hydroxyl. Such a formula is termed a *rational formula**.

Occasionally, however, owing to the atomic arrangement of a compound not being known, its formula cannot be written according to this rule; and in order to prevent such more or less *empirical formulæ* from being mistaken for rational formulæ, the first symbol of a rational formula will always be printed in thick type whenever the element has more than one bond. It deserves also to be mentioned that, as a rule, the element having the greatest number of bonds will occupy this prominent position. Thus,

Sulphuric acid...................... SO_2Ho_2.

Water .... OH_2.

Nitric acid .... NO_2Ho.

Microcosmic salt.................. $POHoAmoNao$.

ATOMICITY OF ELEMENTS.—It has been already stated that the atomic weight of an element is the smallest proportion by weight in which that element enters into or is expelled from a chemical compound. The atoms of the various elements, the relative weights of which are thus expressed, exhibit very different values in chemical reactions. Thus an atom of zinc is equivalent to two atoms of hydrogen; for when zinc is brought into contact with steam at a high temperature, one atom of zinc expels from the steam two atoms of hydrogen and occupies their place—thus,

$$OH_2 \; + \; Zn \; = \; OZn \; + \; H_2.$$
Water.   Zincic oxide.

Again, when zincic oxide is brought into contact with hydrochloric acid, the place of the zinc becomes once more occupied

* For further information on this subject see ATOMICITY OF ELEMENTS below.

by hydrogen, but two atoms of hydrogen are found to be necessary to take the place of one atom of zinc:

$$\mathbf{O}\text{Zn} \quad + \quad 2\text{HCl} \quad = \quad \mathbf{ZnCl_2} \quad + \quad \cdot\mathbf{OH_2}.$$

Zincic oxide.   Hydrochloric acid.   Zincic chloride.   Water.

In like manner one atom of boron can be substituted for three atoms of hydrogen, one atom of carbon for four, one of nitrogen for five, and one atom of sulphur for no less than six atoms of hydrogen.

To give a concrete expression to these facts, the atom of hydrogen may be represented as having only one point of attachment or *bond* by which it can be united with any other element, zinc as having two such bonds, boron three, and so on. Thus the atoms of these elements may be graphically represented in the following manner:—

Hydrogen ....

Zinc ....

Boron ....

Carbon ....

Nitrogen............................

Sulphur ....

In symbolic notation, the same idea is conveyed by the use of dashes and Roman numerals placed above and to the right of the symbol of the element; thus,

Hydrogen......... H',   Carbon........... C^{iv},
Zinc .... Zn'',   Nitrogen .... N^v,
Boron .... B''',   Sulphur .... S^{vi}.

No element, either alone or in combination, can exist with any of its bonds disconnected; hence *the molecules of all ele-*

ments with an odd number of bonds are generally diatomic, and always polyatomic; that is, they contain two or more atoms of the element united together.   Thus,

	Symbolic.	Graphic.
Hydrogen	H_2	
Chlorine	Cl_2................	
Nitrogen	N_2^v	
Phosphorus	P_4^v.................	

An element with an even number of bonds can exist as a monatomic molecule, its own bonds satisfying each other. Thus,

	Symbolic.	Graphic.
Mercury	Hg''	
Cadmium..............	Cd''	
Zinc....................	Zn''	

It is nevertheless obvious that such an element may also exist as a polyatomic molecule.   Oxygen furnishes us with an example of this; for, in its ordinary condition it is a diatomic molecule, and in the allotropic form of ozone, a triatomic molecule:

	Symbolic.	Graphic.
Oxygen	O_2''..............	
Ozone..................	O_3''..............	

This combining value of the elementary atoms is usually termed their *atomicity* or *atom-fixing power*.   An element with one bond is termed a *monad*, with two bonds a *dyad*, with three a *triad*, with four a *tetrad*, with five a *pentad*, and with six a *hexad*. Elements with an odd number of bonds are termed *perissads*, whilst those with an even number are named *artiads*.

In order to avoid the unnecessary use of atomicity-marks in symbolic notation, I shall never attach them to a monad or to oxygen, which, it must be remembered, is always a dyad. Neither will the atomicity-coefficient be attached to the tetrad element carbon, in the formulæ of organic bodies, unless this element plays the part of a dyad—an occurrence of extreme rarity. When not otherwise marked, therefore, carbon must always be understood to be a tetrad.

It will also, as a rule, be unnecessary to mark the atomicity of the elements which are expressed by symbols in thick type, because their atomicity is clearly indicated by the sum of the atomicities of the elements or compound radicals placed to their right, or connected with them perpendicularly by a bracket. Thus in the formula

$$\left\{ \begin{array}{l} \mathbf{C}Cl_3 \\ \mathbf{C}Cl_3 \end{array} \right.,$$

each atom of carbon is united with three atoms of the monad chlorine, whilst the bracket indicates that the two atoms of carbon are also united, thus stamping $\mathbf{C}$ as a tetrad element.

From what has just been said with regard to carbon, it is evident that the atomicity of an element is, apparently at least, not a fixed and invariable quantity : thus nitrogen is sometimes equivalent to five atoms of hydrogen, as in ammonic chloride, ($N^v H_4 Cl$), sometimes to three atoms, as in ammonia ($N''' H_3$), and sometimes to only one atom, as in nitrous oxide ($N_2 O$). But it is found that this variation in atomicity always takes place by the disappearance or development of an even number of bonds : thus nitrogen is either a pentad, a triad, or a monad ; phosphorus and arsenic, either pentads or triads ; carbon and tin, either tetrads or dyads ; and sulphur, selenium, and tellurium, either hexads, tetrads, or dyads.

These remarkable facts can be explained by a very simple

and obvious assumption, viz. that *one or more pairs of bonds belonging to one atom of the same element can unite and, having saturated each other, become, as it were, latent.* Thus the pentad nitrogen becomes a triad when one pair of its bonds becomes latent, and a monad when two pairs, by combination with each other, are, in like manner, rendered latent,—conditions which may be graphically represented thus:—

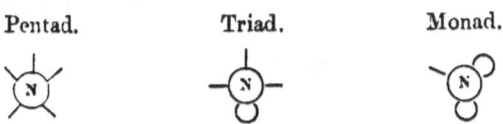

| Pentad. | Triad. | Monad. |

And in the case of sulphur:

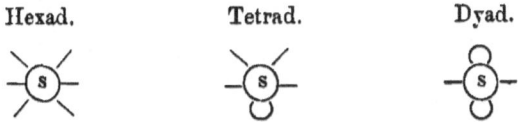

| Hexad. | Tetrad. | Dyad. |

Adopting this hypothesis, it will be convenient to distinguish the maximum number of bonds of an element as its *absolute atomicity*, the number of bonds united together as its *latent atomicity*, and the number of bonds actually engaged in linking it with the other elements of a compound as its *active atomicity*. The sum of the active and latent atomicity of any element must evidently always be equal to the absolute atomicity. Thus in sulphuric acid ($S^{vi}O_2Ho_2$) the absolute and active atomicities are both $= vi$, therefore the latent atomicity $= 0$. In sulphurous acid ($''S^{iv}OHo_2$) the active atomicity $= iv$, and consequently the latent $= vi - iv = ii$; whilst in sulphuretted hydrogen ($^{iv}S''H_2$) the active and latent atomicities are respectively ii and iv.

The apparent exceptions to this hypothesis disappear on investigation: thus iron, which is a dyad in ferrous compounds (as $FeCl_2$), a tetrad in cubical pyrites (FeS_2''), and a hexad in ferric acid (FeO_2Ho_2), is apparently a triad in ferric chloride ($FeCl_3$); but the vapour-density of ferric chloride shows that its formula must be doubled—that, in fact, the two atoms of the hypothetical molecule of iron (Fe_2) have not been completely separated. The formulæ of the ferrous and ferric chlorides and of ferric acid then become

Symbolic.   Graphic.

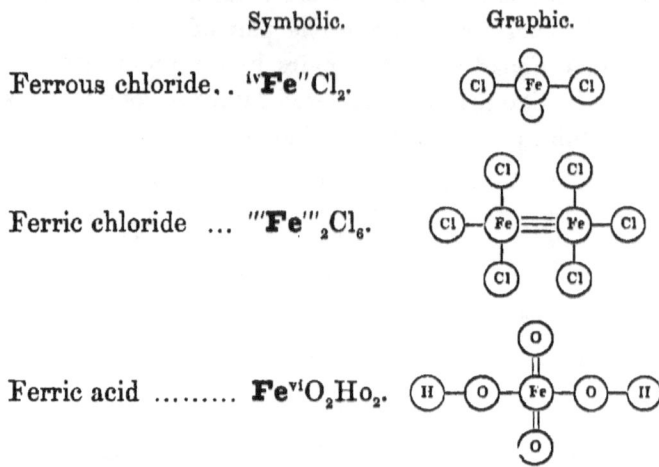

Ferrous chloride.. ^{iv}**Fe**$''$Cl$_2$.

Ferric chloride ... $'''$**Fe**$'''_2$Cl$_6$.

Ferric acid .... **Fe**^{vi}O$_2$Ho$_2$.

Again, mercury is apparently a monad in mercurous chloride (calomel, HgCl) and a dyad in mercuric chloride (corrosive sublimate, Hg$''$Cl$_2$); but there are strong reasons for believing that the formula of calomel ought to be doubled, in which case mercury would assume the dyad form in both compounds:

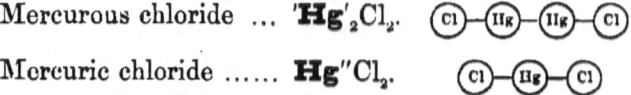

Mercurous chloride ... $'$**Hg**$'_2$Cl$_2$.

Mercuric chloride ...... **Hg**$''$Cl$_2$.

It will be remarked that the number of bonds supposed to be combined with each other in the atom of iron in ferrous chloride is expressed in the above symbol by the atomicity numeral IV placed to the left of the symbol, whilst the analogous union of three bonds of each atom of iron in ferric chloride is expressed by the three dashes$'''$ to the left of the symbol **Fe**$_2$. I shall not, however, use these coefficients of latent atomicity in the case of the single atom of an element, the student being supposed to have made himself acquainted with the absolute atomicity of every element as expressed in the Table at page 32. For a similar reason, it will also rarely be necessary to express the same idea in graphic notation : thus, for instance, ammonia will be drawn

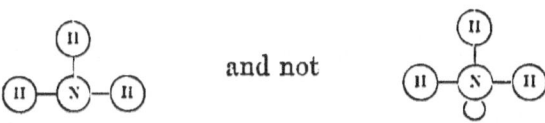

and not

It will be necessary, however, to employ these coefficients in symbolic formulæ, where two or more atoms of the same element are joined together under such circumstances, that the number of bonds uniting them cannot be found by subtracting the coefficient of active atomicity from the absolute atomicity of the element; as in hydric persulphide ($'S'_2H_2$), for instance, which might otherwise be viewed as $'''S'_2H_2$, or $'S'_2H_2$.

In rare cases, in which oxygen links together two elements or radicals in the same line, a hyphen is placed before and after the symbol O, thus :—

$$\left\{ \begin{array}{l} \mathbf{C}H_2\text{-O-}\mathbf{C}McO \\ \mathbf{C}H_2\text{-O-}\mathbf{C}McO \end{array} \right.$$

Diacetic glycol.

GRAPHIC NOTATION.—This mode of notation, although far too cumbrous for general use, is invaluable for clearly showing the arrangement of the individual atoms of a chemical compound. It is true that it expresses nothing more than the symbolic notation of the same compound, if the latter be written and understood as above described; nevertheless the graphic form affords most important assistance, both in fixing upon the mind the true meaning of symbolic formulæ, and also in making comparatively easy of comprehension the internal arrangement of the very complex molecules frequently met with both in mineral and organic compounds. It is also of especial value in rendering evident the causes of isomerism in organic bodies.

Graphic notation, like the above method of symbolic notation, is founded almost entirely upon the doctrine of atomicity, and consists in representing, graphically, the mode in which every bond in a chemical compound is disposed of. Inasmuch, however, as the principles involved are precisely the same as those already described under the heads of SYMBOLIC NOTATION and ATOMICITY OF ELEMENTS, it is unnecessary here to do more than give the following comparative examples of symbolic and graphic formulæ :—

	Symbolic.	Graphic.
Water	OH_2.	
Nitric acid	NO_2Ho.	
Ammonic chloride	NH_4Cl.	
Sulphuric anhydride	SO_3.	
Sulphuric acid	SO_2Ho_2.	
Carbonic anhydride.........	CO_2.	
Potassic carbonate	$COKo_2$.	

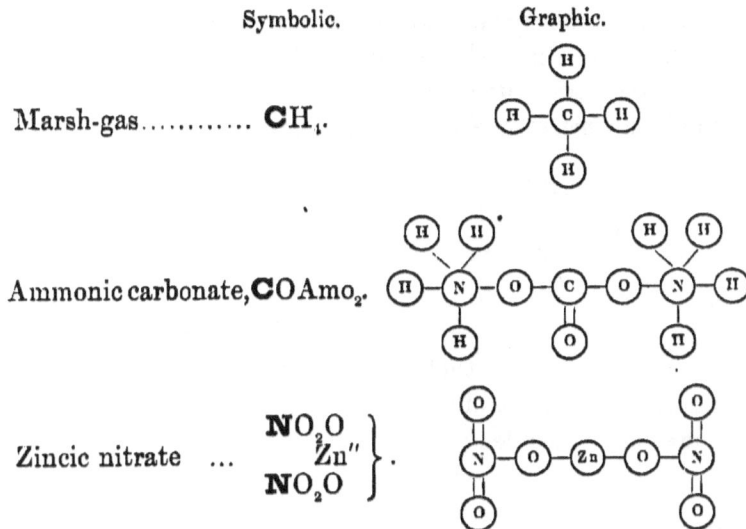

Symbolic.   Graphic.

Marsh-gas............ CH_4.

Ammonic carbonate, $COAmo_2$.

Zincic nitrate ...
$$\left.\begin{matrix} NO_2O \\ Zn'' \\ NO_2O \end{matrix}\right\}.$$

It must be carefully borne in mind that these graphic for-mulæ are intended to represent neither the shape of the mole-cules, nor the relative position of the constituent atoms. The lines connecting the different atoms of a compound, and which might with equal propriety be drawn in any other direction, provided they connected together the same elements, serve only to show the definite disposal of the bonds: thus the for-mula for nitric acid indicates that two of the three constituent atoms of oxygen are com-bined with nitrogen alone, and are conse-quently united to that element by both their bonds, whilst the third oxygen atom is com-bined both with nitrogen and hydrogen.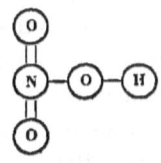

The lines connecting the different atoms of a compound are but crude symbols of the bonds of union between them; and it is scarcely necessary to remark that no such material con-nexions exist, the bonds which actually hold together the atoms of a compound being in all probability, as regards their nature, much more like those which connect the members of our solar system.

It may also be here mentioned that graphic, like symbolic

formulæ, are purely statical representations of chemical compounds, they take no cognizance of the amount of potential
energy associated with the different elements.   Thus in the
formulæ for marsh-gas and carbonic anhydride,

Marsh-gas.   Carbonic anhydride.

there is no indication that the molecule of the first compound
contains a vast store of force, whilst the last is comparatively
a powerless molecule.

CHAPTER IV.

COMPOUND RADICALS.

THE term compound radical may be applied to any group of
two or more atoms, which takes the place and performs the
functions of an element in a chemical compound.   In practice,
however, it is only applied to any such group when the latter
is met with in numerous chemical compounds.

An element is a *simple radical*, and enters into combination
in the following manner, a, b, c, and d being monad elements,
a'' a dyad, a''' a triad, and a^{iv} a tetrad element :—

$$a + b = ab,$$
$$a'' + 2b = a''b_2,$$
$$a''' + 3b = a'''b_3,$$
$$\&c. \qquad \&c.$$

A group of elements replacing a, a'', or a''' in the above equations is a *compound radical*, as in the following examples.

$$(a''b) \quad + b = (a''b)b,$$
$$(a'''b)'' \quad +2b = (a'''b)''b_2,$$
$$(a'''bc) \quad + b = (a'''bc)b,$$
$$(a^{iv}b)''' \quad +3b = (a^{iv}b)'''b_3,$$
$$(a^{iv}bc)'' \quad +2b = (a^{iv}bc)''b_2,$$
$$(a^{iv}bcd) \quad + b = (a^{iv}bcd)b.$$

The group of elements $(a''b)$ constitutes a compound monad radical equivalent to one atom of hydrogen or chlorine. The group $(a'''b)''$ is a compound dyad radical, &c. It is therefore evident that a polyad element is essential to every compound radical; in fact *a compound radical consists of one or more atoms of a polyad element in which one or more bonds are unsatisfied; and it is either a monad, dyad, triad, &c. radical, according to the number of monad atoms required to satisfy its active atomicity.* Such a radical, when a monad, triad, or pentad, cannot exist as a separate atom; like hydrogen or nitrogen, when isolated, it combines with itself, forming a diatomic molecule. It is only by the union of two atoms that the vacated bonds can in these cases be satisfied.

From the above definition of a compound radical, it is evident that an almost infinite number of such bodies must exist; for in the compounds of every polyad element it is only necessary to vacate successive bonds to create each time a new compound radical. Thus marsh-gas (CH_4) minus one atom of hydrogen gives the compound radical methyl (CH_3); minus two atoms of hydrogen, it forms methylene $(CH_2)''$, and by the abstraction of three hydrogen atoms it is transformed into the triad radical formyl $(CH)'''$; but, except in a few cases, it is not advantageous thus to incorporate, as it were, compound radicals, which, instead of simplifying notation and nomenclature, would, if thus multiplied, only embarrass them. No compound radical, therefore, ought to receive a recognition as such unless it can be shown to enter into the composition of a large number of compounds.

The following are the names and formulæ of the chief inorganic compound radicals recognized in the notation of this book :—

	Molecular formulæ.	Atomic formulæ.	Abbreviated atomic formulæ.
Hydroxyl	$(HO)_2$	HO	Ho.
Hydrosulphyl......	$(HS)_2$	HS	Hs.
Ammonium	$(NH_4)_2$	NH_4	Am
Ammonoxyl	$(NH_4O)_2$	NH_4O	Amo.
Amidogen	$(NH_2)_2$	NH_2	Ad.

In addition to these, certain compounds which metals form with oxygen are also regarded as compound radicals—for instance,

	Molecular formulæ.	Atomic formulæ.	Abbreviated atomic formulæ.
Potassoxyl	$(KO)_2$	KO	Ko.
Zincoxyl	(ZnO_2)	O Zn'' O	Zno''.

The essential character of these last compound radicals is that the whole of the oxygen they contain is united with the metal by one bond only of each oxygen atom, as seen in the following graphic formulæ :—

Hydroxyl ....

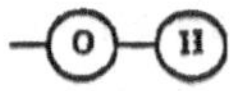

Potassoxyl ....

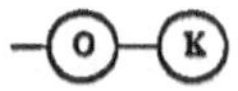

Zincoxyl ....

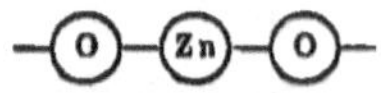

The metal thus becomes linked to other elements by these dyad atoms of oxygen. The functions of such compound radicals will be sufficiently evident from the following examples of compounds into which they enter, and in which their position is marked by dotted lines.

Nitric acid ....

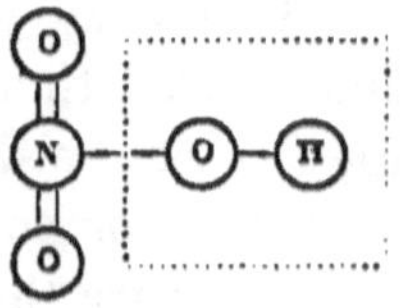

Potassic sulphate ....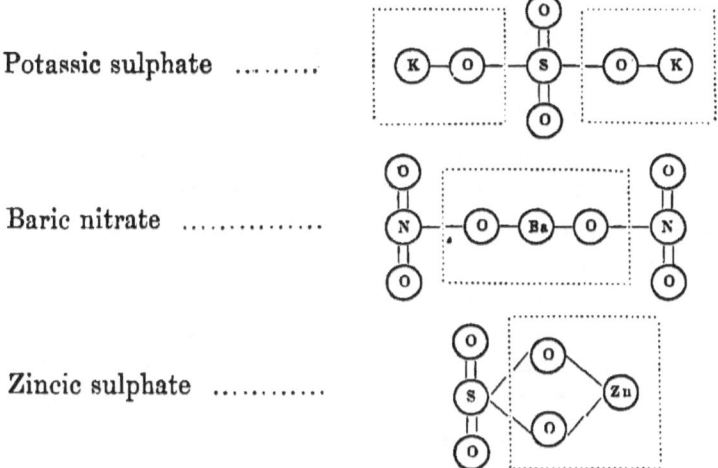

Baric nitrate ....

Zincic sulphate ....

It is not necessary to dignify all these metallic compound radicals with names; the chief point of importance about them is their abbreviated notation, in which the small letter o is attached to the symbol of the metal, the atomicity of the radical being marked in the usual manner. It must be borne in mind that the number of atoms of oxygen in any radical of this class depends upon its atomicity: thus a monad contains only one atom of oxygen, a dyad two, and a triad always three atoms of oxygen. The use of any but monad and dyad metallic compound radicals is very rare.

CHAPTER V.

ATOMIC AND MOLECULAR COMBINATION.

In all the cases of chemical combination considered in the above Chapters, a union of atoms has been invariably contemplated. This atomic union is generally attended by the breaking up of previously existing molecules—two such molecules, by the mutual exchange of their atomic constituents, producing two new and perfectly distinct molecules. Thus when chlorine unites with hydrogen to form hydrochloric acid, a molecule of chlorine and one of hydrogen yield up their constituent atoms, forming two molecules of hydrochloric acid,

$$Cl_2 + H_2 = 2HCl.$$

In comparatively rare cases, two molecules combine to form only one new molecule; thus a molecule of carbonic oxide and one of chlorine combine to form one molecule of carbonic oxydichloride or phosgene gas: but the union is even here essentially atomic; for after combination both the oxygen and chlorine are directly united with the atom of carbon:

$$\mathbf{C}''O \ + \ Cl_2 \ = \ \mathbf{C}^{iv}OCl_2.$$

Carbonic oxide.   Chlorine.   Phosgene gas.

Chemists are, however, compelled to admit an entirely different kind of union, which not unfrequently occurs, and which may be appropriately termed *molecular union* or *molecular combination*. In the formation of such compounds, no change takes place in the active atomicity of any of the molecules. It is this kind of combination which holds together salts and their water of crystallization, as, for instance,

Sodic chloride crystallized at −10° C.......$NaCl, 2OH_2$.
Sodic bromide crystallized below +30°C....$NaBr, 2OH_2$.
Sodic iodide crystallized below +50° C....$NaI, \ 2OH_2$.
Alum ....$S_4O_8('\mathbf{Al}_2'''O_6)^{vi}Ko_2, 24OH_2$.

The researches of Tyndall upon the absorption of radiant heat by different vapours, render it more than probable that aqueous vapour does not consist of an assemblage of single and separate molecules of the compound OH_2, but of groups of these molecules of great complexity, united without contraction of volume.

Numerous other instances of molecular combination might be adduced; but it is only necessary here to point out that such molecular unions will be distinguished from atomic combinations by the use of the comma, as in the above and following examples:

Tetramethylammonic tri-iodide.........$NMe_4 I, I_2$.
Tetramethylammonic pentiodide ......$NMe_4 I, 2I_2$.
Tetramethylammonic iodo-dichloride .$NMe_4 I, Cl_2$.

In all cases molecular combination seems to be of a much more feeble character than atomic union; for, in the first place, such bodies are generally decomposed with facility; and secondly, the properties of their constituent molecules are markedly perceptible in the compounds. Thus the above so-called periodides of the organic bases present in appearance great resemblance to iodine.

CHAPTER VI.

CLASSIFICATION OF ELEMENTS.

It has already been mentioned that the elements may be divided into two great classes, the metals and the non-metals or metalloids. A second division into positive or basylous and negative or chlorous elements has also been explained. A third and still more important classification is founded upon the atomicity of the elements. In the following classified Table all three methods are embodied, the metalloids being printed in red type, and the metals in black, whilst the positive or basylous elements are printed in Roman characters, and the negative or chlorous in *italics*. In addition, the different classes are also divided into sections, consisting of elements closely related in their chemical characters.

Monads.	Dyads.	Triads.	Tetrads.	Pentads.	Hexads.
1st Section.	1st Section.	1st Section.	1st Section.	1st Section.	1st Section.
Hydrogen.	*Oxygen.*	Boron.	Carbon.	Nitrogen.	*Sulphur.*
			Silicon.	Phosphorus.	*Selenium.*
			Tin.	Arsenic.	*Tellurium.*
2nd Section.	2nd Section.	2nd Section.	Titanium.	Antimony.	
Fluorine.	Barium.	Gold.		Bismuth.	2nd Section.
Chlorine.	Strontium.				
Bromine.	Calcium.		2nd Section.		Tungsten.
Iodine.	Magnesium.		Thorinum.		Vanadium.
	Zinc.		Niobium.		Molybdenum.
			Tantalum.		
3rd Section.	3rd Section.		Zirconium.		3rd Section.
Cæsium.	Didymium.		Aluminium.		Osmium.
Rubidium.	Lanthanum.				Iridium.
Potassium.	Yttrium.		3rd Section.		Ruthenium.
Sodium.	Glucinum.		Platinum.		Rhodium.
Lithium.			Palladium.		
					4th Section.
4th Section.	4th Section.		4th Section.		Chromium.
Thallium.	Cadmium.		Lead.		Manganese.
Silver.	Mercury.				Iron.
	Copper.				Cobalt.
					Nickel.
					Uranium.
					Cerium.

CHAPTER VII.

WEIGHTS AND MEASURES.

THE weights and measures employed in this book are chiefly those of the French decimal system. The following Tables, published by Messrs. De la Rue and Co., will enable the student to convert these into their English equivalents whenever this may be necessary.

French Measures of Length.

	In English inches.	In English feet=12 inches.	In English yards=3 feet.	In English fathoms=6 feet.	In English miles= 1760 yards.
Millimètre	0·03937	0·003281	0·0010936	0·0005468	0·0000005
Centimètre	0·39371	0·032809	0·0109363	0·0054682	0·0000062
Décimètre	3·93708	0·328090	0·1093633	0·0546816	0·0000621
Mètre	39·37079	3·280899	1·0936331	0·5468165	0·0006214
Décamètre	393·70790	32·808992	10·9363310	5·4681655	0·0062138
Hectomètre......	3937·07900	328·089920	109·3633100	54·6816550	0·0621382
Kilomètre	39370·79000	3280·899200	1093·6331000	546·8165500	0·6213824
Myriomètre......	393707·90000	32808·992000	10936·3310000	5468·1655000	6·2138244

1 inch = 2·539954 centimètres.   1 yard = 0·9143835 mètre.
1 foot = 3·0479449 décimètres.   1 mile = 1·6093149 kilomètre.

French Measures of Surface.

	In English square feet.	In English square yards =9 square feet.	In English poles= 272·25 sq. feet.	In English roods= 10890 sq. feet.	In English acres= 43560 sq. feet.
Centiare or sq. mètre	10·764299	1·196033	0·0395383	0·0009885	0·0002471
Are or 100 sq. mètres	1076·429934	119·603326	3·9538290	0·0988457	0·0247114
Hectare or 10,000 square mètres...	107642·993418	11960·332602	395·3828959	9·8845724	2·4711431

1 square inch = 6·4513669 square centimètres.
1 square foot = 9·2899683 square décimètres.
1 square yard = 0·83609715 square mètre or centiare.
1 acre   = 0·40467102 hectare.

French Measures of Capacity.

	In cubic inches.	In cubic feet= 1728 cubic inches.	In pints 34·65923 cubic inches.	In gallons =8 pints =277·27384 cubic inches.	In bushels =8 gallons =2218·19075 cubic inches.
Millilitre or cubic centimètre	0·06103	0·000035	0·00176	0·0002201	0·0000275
Centilitre or 10 cubic centimètres	0·61027	0·000353	0·01761	0·0022010	0·0002751
Décilitre or 100 cubic centimètres	6·10271	0·003532	0·17608	0·0220097	0·0027512
Litre or cubic décimètre	61·02705	0·035317	1·76077	0·2200967	0·0275121
Décalitre or centistère...	610·27052	0·353166	17·60773	2·2009668	0·2751208
Hectolitre or décistère..	6102·70515	3·531658	176·07734	22·0096677	2·7512085
Kilolitre, or Stère, or cubic mètre	61027·05152	35·316581	1760·77341	220·0966767	27·5120846
Myriolitre or decastère..	610270·51519	353·165807	17607·73414	2200·9667675	275·1208459

1 cubic inch = 16·386176 cubic centimètres.
1 cubic foot = 28·315312 cubic décimètres, or litres.
1 gallon   = 4·543358 litres.

French Measures of Weight.

	In English grains.	In troy ounces= 480 grains.	In avoirdu- pois lbs. = 7000 grains.	In cwts. = 112 lbs. = 784000 grs.	Tons= 20 cwts. = 15680000 grs.
Milligramme	0·01543	0·000032	0·0000022	0·0000000	0·0000000
Centigramme	0·15432	0·000322	0·0000220	0·0000002	0·0000000
Décigramme	1·54323	0·003215	0·0002205	0·0000020	0·0000001
Gramme	15·43235	0·032151	0·0022046	0·0000197	0·0000010
Décagramme	154·32349	0·321507	0·0220462	0·0001968	0·0000098
Hectogramme.........	1543·23488	3·215073	0·2204621	0·0019684	0·0000984
Kilogramme	15432·34880	32·150727	2·2046213	0·0196841	0·0009842
Myriogramme	154323·48800	321·507267	22·0462126	0·1968412	0·0098421

1 grain =0·064799 gramme.	1 lb. avoir.=0·453593 kilogr.
1 troy oz.=31·103496 grammes.	1 cwt. =50·802377 kilogrs.

Temperatures are expressed upon the Centigrade scale, and
barometric measurements are given in millimetres.

For the ready conversion of gaseous volumes into weights, I
have adopted the *crith*, or standard multiple proposed by Dr.
Hofmann. The crith is the weight of one litre or cubic deci-
metre of hydrogen at 0° C. and at a pressure of 760 millimetres
of mercury. The following is Dr. Hofmann's description of
the value and applications of this unit.

" The actual weight of this cube of hydrogen, at the standard
temperature and pressure mentioned, is 0·0896 gramme; a
figure which I earnestly beg you to inscribe, as with a sharp gra-
ving tool, upon your memory. There is probably no figure in
chemical science more important than this one to be borne
in mind, and to be kept ever in readiness for use in calculation
at a moment's notice. For this litre-weight of hydrogen
=0·0896 gramme (I purposely repeat it) is the standard
multiple, or coefficient, by means of which the weight of one
litre of any other gas, simple or compound, is computed. Again,
therefore, I say, do not let slip this figure—0·0896 gramme.
So important, indeed, is this standard weight unit, that some
name—the simpler and briefer the better—is needed to denote
it. For this purpose I venture to suggest the term *crith*, derived from the Greek word κριθή, signifying a barley-corn,
and figuratively employed to imply a small weight. The weight
of 1 litre of hydrogen being called 1 crith, the volume-weight

of other gases, referred to hydrogen as a standard, may be expressed in terms of this unit.

" For example, the relative volume-weight of chlorine being 35·5, that of oxygen 16, that of nitrogen 14, the actual weight of 1 litre of each of these elementary gases, at 0° C. and 0^m·76 pressure, may be called respectively 35·5 *criths*, 16 *criths*, and 14 *criths*.

" So, again, with reference to the compound gases; the relative volume-weight of each is equal to half the weight of its product-volume. Hydrochloric acid (HCl), for example, consists of 1 vol. of hydrogen + 1 vol. of chlorine = 2 volumes; or, by weight, 1 + 35·5 = 36·5 units; whence it follows that the relative volume-weight of hydrochloric acid gas is $\frac{36\cdot5}{2}$ = 18·25 units; which last figure therefore expresses the number of *criths* which one litre of hydrochloric acid gas weighs at 0° C. temperature and 0^m·76 pressure; and the crith being (as I trust you already bear in mind) 0·0896 gramme, we have

$$18\cdot25 \times 0\cdot0896 = 1\cdot6352$$

as the actual weight in grammes of hydrochloric acid gas.

" So, once more, as the product-volume of water-gas (H_2O) (taken at the above temperature and pressure) contains 2 vols. of hydrogen + 1 vol. of oxygen, and therefore weighs 2 + 16 = 18 units, the single volume of water-gas weighs $\frac{18}{2}$ = 9 units; or, substituting as before the concrete for the abstract value, 1 litre of water-gas weighs 9 *criths*; that is to say, 9 × 0·0896 gramme = 0·8064 gramme.

" In like manner the product-volume of sulphuretted hydrogen (H_2S) = 2 litres of hydrogen, weighing 2 criths, + 1 litre of sulphur-gas, weighing 32 criths, together 2 + 32 = 34 criths, which, divided by 2, gives $\frac{34}{2}$ = 17 criths = 17 × 0·0896 gramme = 1·5232 gramme = the weight of 1 litre of sulphuretted hydrogen at standard temperature and pressure.

" And so, lastly, of ammonia (NH_3); it contains in 2 litres 3 litres of hydrogen, weighing 3 criths, and 1 litre of nitrogen, weighing 14 criths; its total product-volume-weight is therefore $3+14=17$ criths, and its single volume or litre weight is consequently

$$\frac{17}{2}=8\text{·}5 \text{ criths}=8\text{·}5 \times 0\text{·}0896 \text{ gramme}=0\text{·}7616 \text{ gramme}.$$

"Thus, by the aid of the hydrogen-litre-weight or *crith* $=0\text{·}0896$ gramme, employed as a common multiple, the actual or concrete weight of 1 litre of any gas, simple or compound, at standard temperature and pressure, may be deduced from the mere abstract figure expressing its volume-weight relatively to hydrogen."

CHAPTER VIII.

MONAD ELEMENTS.

SECTION I.

HYDROGEN, H_2.

Atomic weight $=1$. *Molecular weight* $=2$. *Molecular volume* ☐☐. *1 litre weighs 1 crith. Atomicity', being the standard of comparison.*

Occurrence.—In combination, as water, in very large quantities in nature. In almost all vegetable and animal substances, and in many minerals. In the free state in the gases of volcanoes. In certain stars and nebulæ?

Preparation.—1. By the action of sodium upon water :—

$$\underset{\text{Water.}}{2OH_2} \quad + \quad \underset{\text{Sodium.}}{Na_2} \quad = \quad \underset{\text{Sodic hydrate.}}{2ONaH} \quad + \quad \underset{\text{Hydrogen.}}{H_2.}$$

2. By the action of sodium upon dry hydrochloric acid :—

$$\underset{\text{Hydrochloric acid.}}{2HCl} \quad + \quad \underset{\text{Sodium.}}{Na_2} \quad = \quad \underset{\text{Sodic chloride.}}{2NaCl} \quad + \quad \underset{\text{Hydrogen.}}{H_2.}$$

3. By the action of zinc, iron, or certain other metals on hydrochloric acid:—

$$2HCl \quad + \quad Zn \quad = \quad \mathbf{ZnCl_2} \quad + \quad H_2.$$

Hydrochloric acid.   Zincic chloride.

4. By the action of zinc or certain other metals on dilute sulphuric acid:—

$$\mathbf{SO_2Ho_2} \quad + \quad Zn \quad = \quad \mathbf{SO_2Zno''} \quad + \quad H_2.$$

Sulphuric acid.   Zincic sulphate.

5. By passing steam over iron heated to redness:—

$$Fe_3 \quad + \quad 4OH_2 \quad = \quad {}^{iv}(\mathbf{Fe_3})^{viii}O_4 \quad + \quad 4H_2.$$

Water.   Triferric tetroxide.

6. By the action of zinc on a boiling solution of potassic hydrate:—

$$2OKH \quad + \quad Zn \quad = \quad \mathbf{ZnKo_2} \quad + \quad H_2.$$

Potassic hydrate.   Potassic zinc oxide.

7. By the electrolysis of water and of some other liquids containing hydrogen.

8. By the action of intense heat upon water.

9. In the destructive distillation of some organic substances.

Section II.

CHLORINE, Cl_2.

Atomic weight $=35\cdot5$. *Molecular weight* $=71$. *Molecular volume* $\square\square$. 1 *litre weighs* $35\cdot5$ *criths. Has not been solidified. Liquefies at* $15°\cdot5$ C., *under a pressure of* 4 *atmospheres. Atomicity'. Evidence of atomicity,* HCl.

Occurrence.—Always in combination—with sodium and other metals in sea-water, and in the solid state in the salt-beds of Cheshire, Worcester, &c. Evolved from volcanoes in the form of hydrochloric acid.

Preparation.—1. By heating certain metallic chlorides, as platinic and auric chlorides:—

$$\mathbf{PtCl_4} \quad = \quad 2Cl_2 \quad + Pt.$$

Platinic chloride.

2. By gently heating a mixture of manganic oxide and hydrochloric acid, when the reaction takes place in two stages :—

$$\underset{\text{Manganic oxide.}}{\mathbf{MnO_2}} + \underset{\text{Hydrochloric acid.}}{4\,\mathrm{HCl}} = \underset{\text{Manganic chloride.}}{\mathbf{MnCl_4}} + \underset{\text{Water.}}{2\mathbf{OH_2}};$$

$$\underset{\text{Manganic chloride.}}{\mathbf{MnCl_4}} = \underset{\text{Manganous chloride.}}{\mathbf{MnCl_2}} + \mathrm{Cl_2}.$$

3. By heating a mixture of sulphuric acid, sodic chloride, and manganic oxide, when the whole of the chlorine present is liberated :—

$$\underset{\text{Manganic oxide.}}{\mathbf{MnO_2}} + \underset{\text{Sulphuric acid.}}{2\,\mathrm{SO_2Ho_2}} + \underset{\text{Sodic chloride.}}{2\,\mathrm{NaCl}} = \underset{\text{Sodic sulphate.}}{\mathbf{SO_2Nao_2}} +$$

$$\underset{\text{Manganous sulphate.}}{\mathbf{SO_2Mno''}} + \underset{\text{Water.}}{2\mathbf{OH_2}} + \mathrm{Cl_2}.$$

If in the second process a mixture of manganic oxide, hydrochloric acid, and sulphuric acid be employed, the whole of the chlorine is evolved :—

$$\underset{\text{Manganic oxide.}}{\mathbf{MnO_2}} + \underset{\text{Sulphuric acid.}}{\mathbf{SO_2Ho_2}} + \underset{\text{Hydrochloric acid.}}{2\,\mathrm{HCl}} =$$

$$\underset{\text{Manganic sulphate.}}{\mathbf{SO_2Mno''}} + \underset{\text{Water.}}{2\mathbf{OH_2}} + \mathrm{Cl_2}.$$

4. By the electrolysis of hydrochloric acid.

Reactions.—1. A mixture of chlorine and hydrogen unite instantly, with explosion, under the influence of sunlight, or of powerful artificial light, or on the application of a burning body to the mixture. A burning jet of hydrogen continues to burn when plunged into chlorine,

$$\mathrm{H_2} + \mathrm{Cl_2} = 2\,\mathrm{HCl}.$$

2. Chlorine has so great an attraction for hydrogen, that it removes the latter from its compounds with carbon. When a rag moistened with turpentine is plunged into chlorine, the chlorine and hydrogen unite, with evolution of heat and light, carbon being liberated :—

$$\underset{\text{Turpentine.}}{\mathrm{C_{10}H_{16}}} + 8\,\mathrm{Cl_2} = \underset{\text{Hydrochloric acid.}}{16\,\mathrm{HCl}} + 10\,\mathrm{C}.$$

HYDROCHLORIC ACID, *Chlorhydric Acid, Muriatic Acid.*
HCl.

Molecular weight $=36\cdot5$. *Molecular volume* ☐. 1 *litre weighs* $18\cdot25$ *criths. Has not been solidified. Condenses at* $10°$ *under a pressure of* 40 *atmospheres.*

Occurrence.—Evolved from volcanoes.

Preparation.—1. From its elements, as above described.

2. By gently heating sodic chloride with sulphuric acid, previously diluted with a small quantity of water :—

$$\mathbf{SO_2Ho_2} \; + \; \mathrm{NaCl} \; = \; \mathbf{SO_2HoNao} \; + \; \mathrm{HCl}:$$
Sulphuric acid. Sodic chloride. Hydric sodic sulphate. Hydrochloric acid.

or $\quad \mathbf{SO_2Ho_2} \; + \; \mathrm{2NaCl} \; = \; \mathbf{SO_2Nao_2} \; + \; \mathrm{2HCl}.$
Sulphuric acid. Sodic chloride. Sodic sulphate. Hydrochloric acid.

Reactions.—Hydrochloric acid may be converted into salts termed chlorides by the action of certain metals as described above, and also by that of the metallic hydrates or oxides :—

$$\mathbf{OKH} \; + \; \mathrm{HCl} \; = \; \mathrm{KCl} \; + \; \mathbf{OH_2}.$$
Potassic hydrate. Hydrochloric acid. Potassic chloride. Water.

$$\mathbf{ZnO} \; + \; \mathrm{2HCl} \; = \; \mathbf{ZnCl_2} \; + \; \mathbf{OH_2}.$$
Zincic oxide. Hydrochloric acid. Zincic chloride. Water.

For the remaining monad elements of this Section, see Chapter XIV.

CHAPTER IX.

DYAD ELEMENTS

SECTION I.

OXYGEN, O_2.

Atomic weight $=16$. *Molecular weight* $=32$. *Molecular volume* ☐. 1 *litre weighs* 16 *criths. Atomicity". Evidence of atomicity* :—

Water .... OH_2.
Potassic hydrate .... OKH.
Argentic oxide .... OAg_2.
Hypochlorous anhydride .... OCl_2.

Occurrence.—In the free state in the atmosphere. In the combined state in water, in most mineral bodies, and in almost all animal and vegetable compounds.

Preparation.—1. If metallic mercury be heated to its boiling-point with access of air, it gradually absorbs oxygen, being converted into mercuric oxide, $Hg''O$. This compound, when more strongly heated, is resolved into its elements,

$$2\mathbf{HgO} \quad = \quad 2Hg \quad + \quad O_2.$$
Mercuric oxide.   Mercury.   Oxygen.

2. By heating native manganic oxide (pyrolusite) a portion of its oxygen is liberated :—

$$3\mathbf{MnO_2} \quad = \quad ^{iv}(\mathbf{Mn_3})^{viii}O_4 \quad + \quad O_2.$$
Manganic oxide.   Trimanganic tetroxide.

3. Oxygen is evolved in nature in a remarkable manner by the decomposition of carbonic anhydride, CO_2, by the green leaves of plants, the vegetable assimilating the carbon, whilst the oxygen escapes into the atmosphere :—

$$\mathbf{CO_2} \quad = \quad C \quad + \quad O_2.$$
Carbonic anhydride.

4. By the action of heat upon potassic chlorate .—

(Atomic) $\quad \begin{cases} \mathbf{O}Cl \\ O \\ \mathbf{O}K \end{cases} \quad = KCl \quad + \quad 3O,$ or

(Molecular) $\quad 2 \begin{cases} \mathbf{O}Cl \\ O \\ \mathbf{O}K \end{cases} \quad = 2KCl \quad + \quad 3O_2.$

Potassic chlorate.   Potassic chloride.   Oxygen.

5. By mixing the potassic chlorate with manganic oxide, the oxygen is evolved at a much lower temperature; the manganic oxide appears to take no part in the reaction.

6. By dropping concentrated sulphuric acid into a red-hot

platinum retort, the acid is decomposed into oxygen, sulphurous anhydride, and water :—

(Atomic) $\quad SO_2Ho_2 \;=\; SO_2 \;+\; OH_2 \;+\; O$, or

(Molecular) $\quad 2SO_2Ho_2 \;=\; 2SO_2 \;+\; 2OH_2 \;+\; O_2$.

Sulphuric acid.   Sulphurous anhydride.   Water.

7. By the electrolysis of water.

8. By the action of heat upon a mixture of manganic oxide and sulphuric acid :—

(Atomic) $\quad MnO_2 + SO_2Ho_2 = SO_2Mno'' + OH_2 + O$, or

(Molecular) $2MnO_2 + 2SO_2Ho_2 = 2SO_2Mno'' + 2OH_2 + O_2$.

Manganic oxide.   Sulphuric acid.   Manganous sulphate.   Water.

9. By heating a mixture of potassic bichromate and sulphuric acid :—

$$2\left\{\begin{array}{l} CrO_2Ko \\ O \\ CrO_2Ko \end{array}\right. \;+\; 8SO_2Ho_2 \;=\; 2SO_2Ko_2$$

Potassic bichromate.   Sulphuric acid.   Potassic sulphate.

$$+ \;\; 2S_3O_6('Cr_2'''O_6)^{vi} \;+\; 3O_2 \;+\; 8OH_2.$$

Chromic sulphate.   Water.

10. By passing steam and chlorine through a red-hot porcelain tube, hydrochloric acid and oxygen are formed :—

$$2OH_2 \;+\; 2Cl_2 \;=\; 4HCl \;+\; O_2.$$

Water.   Chlorine.   Hydrochloric acid.

Reaction.—A mixture of two volumes of hydrogen and one volume of oxygen explodes at a red heat, water being produced. The same compound is formed when hydrogen is burnt in oxygen or oxygen in hydrogen :—

(Molecular) $\quad 2H_2 \;+\; O_2 \;=\; 2OH_2$.

ALLOTROPIC OXYGEN or OZONE, O_3.

Molecular weight $=48$. *Molecular volume* ▢. 1 *litre weighs* 24 *criths.*

Preparation.—1. When electric sparks are passed through air or oxygen, a peculiar odour, which is due to ozone, is observed.

2. By placing phosphorus in moist air at about the ordinary temperature for a few hours.

3. By passing an electric current through dilute sulphuric or chromic acid.

Thus obtained, ozone is always mixed with a large proportion of air or oxygen.

Properties.—Powerfully oxidizing. It oxidizes the metals silver and mercury, and organic matters at ordinary temperature. When oxygen is converted into ozone, contraction of volume takes place ; and when the ozone is heated to 290°, it is retransformed into the original volume of ordinary oxygen,—indicating that the molecule of ozone contains more atoms than the molecule of ordinary oxygen.

In most cases of oxidation by ozone no diminution of the volume of gas takes place, the additional atoms previously introduced into the molecules of oxygen being removed, and ordinary oxygen becoming free. But it has been recently shown by Soret, that oil of turpentine absorbs the whole molecule of the ozone, leaving untouched the oxygen which was previously present in the state of admixture. By observing the contraction during the production of the ozone and the diminution of volume produced by absorbing it with oil of turpentine, the density of ozone may be readily calculated, and consequently its atomic constitution.

By this means the specific gravity of ozone has been shown to be 24, the molecular weight being therefore 48, which is the weight of 3 atoms of oxygen.

In ordinary oxygen, the molecule is composed of two atoms of oxygen, and is represented by

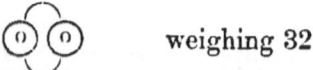 weighing 32.

In ozone the molecule contains 3 atoms of oxygen, and is represented by

weighing 48.

WATER, *Hydric Oxide.*

(H)–(O)–(H) OH_2.

Molecular weight $=18$.   *Molecular volume* ☐.   1 *litre of water-vapour weighs* 9 *criths. Fuses at* 0°. *Boils at* 100°.

Occurrence.—Most abundantly in nature.

Formation.—1. By the direct union of hydrogen and oxygen, as above.

2. As a secondary product in numberless chemical reactions, as, for instance, in the action of hydrochloric acid on potassic hydrate :—

$$OKH \quad + \quad HCl \quad = \quad OH_2 \quad + \quad KCl.$$
Potassic hydrate.   Hydrochloric acid.   Water.   Potassic chloride.

Reactions.—1. By its action many metallic oxides are converted into hydrates :—

$$OK_2 \quad + \quad OH_2 \quad = \quad 2OKH.$$
Potassic oxide.   Water.   Potassic hydrate.

$$BaO \quad + \quad OH_2 \quad = \quad BaHo_2.$$
Baric oxide.   Water.   Baric hydrate.

2. By its action on anhydrides it transforms them into acids :—

$$N_2O_5 \quad + \quad OH_2 \quad = \quad 2NO_2Ho.$$
Nitric anhydride.   Water.   Nitric acid.

$$SO_3 \quad + \quad OH_2 \quad = \quad SO_2Ho_2.$$
Sulphuric anhydride.   Water.   Sulphuric acid.

$$P_2O_5 \quad + \quad 3OH_2 \quad = \quad 2POHo_3.$$
Phosphoric anhydride.   Water.   Phosphoric acid.

3. It also unites molecularly with many compounds as water of crystallization (see Chapter V.), as in the following instances :—

$BaCl_2, 2OH_2$ .... Baric chloride.
$SO_2Nao_2, 1OOH_2$ .... Sodic sulphate.
$S_4O_8Ko_2('Al_2'''O_6)^{vi}, 24OH_2$ ... Alum.

HYDROXYL, *Hydric Peroxide.*

$$\text{(H)}-\text{(O)}\dashv\text{(O)}-\text{(H)} \qquad H_2O_2 \text{ or } (HO)_2 \text{ or } Ho_2 \text{ or } \begin{cases} OH \\ OH \end{cases}$$

Probable molecular weight = 34.

Preparation.—By passing a current of carbonic anhydride through water in which baric peroxide is suspended :—

$$\begin{cases} O \\ O \end{cases}Ba'' \;+\; CO_2 \;+\; OH_2 \;=\; COBao'' \;+\; \begin{cases} OH \\ OH \end{cases}$$

Baric peroxide.   Carbonic anhydride   Water.   Baric carbonate.   Hydroxyl.

Reactions.—1. By heat it is decomposed into water and oxygen :—

$$2 \begin{cases} OH \\ OH \end{cases} = \; 2OH_2 \;+\; O_2.$$

Hydroxyl.   Water.   Oxygen.

2. Hydroxyl is transformed into water by the action of nascent hydrogen: if hydroxyl be introduced into an apparatus generating hydrogen, the gas ceases to be evolved :—

$$SO_2Ho_2 \;+\; Zn \;+\; \begin{cases} OH \\ OH \end{cases} = \; SO_2Zno'' \;+\; 2OH_2.$$

Sulphuric acid.   Hydroxyl.   Zincic sulphate.   Water.

3. Hydroxyl liberates iodine from potassic iodide :—

$$2KI \;+\; \begin{cases} OH \\ OH \end{cases} = \; 2OKH \;+\; I_2.$$

Potassic iodide.   Hydroxyl.   Potassic hydrate.   Iodine.

4. Hydroxyl is a powerful oxidizing agent; it converts, for instance, plumbic sulphide into plumbic sulphate :

$$PbS'' \;+\; 4 \begin{cases} OH \\ OH \end{cases} = \; SO_2Pbo'' \;+\; 4OH_2.$$

Plumbic sulphide.   Hydroxyl.   Plumbic sulphate.   Water.

OXIDES AND ACIDS OF CHLORINE.

Oxygen forms many compounds with chlorine and with chlorine and hydroxyl; but none of them can be produced by direct combination. The following list contains all that are known:—

Hypochlorous anhydride...... } $\mathbf{O}Cl_2$.

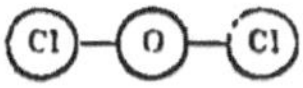

Cl—O—Cl

Chloric oxide?... { $\mathbf{O}Cl$ / $\mathbf{O}Cl$.

Cl—O—O—Cl

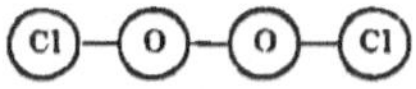

Chlorous anhydride........... { $\mathbf{O}Cl$ / O / $\mathbf{O}Cl$

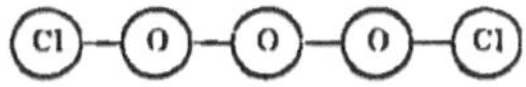

Cl—O—O—O—Cl

Chloric peroxide { $\mathbf{O}Cl$ / O / O / $\mathbf{O}Cl$

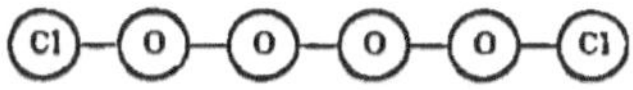

Cl—O—O—O—O—Cl

Chloric hyperoxide?......... { $\mathbf{O}Cl$ / O / O / O / O / $\mathbf{O}Cl$

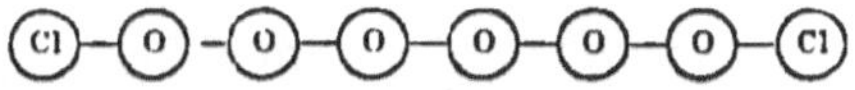

Cl—O—O—O—O—O—O—Cl

Hypochlorous acid, $\mathbf{O}ClH$, or $ClHo$.

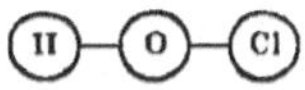

H—O—Cl

Chlorous acid ... $\mathbf{O}ClHo$, or { $\mathbf{O}Cl$ / $\mathbf{O}H$.

H—O—O—Cl

Chloric acid ... { $\mathbf{O}Cl$ / $\mathbf{O}Ho$, or { $\mathbf{O}Cl$ / O / $\mathbf{O}H$.

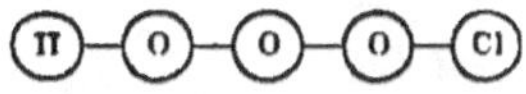

H—O—O—O—Cl

Perchloric acid { $\mathbf{O}Cl$ / O / $\mathbf{O}Ho$, or { $\mathbf{O}Cl$ / O / O / $\mathbf{O}H$.

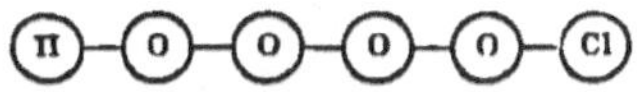

H—O—O—O—O—Cl

HYPOCHLOROUS ANHYDRIDE.

$$\mathbf{O}Cl_2.$$

Molecular weight $=87$.   *Molecular volume* ☐☐.   1 *litre of hypochlorous anhydride vapour weighs* 43·5 *criths.   Boils at about* 20°.

Preparation.—By passing chlorine over mercuric oxide at a low temperature :—

$$2\mathbf{Hg}O \quad + \quad 2Cl_2 \quad = \quad \left\{ \begin{array}{l} \mathbf{Hg}Cl \\ O \\ \mathbf{Hg}Cl \end{array} \right. \quad + \quad \mathbf{O}Cl_2.$$

Mercuric oxide.   Mercuric oxychloride.   Hypochlorous anhydride.

CHLOROUS ANHYDRIDE.

$$\left\{ \begin{array}{l} \mathbf{O}Cl \\ O \\ \mathbf{O}Cl \end{array} \right..$$

Molecular weight $=119$.   *Molecular volume anomalous* ☐☐.
   1 *litre weighs* 79·4 *criths.*

Preparation.—By gently heating in a water-bath a mixture of potassic chlorate, nitric acid, and arsenious acid.   Four different reactions are to be distinguished in this operation :—

1. $\left\{ \begin{array}{l} \mathbf{O}Cl \\ \mathbf{O}Ko \end{array} \right. + \mathbf{NO}_2Ho = \left\{ \begin{array}{l} \mathbf{O}Cl \\ \mathbf{O}Ho \end{array} \right. + \mathbf{NO}_2Ko$;

   Potassic chlorate.   Nitric acid.   Chloric acid.   Potassic nitrate.

2. $\mathbf{As}Ho_3 + \mathbf{NO}_2Ho = \mathbf{NOHo} + \mathbf{As}OHo_3$;

   Arsenious acid.   Nitric acid.   Nitrous acid.   Arsenic acid.

3. $\left\{ \begin{array}{l} \mathbf{O}Cl \\ \mathbf{O}Ho \end{array} \right. + \mathbf{NOHo} = \mathbf{O}ClHo + \mathbf{NO}_2Ho$;

   Chloric acid.   Nitrous acid.   Chlorous acid.   Nitric acid.

4. $\quad 2\mathbf{O}ClHo = \left\{ \begin{array}{l} \mathbf{O}Cl \\ O \\ \mathbf{O}Cl \end{array} \right. + \mathbf{O}H_2.$

   Chlorous acid.   Chlorous anhydride.   Water.

CHLORIC PEROXIDE.

$$\left\{ \begin{array}{l} \mathbf{O}Cl \\ O \\ O \\ \mathbf{O}Cl \end{array} \right. \cdot$$

Molecular weight $= 135.$ *Boils at* $20°$.

Preparation.—By the action of sulphuric acid on potassic chlorate :—

$$3 \left\{ \begin{array}{l} \mathbf{O}Cl \\ \mathbf{O}Ko \end{array} \right. + 2\mathbf{S}O_2Ho_2 = \left\{ \begin{array}{l} \mathbf{O}Cl \\ O \\ \mathbf{O}Ko \end{array} \right. + 2SO_2HoKo + \mathbf{O}H_2 + \left\{ \begin{array}{l} \mathbf{O}Cl \\ O \\ O \\ \mathbf{O}Cl \end{array} \right.$$

Potassic   Sulphuric   Potassic   Hydric potassic   Water.   Chloric
chlorate.   acid.   perchlorate.   sulphate.   peroxide.

HYPOCHLOROUS ACID.

$\mathbf{O}ClH$, or $ClHo$.

Molecular weight $= 52\cdot5$.

Preparation.—1. By the action of water on hypochlorous anhydride :—

$$\mathbf{O}Cl_2 \quad + \quad \mathbf{O}H_2 \quad = \quad 2ClHo.$$

Hypochlorous anhydride.   Water.   Hypochlorous acid.

2. By the action of chlorine upon mercuric oxide in the presence of water :—

$$2\mathbf{Hg}O \quad + \quad \mathbf{O}H_2 \quad + \quad 2Cl_2 \quad = \quad \left\{ \begin{array}{l} \mathbf{Hg}Cl \\ O \\ \mathbf{Hg}Cl \end{array} \right. + \quad 2ClHo.$$

Mercuric oxide.   Water.   Chlorine.   Mercuric   Hypochlorous
            oxychloride.   acid.

Reactions.—1. By the action of hydrochloric acid, chlorine is evolved from both the hydrochloric acid and hypochlorous acid :—

$$ClHo \quad + \quad HCl \quad = \quad Cl_2 \quad + \quad \mathbf{O}H_2.$$

Hypochlorous acid.   Hydrochloric acid.   Chlorine.   Water.

2. By the action of argentic oxide, oxygen is evolved from both compounds:—

$$\mathbf{O}Ag_2 \ + \ 2ClHo \ = \ 2AgCl \ + \ \mathbf{O}H_2 \ + \ O_2$$

Argentic oxide.   Hypochlorous acid.   Argentic chloride.   Water.

3. By the action of hypochlorous acid, metallic oxides or hydrates are converted into hypochlorites:—

$$\mathbf{O}KH \ + \ ClHo \ = \ ClKo \ + \ \mathbf{O}H_2.$$

Potassic hydrate.   Hypochlorous acid.   Potassic hypochlorite.   Water.

It was formerly supposed that hypochlorites, together with chlorides, were formed when chlorine acted upon certain metallic oxides and hydrates :—

$$2Cl_2 \ + \ 2\mathbf{Ca}Ho_2 \ = \ \mathbf{Ca}Cl_2 \ + \ Cao''Cl_2 \ + \ 2\mathbf{O}H_2.$$

Calcic hydrate.   Calcic chloride.   Calcic hypochlorite.   Water.

But the so-called chloride of lime or bleaching powder does not contain calcic chloride, and the true reaction appears to be

$$\mathbf{Ca}Ho_2 \ + \ Cl_2 \ = \ \mathbf{Ca}(OCl)Cl \ + \ \mathbf{O}H_2.$$

Calcic hydrate.   Bleaching powder *.   Water.

By the action of acids this compound yields free chlorine :—

$$\mathbf{Ca}(OCl)Cl \ + \ \mathbf{S}O_2Ho_2 \ = \ \mathbf{S}O_2Cao'' \ + \ \mathbf{O}H_2 \ + \ Cl_2.$$

Bleaching powder.   Sulphuric acid.   Calcic sulphate.   Water.

CHLOROUS ACID.

$$\mathbf{O}ClHo \ \ \text{or} \ \left\{ \begin{array}{l} \mathbf{O}Cl \\ \mathbf{O}H. \end{array} \right.$$

Molecular weight $= 68\cdot5$.

Preparation.—By the action of water upon chlorous anhydride :—

$$\left\{ \begin{array}{l} \mathbf{O}Cl \\ O \\ \mathbf{O}Cl \end{array} \right. \ + \ \mathbf{O}H_2 \ = \ 2 \left\{ \begin{array}{l} \mathbf{O}Cl \\ \mathbf{O}H. \end{array} \right.$$

Chlorous anhydride.   Water.   Chlorous acid.

* (Cl)—(Ca)—(o)—(Cl).

CHLORIC ACID.

$$\begin{cases} \mathbf{O}Cl \\ \mathbf{O}Ho \end{cases} \text{ or } \begin{cases} \mathbf{O}Cl \\ O \\ \mathbf{O}H \end{cases}.$$

Molecular weight $= 84\cdot5$.

Preparation.—By the action of dilute sulphuric acid upon baric chlorate :—

$$\begin{cases} \mathbf{O}Cl \\ O \\ Bao'' \\ O \\ \mathbf{O}Cl \end{cases} + \mathbf{S}O_2Ho_2 = 2\begin{cases} \mathbf{O}Cl \\ \mathbf{O}Ho \end{cases} + \mathbf{S}O_2Bao''.$$

Baric chlorate.　　Sulphuric acid.　　Chloric acid.　　Baric sulphate.

Decomposition.—By boiling, it is decomposed into perchloric acid, water, chlorine, and oxygen :—

$$3\begin{cases} \mathbf{O}Cl \\ \mathbf{O}Ho \end{cases} = \begin{cases} \mathbf{O}Cl \\ O \\ \mathbf{O}Ho \end{cases} + \mathbf{O}H_2 + Cl_2 + 2O_2.$$

Chloric acid.　　　Perchloric acid.　　　Water.

Preparation of Chlorates.—1. Potassic chlorate may be prepared by the action of chlorine upon a concentrated solution of potassic hydrate :—

$$6\mathbf{O}KH + 3Cl_2 = 5KCl + \begin{cases} \mathbf{O}Cl \\ \mathbf{O}Ko \end{cases} + 3\mathbf{O}H_2.$$

Potassic　　　Chlorine.　　Potassic　　Potassic　　　Water.
hydrate.　　　　　　　　chloride.　　chlorate.

2. Calcic chlorate is made by passing chlorine through boiling milk of lime :—

$$6\mathbf{Ca}Ho_2 + 6Cl_2 = \begin{cases} \mathbf{O}Cl \\ O \\ Cao'' \\ O \\ \mathbf{O}Cl \end{cases} + 5\mathbf{Ca}Cl_2 + 6\mathbf{O}H_2.$$

Calcic hydrate.　　　　　Calcic chlorate.　　Calcic chloride.　　Water.

By the addition of potassic chloride to the calcic chlorate,

D

potassic chlorate is formed; the latter is then separated from the calcic chloride by crystallization :—

$$\left\{\begin{matrix} \mathbf{O}Cl \\ O \\ Cao'' \\ O \\ \mathbf{O}Cl \end{matrix}\right. \quad + \quad 2KCl \quad = \quad 2\left\{\begin{matrix} \mathbf{O}Cl \\ \mathbf{O}Ko \end{matrix}\right. \quad + \quad \mathbf{Ca}Cl_2.$$

Calcic chlorate.   Potassic chloride.   Potassic chlorate.   Calcic chloride.

PERCHLORIC ACID.

$$\left\{\begin{matrix} \mathbf{O}Cl \\ O \\ \mathbf{O}Ho \end{matrix}\right. \ or \ \left\{\begin{matrix} \mathbf{O}Cl \\ O \\ O \\ \mathbf{O}H \end{matrix}\right. .$$

Molecular weight $= 100\cdot5$.

Preparation.—Potassic perchlorate is distilled with about three times its weight of sulphuric acid :—

$$2\left\{\begin{matrix} \mathbf{O}Cl \\ O \\ \mathbf{O}Ko \end{matrix}\right. \quad + \quad \mathbf{S}O_2Ho_2 \quad = \quad 2\left\{\begin{matrix} \mathbf{O}Cl \\ O \\ \mathbf{O}Ho \end{matrix}\right. \quad + \quad \mathbf{S}O_2Ko_2.$$

Potassic perchlorate.   Sulphuric acid.   Perchloric acid.   Potassic sulphate.

The impure perchloric acid is then carefully rectified, when pure perchloric acid passes over as an oily liquid towards the end of the operation.

It forms with water a white crystalline hydrate.

Preparation of Potassic Perchlorate.—1. Potassic chlorate is heated gradually, and the process arrested when one-third of the oxygen present has been evolved; the residue then contains potassic chloride and perchlorate :—

$$2\left\{\begin{matrix} \mathbf{O}Cl \\ \mathbf{O}Ko \end{matrix}\right. \quad = \quad KCl \quad + \quad \left\{\begin{matrix} \mathbf{O}Cl \\ O \\ \mathbf{O}Ko \end{matrix}\right. \quad + \quad O_2.$$

Potassic chlorate.   Potassic chloride.   Potassic perchlorate.

By crystallization the two salts are separated.

2. When potassic chlorate is gradually introduced into boiling nitric acid, chlorine and oxygen are evolved, potassic nitrate and perchlorate being formed :—

$$3\left\{ \begin{matrix} \mathbf{O}\mathrm{Cl} \\ \mathbf{O}\mathrm{Ko} \end{matrix} \right. + 2\mathbf{N}\mathrm{O_2Ho} = 2\mathbf{N}\mathrm{O_2Ko} + \mathbf{O}\mathrm{H_2} + \left\{ \begin{matrix} \mathbf{O}\mathrm{Cl} \\ \mathrm{O} \\ \mathbf{O}\mathrm{Ko} \end{matrix} \right. + \mathrm{Cl_2} + 2\mathrm{O_2}.$$

Potassic   Nitric acid.   Potassic nitrate.   Water.   Potassic
chlorate.   perchlorate.

These salts are then separated by crystallization.

CHAPTER X.

TRIAD ELEMENTS.

Section I.

BORON, B_2.

Atomic weight $= 11$. *Probable molecular weight* $= 22$. *Sp. gr.,* *diamond variety,* $2\cdot68$. *Atomicity* ‴. *Evidence of atomicity* :—

      Boric chloride .... $\mathbf{B}'''\mathrm{Cl_3}$.
      Boric fluoride.............. $\mathbf{B}'''\mathrm{F_3}$.
      Boric ethide .... $\mathbf{B}'''\mathrm{Et_3}$.

Occurrence.—Found only in combination with oxygen.
Preparation :—
a. Amorphous boron.—1. By igniting boric anhydride with sodium :—

$$\mathbf{B}_2\mathrm{O_3} \ + \ 3\mathrm{Na_2} \ = \ 3\mathbf{O}\mathrm{Na_2} \ + \ \mathrm{B_2}.$$
Boric anhydride.   Sodium.   Sodic oxide.

2. By passing boric chloride over heated potassium :—

$$2\mathbf{B}\mathrm{Cl_3} \ + \ 3\mathrm{K_2} \ = \ 6\mathrm{KCl} \ + \ \mathrm{B_2}.$$
Boric chloride.   Potassic chloride.

β. Graphitoidal boron.—By passing boric chloride over fused aluminium :—

$$Al_2 \;+\; 2BCl_3 \;=\; 'Al_2'''Cl_6 \;+\; B_2.$$
Boric chloride.   Aluminic chloride.

γ. Diamond boron.—By fusing boric anhydride with aluminium :—

$$Al_2 \;+\; B_2O_3 \;=\; 'Al_2'''O_3 \;+\; B_2.$$
Boric anhydride.   Aluminic oxide.

Reactions :—

a. Amorphous boron.—1. Decomposes hot sulphuric acid :—

$$B_2 \;+\; 3SO_2Ho_2 \;=\; B_2O_3 \;+\; 3OH_2 \;+\; 3SO_2.$$
Sulphuric acid.   Boric anhydride.   Water.   Sulphurous anhydride.

2. Decomposes nitric acid :—

$$B_2 \;+\; 6NO_2Ho \;=\; 2BHo_3 \;+\; 3'N_2^{iv}O_4.$$
Nitric acid.   Boric acid.   Nitric peroxide.

3. Decomposes alkaline carbonates, sulphates, and nitrates :—

$$B_2 \;+\; 3CONao_2 \;=\; 2BNao_3 \;+\; 3C''O.$$
Sodic carbonate.   Trisodic borate.   Carbonic oxide.

$$B_2 \;+\; 3SO_2Ko_2 \;=\; 2BKo_3 \;+\; 3SO_2.$$
Potassic sulphate.   Tripotassic borate.   Sulphurous anhydride.

$$B_2 \;+\; 6NO_2Ko \;=\; 2BKo_3 \;+\; 3'N_2^{iv}O_4.$$
Potassic nitrate.   Tripotassic borate.   Nitric peroxide.

4. Boron is one of the very few elements which unite directly with nitrogen :—

$$B_2 \;+\; N_2 \;=\; 2B'''N'''.$$
Boric nitride.

γ. Diamond boron.—1. When fused with hydric potassic sulphate, boric anhydride is formed :—

$$6SO_2HoKo \;+\; B_2 = B_2O_3 \;+\; 3SO_2Ko_2 \;+\; 3OH_2 \;+\; 3SO_2.$$
Hydric potassic   Boric   Potassic sul-   Water.   Sulphurous
sulphate.   anhydride.   phate.   anhydride.

No compound of boron with hydrogen has been obtained ; but the chloride, bromide, and fluoride are known.

BORIC CHLORIDE.
BCl_3.

Molecular weight $=117\cdot5$. *Molecular volume* ☐. 1 *litre of boric chloride vapour weighs* $58\cdot75$ *criths.* *Sp. gr.* $1\cdot35$ *at* $7°$. *Boils at* $17°$.

Preparation.—By passing chlorine over a mixture of boric anhydride and charcoal heated to redness :—

$$\underset{\text{Boric anhydride.}}{B_2O_3} + 3Cl_2 + C_3 = \underset{\text{Boric chloride.}}{2BCl_3} + \underset{\text{Carbonic oxide.}}{3CO}.$$

Reaction.—In contact with water it forms hydrochloric and boric acids :—

$$\underset{\text{Boric chloride.}}{BCl_3} + \underset{\text{Water.}}{3OH_2} = \underset{\text{Hydrochloric acid.}}{3HCl} + \underset{\text{Boric acid.}}{BHo_3}.$$

BORIC BROMIDE.
BBr_3.

Molecular weight $=251$. *Molecular volume* ☐. 1 *litre of boric bromide vapour weighs* $125\cdot5$ *criths.* *Sp.gr. of liquid* $=2\cdot69$. *Boils at* $90°$ C.

Prepared and decomposed in exactly the same way as the chloride.

BORIC FLUORIDE.
BF_3.

Molecular weight $=68$. *Molecular volume* ☐. 1 *litre weighs* 34 *criths.*

Preparation.—1. By strongly heating boric anhydride with calcic fluoride :—

$$\underset{\text{Boric anhydride.}}{2B_2O_3} + \underset{\text{Calcic fluoride.}}{3CaF_2} = \underset{\text{Calcic borate.}}{B_2Cao_3''} + \underset{\text{Boric fluoride.}}{2BF_3}.$$

2. By heating together boric anhydride with calcic fluoride and sulphuric acid :—

$$\mathbf{B_2O_3} + 3\mathbf{CaF_2} + 3SO_2Ho_2 = 3SOHo_2Cao'' + 2\mathbf{BF_3}.$$

Boric   Calcic   Sulphuric   Dihydric calcic   Boric fluo-
anhydride.   fluoride.   acid.   sulphate *.   ride.

Reaction.—By contact with water boric fluoride forms a peculiar acid, the hydrofluoboric acid, the constitution of which is not well understood :—

$$4\mathbf{BF_3} + 3\mathbf{OH_2} = 3(BF_3, HF) + \mathbf{BHo_3}.$$

Boric fluoride.   Water.   Hydrofluoboric acid.   Boric acid.

This acid acts upon metallic hydrates, forming salts :—

$$BF_3, HF + \mathbf{OKH} = BF_3, KF + \mathbf{OH_2}.$$

Hydrofluoboric   Potassic   Potassic   Water.
acid.   hydrate.   borofluoride.

Possibly the boron in these compounds is pentadic; thus $\mathbf{B'HF_4}$ and $\mathbf{B'KF_4}$.

BORIC ANHYDRIDE AND ACIDS.

Boric anhydride .... $\mathbf{B_2O_3}$.
Monobasic boric acid ⎫
Metaboric acid ...... ⎬ .... $\mathbf{BOHo}$.
               ⎭
Tribasic boric acid ... ⎫
Boric acid .... ⎬ .... $\mathbf{BHo_3}$.
               ⎭

BORIC ANHYDRIDE, *Boracic anhydride.*
$\mathbf{B_2O_3}$.

Molecular weight = 70.   *Sp. gr.* 1·83.

Preparation.—By fusing boric acid at a red heat :—

$$2\mathbf{BHo_3} = \mathbf{B_2O_3} + 3\mathbf{OH_2}.$$

Boric acid.   Boric anhydride.   Water.

* See sulphuric acid, Chap. XIII. page 81.

BORIC ACID, *Boracic Acid, Orthoboric Acid.*

$\mathbf{B}Ho_3$.

Molecular weight $=62$.   *Sp. gr.* $1\cdot479$.

Occurrence.—Contained in the steam which escapes from the *suffioni* in some parts of Tuscany.

Preparation.—By the addition of hydrochloric acid to a hot saturated solution of borax, when the acid crystallizes out on cooling:—

$$\mathbf{B}_4O_5Nao_2 \quad + \quad 2HCl \quad + \quad 5\mathbf{O}H_2 \quad = \quad 4\mathbf{B}Ho_3 \quad + \quad 2NaCl.$$

Borax. Hydrochloric acid. Water. Boric acid. Sodic chloride.

Reactions.—1. At the temperature of 100° it loses water, being converted into metaboric acid:—

$$\mathbf{B}H o_3 \quad = \quad \mathbf{B}OHo \quad + \quad \mathbf{O}H_2.$$

Boric acid. Metaboric acid. Water.

2. By the action of metallic hydrates, oxides, or carbonates, borates are formed.

The mineral *tincal* contains borax, an abnormal sodic borate, $\mathbf{B}_4O_5Nao_2, 10\mathbf{O}H_2$.

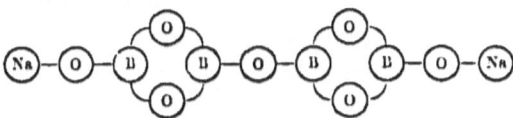

A *trimagnesic octoborate* is known as the mineral *boracite*:—

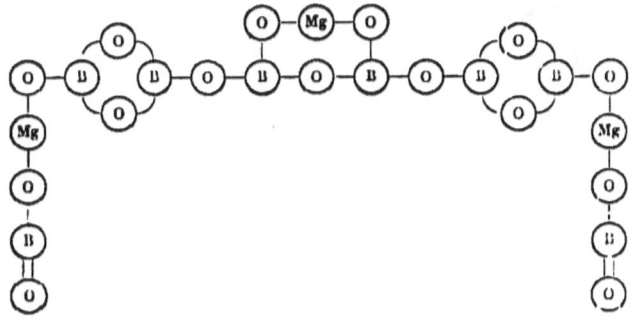

BORIC SULPHIDE.
$\mathbf{B_2S_3}''$.

Molecular weight $=118$.

Preparation.—By passing carbonic disulphide over a mixture of carbon and boric anhydride heated to bright redness :—

$$2\mathbf{B_2O_3} \quad + \quad 3\mathbf{CS_2}'' \quad + \quad 3\mathbf{C} \quad = \quad 2\mathbf{B_2S_3}'' \quad + \quad 6\mathbf{C}''O.$$

Boric anhydride.	Carbonic disulphide.		Boric sulphide.	Carbonic oxide.

Reaction.—Boric sulphide is readily decomposed by water, giving sulphuretted hydrogen and boric acid :—

$$\mathbf{B_2S_3}'' \quad + \quad 6\mathbf{OH_2} \quad = \quad 3\mathbf{SH_2} \quad + \quad 2\mathbf{B}Ho_3.$$

Boric sulphide.	Water.	Sulphuretted hydrogen.	Boric acid.

BORIC NITRIDE.
$\mathbf{B}N'''$.

Molecular weight $=25$.

Preparation.—1. By heating boron in nitrogen (see p. 52).
2. By heating together borax and ammonic chloride :—

$$\mathbf{B_4O_5Nao_2} \quad + \quad 4\mathbf{NH_4Cl} \quad = \quad 4\mathbf{B}N''' \quad + \quad 2NaCl +$$

Borax.	Ammonic chloride.	Boric nitride.	Sodic chloride.

$$7\mathbf{OH_2} \quad + \quad 2HCl.$$

Water.	Hydrochloric acid.

Reaction.—When fused with potassic hydrate, boric nitride yields tripotassic borate and ammonia :—

$$\mathbf{B}N''' \quad + \quad 3\mathbf{OKH} \quad = \quad \mathbf{B}Ko_3 \quad + \quad \mathbf{NH_3}.$$

Boric nitride.	Potassic hydrate.	Tripotassic borate.	Ammonia.

CHAPTER XI.

TETRAD ELEMENTS.

SECTION I.

CARBON, C.

Atomic weight $=12$. *Atomicity* $''$ *and* ^{iv}. *Evidence of atomicity* :—

Carbonic oxide .... $C''O$.
Carbonic tetrachloride ... $C^{iv}Cl_4$.
Marsh-gas .... $C^{iv}H_4$.
Chloroform .... $C^{iv}HCl_3$.

Occurrence.—In large quantities in nature, but chiefly in combination.

Manufacture.—By the carbonization of animal and vegetable matters.

COMPOUNDS OF CARBON WITH OXYGEN.

CARBONIC ANHYDRIDE.

CO_2.

Molecular weight $= 44$. *Molecular volume* ☐☐. 1 *litre weighs* 22 *criths. Fuses at* $-57°$. *Boils below its melting-point.*

Occurrence.—In the atmosphere, and dissolved in water.

Formation.—By the combustion of carbon and of carbonaceous substances in air or oxygen. In respiration, decay, putrefaction, and fermentation. During the formation of coal. Evolved from volcanoes.

Preparation.—1. By burning carbon in air or oxygen :—

$$C \ + \ O_2 \ = \ CO_2.$$
Carbonic
anhydride.

D 5

2. By the action of acids upon metallic carbonates :—

$$\mathbf{C}OKo_2 \;+\; \mathbf{S}O_2Ho_2 \;=\; \mathbf{C}O_2 \;+\; OH_2 \;+\; SO_2Ko_2.$$

Potassic   Sulphuric acid.   Carbonic   Water.   Potassic
carbonate.   anhydride.   sulphate.

$$\mathbf{C}OKoHo \;+\; \mathbf{N}O_2Ho \;=\; \mathbf{C}O_2 \;+\; OH_2 \;+\; NO_2Ko.$$

Hydric potassic   Nitric acid.   Carbonic   Water.   Potassic
carbonate.   anhydride.   nitrate.

$$\mathbf{C}OCao'' \;+\; 2HCl \;=\; \mathbf{C}O_2 \;+\; OH_2 \;+\; CaCl_2.$$

Calcic   Hydrochloric   Carbonic   Water.   Calcic
carbonate.   acid.   anhydride.   chloride.

Reactions.—1. Carbonic anhydride is decomposed by heated potassium :—

$$3\mathbf{C}O_2 \;+\; 2K_2 \;=\; 2\mathbf{C}OKo_2 \;+\; C.$$

Carbonic anhydride.   Potassic carbonate.

2. It acts upon metallic hydrates, forming carbonates :—

$$\mathbf{C}O_2 \;+\; 2KHo \;=\; \mathbf{C}OKo_2 \;+\; OH_2.$$

Carbonic anhydride.   Potassic hydrate. Potassic carbonate.   Water.

$$\mathbf{C}O_2 \;+\; CaHo_2 \;=\; \mathbf{C}OCao'' \;+\; OH_2.$$

Carbonic   Calcic   Calcic   Water.
anhydride.   hydrate.   carbonate.

Carbonic acid, $\mathbf{C}OHo_2$, is not known.

CARBONIC OXIDE,

$\mathbf{C}O$.

Molecular weight $=28$. *Molecular volume* ⬚. 1 *litre weighs* 14 *criths.*

Formation.—In the combustion of carbon or carbonaceous matter with a limited supply of air. In destructive distillation of many organic substances containing oxygen.

Preparation.—1. By passing carbonic anhydride over red-hot charcoal :—

$$\mathbf{C}O_2 \;+\; C \;=\; 2\mathbf{C}O.$$

Carbonic anhydride.   Carbonic oxide.

2. By passing carbonic anhydride over red-hot iron :—

$$4CO_2 + Fe_3 = {}^{iv}(Fe_3)^{viii}O_4 + 4CO.$$
Carbonic anhydride.　　　　Triferric tetroxide.　Carbonic oxide.

3. By heating iron or carbon with a carbonate :—

$$COCao'' + C = CaO + 2CO.$$
Calcic carbonate.　　　Lime.　Carbonic oxide.

4. By heating oxalic acid with sulphuric acid (by which water is removed from the former), and then separating the carbonic anhydride by washing with sodic hydrate :—

$$\begin{cases} COHo \\ COHo \end{cases} = OH_2 + CO + CO_2.$$
Oxalic acid.　　Water.　　Carbonic　　Carbonic
　　　　　　　　　　　　oxide.　　anhydride.

5. By heating formic acid or a formate with sulphuric acid :—

$$\begin{cases} H \\ COHo \end{cases} = OH_2 + CO.$$
Formic acid.　　　Water.　　Carbonic oxide.

6. By heating potassic ferrocyanide with sulphuric acid :—

$$Fe''C_6N_6K_4 + 6OH_2 + 6SO_2Ho_2 = 6CO$$
Potassic ferrocyanide.　　Water.　　Sulphuric acid.　　Carbonic oxide.

$$+ 2SO_2Ko_2 + SO_2Feo'' + 3SO_2(NH_4O)_2.$$
Potassic sulphate.　　Ferrous sulphate.　　Ammonic sulphate

Reactions.—1. It burns in air and oxygen, producing carbonic anhydride :—

$$CO + O = CO_2.$$
Carbonic　　　Carbonic
oxide.　　　anhydride.

2. Carbonic oxide and chlorine unite under the influence of light (p. 30), forming *carbonic oxydichloride* or *phosgene gas*, $COCl_2$.

The compounds of carbon with chlorine, nitrogen, and hydrogen will be studied in connexion with organic compounds.

CHAPTER XII.

PENTAD ELEMENTS.

SECTION I.

NITROGEN, *Azote,* N_2.

Atomic weight $=14$. *Molecular weight* $= 28$. *Molecular volume*
☐☐. 1 *litre weighs* 14 *criths.* *Atomicity* ^{v}, *which, by
the mutual saturation of pairs of bonds, becomes reduced to*
$'''$ *or to* $'$ (see p. 20). *Evidence of atomicity*:—

Nitrous oxide .... ON_2.
Ammonia .... $N'''H_3$
Ammonic chloride .... N^vH_4Cl.

Occurrence.—In the free state in the atmosphere. In some
nebulæ? In combination, in animal and vegetable bodies.

Preparation.—1. By burning phosphorus in air, whereby
the oxygen is removed from the latter.

2. By passing air over ignited copper, when the oxygen
unites with the copper.

3. By heating ammonic nitrite, or a mixture of ammonic
chloride with potassic or sodic nitrite :—

$$N'''O(N^vH_4O) \quad = \quad N_2 \quad + \quad 2OH_2.$$
Ammonic nitrite.   Water.

$$NH_4Cl \quad + \quad NONao \quad = \quad NaCl \quad + \quad N_2 \quad + \quad 2OH_2.$$
Ammonic chloride.   Sodic nitrite.   Sodic chloride.   Water.

4. By passing chlorine through an excess of solution of
ammonia :—

$$8NH_3 \quad + \quad 3Cl_2 \quad = \quad 6NH_4Cl \quad + \quad N_2.$$
Ammonia.   Ammonic chloride.

OXIDES AND OX-ACIDS OF NITROGEN.

Nitrous oxide............... **ON$_2$**.

Nitric oxide *........... $\begin{cases} \textbf{NO} \\ \textbf{NO} \cdot \end{cases}$

Nitrous anhydride...... $\begin{cases} \textbf{NO} \\ \textbf{O} \\ \textbf{NO} \end{cases}$.

Nitric peroxide .... $\begin{cases} \textbf{NO}_2. \\ \textbf{NO}_2 \cdot \end{cases}$

Nitric anhydride ...... $\begin{cases} \textbf{NO}_2 \\ \textbf{O} \\ \textbf{NO}_2 \end{cases}$.

Nitrous acid .... **NOHo**.

Nitric acid.................. **NO$_2$Ho**.

NITRIC ACID, *Aquafortis.*
NO$_2$Ho.

Molecular weight $=63$. *Molecular volume* ☐☐. 1 *litre of nitric acid vapour weighs* 31·5 *criths. Fuses at* —50°. *Boils at* 84°·5.

* This compound is anomalous; for its molecule, deduced from the specific gravity, is represented by NO. The dissociation which in the case of $\begin{cases} \textbf{NO}_2 \\ \textbf{NO}_2 \end{cases}$ is very imperfect at 0° C., but almost complete at 100° C., is probably nearly complete in the case of N_2O_2 at the lowest temperature to which this gas has hitherto been exposed.

Production.—1. By the slow oxidation of nitrogenized organic matter in the presence of powerful bases.

2. By the passage of electric sparks through moist air.

Manufacture.—By distilling potassic nitrate (nitre) or sodic nitrate (cubic nitre) with concentrated sulphuric acid:—

$$\mathbf{NO_2Ko} \;+\; \mathbf{SO_2Ho_2} \;=\; \mathbf{SO_2HoKo} \;+\; \mathbf{NO_2Ho}.$$

Potassic nitrate.   Sulphuric acid.   Hydric potassic sulphate.   Nitric acid.

By employing two molecules of potassic nitrate and one of sulphuric acid a saving of sulphuric acid is effected, but a higher temperature is required, which destroys some of the nitric acid. The reaction takes place in two stages:—

$$1. \; 2\mathbf{NO_2Ko} \;+\; \mathbf{SO_2Ho_2} \;=\; \mathbf{SO_2HoKo}$$

Potassic nitrate.   Sulphuric acid.   Hydric potassic sulphate.

$$+ \; \mathbf{NO_2Ko} \;+\; \mathbf{NO_2Ho} :$$

Potassic nitrate.   Nitric acid.

$$2. \; \mathbf{SO_2HoKo} \;+\; \mathbf{NO_2Ko} \;=\; \mathbf{SO_2Ko_2} \;+\; \mathbf{NO_2Ho}.$$

Hydric potassic sulphate.   Potassic nitrate.   Potassic sulphate.   Nitric acid.

Decompositions.—1. The decomposition which the nitric acid undergoes by heat is expressed in the following equation:—

$$4\mathbf{NO_2Ho} \;=\; 2\mathbf{OH_2} \;+\; 2'\mathbf{N_2^{iv}O_4} \;+\; \mathbf{O_2}.$$

Nitric acid.   Water.   Nitric peroxide.   Oxygen.

2. By the action of metallic oxides or hydrates, nitric acid produces nitrates:—

$$\mathbf{OKH} \;+\; \mathbf{NO_2Ho} \;=\; \mathbf{NO_2Ko} \;+\; \mathbf{OH_2}.$$

Potassic hydrate.   Nitric acid.   Potassic nitrate.   Water.

$$\mathbf{PbO} \;+\; 2\mathbf{NO_2Ho} \;=\; \left\{ \begin{array}{l} \mathbf{NO_2} \\ \mathrm{Pbo}'' \\ \mathbf{NO_2} \end{array} \right. \;+\; \mathbf{OH_2}.$$

Plumbic oxide.   Nitric acid.   Plumbic nitrate.   Water.

NITRIC ANHYDRIDE.
N_2O_5.

Probable molecular weight $=108$. *Probable molecular volume*
☐☐. *Fuses at* $29°\cdot5$. *Boils at* $45°$.

Preparation.—By passing dry chlorine over argentic nitrate:—

$$4NO_2Ago \quad + \quad 2Cl_2 \quad = \quad 4AgCl \quad + \quad 2N_2O_5 \quad + \quad O_2.$$
Argentic nitrate.　　　　　　　　　Argentic　　Nitric
　　　　　　　　　　　　　　　　chloride.　anhydride.

Reaction.—By the action of water it forms nitric acid:—

$$N_2O_5 \quad + \quad OH_2 \quad = \quad 2NO_2Ho.$$
Nitric anhydride.　　　Water.　　　　Nitric acid.

NITROUS ANHYDRIDE.
N_2O_3.

Probable molecular weight $=76$. *Probable molecular volume*
☐☐.

Preparation.—1. By heating together nitric acid and starch.
2. By gently heating nitric acid with arsenious anhydride:—

$$As_2O_3 \quad + \quad 2NO_2Ho \quad = \quad As_2O_5 \quad + \quad N_2O_3 \quad + \quad OH_2.$$
Arsenious　　　　Nitric acid.　　　Arsenic　　　Nitrous　　Water.
anhydride.　　　　　　　　　anhydride.　anhydride.

3. By the action of nitric acid on silver:—

$$6NO_2Ho \quad + \quad 2Ag_2 \quad = \quad 4NO_2Ago \quad + \quad N_2O_3 \quad + \quad 3OH_2.$$
Nitric acid.　　　　　　　Argentic nitrate.　Nitrous anhydride.　Water.

NITROUS ACID.
NOHo.

Molecular weight $=47$.

Preparation.—By mixing liquefied nitrous anhydride with a small quantity of water:—

$$N_2O_3 \quad + \quad OH_2 \quad = \quad 2NOHo.$$
Nitrous anhydride.　　Water.　　　Nitrous acid.

Decompositions.—1. In the presence of much water, nitric acid and nitric oxide are formed :—

$$6\mathbf{N}OHo \quad = \quad 2\mathbf{N}O_2Ho \quad + \quad 2'\mathbf{N}''_2O_2 \quad + \quad 2\mathbf{O}H_2.$$

Nitrous acid.   Nitric acid.   Nitric oxide.   Water.

2. Nitrous acid acts as a reducing agent under some circumstances :—

$$2\mathbf{N}OHo \quad + \quad O_2 \quad = \quad 2\mathbf{N}O_2Ho \; ;$$

Nitrous acid.   Nitric acid.

and as an oxidizing agent under others :—

$$4\mathbf{N}OHo \quad = \quad 2'\mathbf{N}''_2O_2 \quad + \quad 2\mathbf{O}H_2 \quad + \quad O_2.$$

Nitrous acid.   Nitric oxide.   Water.   Oxygen.

3. By the action of metallic oxides or hydrates, nitrous acid forms nitrites :—

$$\mathbf{O}KH \quad + \quad \mathbf{N}OHo \quad = \quad \mathbf{N}OKo \quad + \quad \mathbf{O}H_2.$$

Potassic hydrate.   Nitrous acid.   Potassic nitrite.   Water.

NITROUS OXIDE, *Laughing Gas.*
$\mathbf{O}N_2.$

Molecular weight $= 44.$ *Molecular volume* ▭.   1 *litre weighs* 22 *criths. Fuses at* $-101°.$ *Boils at* $-88°.$

Preparation.—1. By the action of dilute nitric acid on zinc :—

$$10\mathbf{N}O_2Ho \quad + \quad Zn_4 \quad = \quad \mathbf{O}N_2 \quad + \quad 4\begin{cases} \mathbf{N}O_2 \\ Zno'' \\ \mathbf{N}O_2 \end{cases} \quad +5\mathbf{O}H_2.$$

Nitric acid.   Nitrous oxide.   Zincic nitrate.   Water.

2. By heating ammonic nitrate :—

$$\mathbf{N}O_2(N'H_4O) \quad = \quad 2\mathbf{O}H_2 \quad + \quad \mathbf{O}N_2.$$

Ammonic nitrate.   Water.   Nitrous oxide.

NITRIC OXIDE.

$$\begin{cases} \mathbf{NO} \\ \mathbf{NO} \end{cases}, \text{ or } \mathbf{'N''}_2 O_2.$$

Molecular weight $=60$. *Molecular volume anomalous* ⊞.
  1 *litre weighs* 15 *criths*.

Preparation.—By the action of nitric acid upon mercury or copper :—

$$3Cu \ + \ 8NO_2Ho \ = \ 3\begin{cases} \mathbf{NO_2} \\ \mathbf{Cu}o'' \\ \mathbf{NO_2} \end{cases} + \ \mathbf{'N''}_2O_2 \ + \ 4OH_2.$$

Nitric acid.   Cupric nitrate.   Nitric oxide.   Water.

Reaction.—Unites directly with oxygen :—

$$2\mathbf{'N''}_2O_2 \ + \ O_2 \ = \ 2\mathbf{N'''}_2O_3:$$

Nitric oxide.   Nitrous anhydride.

$$\mathbf{'N''}_2O_2 \ + \ O_2 \ = \ \mathbf{'N''}_2O_4.$$

Nitric oxide.   Nitric peroxide.

NITRIC PEROXIDE.

$$\begin{cases} \mathbf{NO_2} \\ \mathbf{NO_2} \end{cases}, \text{ or } \mathbf{'N''}_2O_4.$$

Molecular weight $=46$ *to* 92. *Molecular volume* ▭ *to* ⊞.
  1 *litre weighs* 23 *to* 46 *criths*.

Preparation.—1. By the union of nitric oxide with oxygen (see above).

  2. By the action of nitric acid upon tin :—

$$Sn_5 \ + \ 20NO_2Ho \ = \ \mathbf{Sn_5}O_5Ho_{10} \ + \ 5OH_2 \ + \ 10\mathbf{'N''}_2O_4.$$

Nitric acid.   Metastannic acid.   Water.   Nitric peroxide.

Decomposition.—By the action of metallic hydrates and oxides it produces nitrites and nitrates :—

$$\mathbf{'N''}_2O_4 \ + \ 2OKH \ = \ \mathbf{NO_2}Ko \ + \ \mathbf{NO}Ko \ + \ OH_2.$$

Nitric peroxide.   Potassic hydrate.   Potassic nitrate.   Potassic nitrite.   Water.

COMPOUNDS CONTAINING NITROGEN, CHLORINE, AND OXYGEN.

NITROUS OXYCHLORIDE, *Chloronitrous Gas.*
NOCl.

Molecular weight $=65\cdot5$. *Molecular volume* $\square$. 1 *litre weighs* $32\cdot75$ *criths. Boils at* $0°$ C.

A mixture of nitric and hydrochloric acids possesses the property of dissolving gold, and is therefore called aqua regia; when heated it evolves chlorine and nitrous oxychloride :—

$$\underset{\text{Nitric acid.}}{NO_2Ho} + \underset{\substack{\text{Hydrochloric}\\\text{acid.}}}{3HCl} = \underset{\substack{\text{Nitrous}\\\text{oxychloride.}}}{NOCl} + \underset{\text{Water.}}{2OH_2} + Cl_2.$$

NITRIC DIOXY-TETRACHLORIDE, *Chloronitric Gas.*
$'N_2^{iv}O_2Cl_4$.

Prepared, together with nitrous oxychloride, by heating a mixture of nitric and hydrochloric acids :—

$$\underset{\text{Nitric acid.}}{2NO_2Ho} + \underset{\text{Hydrochloric acid.}}{6HCl} = \underset{\text{Chloronitric gas.}}{'N_2^{iv}O_2Cl_4} + \underset{\text{Water.}}{4OH_2} + Cl_2.$$

NITRIC DIOXYCHLORIDE, *Chloropernitric Gas.*
NO_2Cl.

Preparation.—By mixing phosphoric oxytrichloride and plumbic nitrate :—

$$3\underset{\text{Plumbic nitrate.}}{\left\{\begin{array}{l}NO_2\\Pbo''\\NO_2\end{array}\right.} + \underset{\substack{\text{Phosphoric}\\\text{oxytrichloride.}}}{2PCl_3O} = \underset{\substack{\text{Triplumbic}\\\text{diphosphate.}}}{P_2O_2Pbo''_3} + \underset{\substack{\text{Nitric}\\\text{dioxychloride.}}}{6NO_2Cl}.$$

COMPOUNDS OF NITROGEN WITH HYDROGEN.
AMMONIA.
NH_3.

Molecular weight $=17$. *Molecular volume* ⊏⊐. 1 *litre weighs* 8·5 *criths. Fuses at* $-75°$. *Boils at* $-38°·5$.

Occurrence.—In the atmosphere in very minute quantities.

Formation.—By the decay of animal and vegetable matters containing nitrogen.

Manufacture.—By the destructive distillation of animal matter, as horn or bones, and of vegetable matter, as coal.

Preparation.—By heating a mixture of lime and ammonic chloride (sal-ammoniac) :—

$$2NH_4Cl \quad + \quad CaO \quad = \quad CaCl_2 \quad + \quad 2NH_3 \quad + \quad OH_2.$$

Ammonic chloride. Lime. Calcic chloride. Ammonia. Water.

Reactions.—1. Decomposed by chlorine (see p. 60).

2. Unites directly with acids, forming the ammonium salts in which the atomicity of nitrogen is ^{v} :—

$$\mathbf{N}'''H_3 \quad + \quad HCl \quad = \quad \mathbf{N}^vH_4Cl.$$

Hydrochloric acid. Ammonic chloride *.

$$\mathbf{N}'''H_3 \quad + \quad \mathbf{N}^vO_2Ho \quad = \quad \mathbf{N}^vO_2(\mathbf{N}^vH_4O).$$

Nitric acid. Ammonic nitrate †.

$$2\mathbf{N}'''H_3 \quad + \quad SO_2Ho_2 \quad = \quad SO_2(\mathbf{N}^vH_4O)_2.$$

Sulphuric acid. Ammonic sulphate ‡.

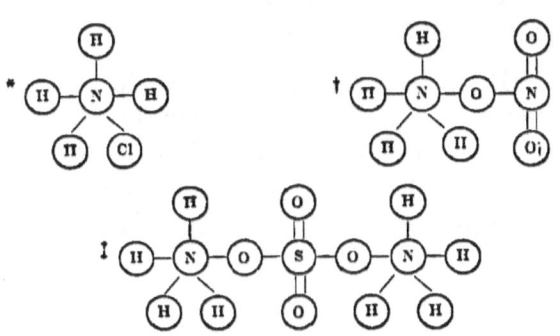

AMMONIUM.

$$\begin{cases} \mathbf{NH_4}, \\ \mathbf{NH_4^{..}} \end{cases}$$

This monad radical has never been obtained in the free state, but its compounds are perfectly analogous, in crystalline form and other properties, to those of potassium. These facts have induced some chemists to consider the group NH_4 as a metal, to which they have given the name ammonium—an hypothesis which is considered to receive support from the production of an unstable amalgam of this radical. All the compounds of mercury with metals are found to possess metallic lustre; and this is also the case with the amalgam of ammonium. It may be prepared by two different processes.

1. If a solution of ammonic chloride be electrolyzed, the negative electrode being mercury and the positive a platinum plate, the mercury is observed to swell up, owing to the formation of a spongy metallic mass.

2. By preparing an amalgam of potassium or sodium, and pouring it into a slightly warmed solution of ammonic chloride, the amalgam is found to swell enormously, potassic or sodic chloride being simultaneously formed :—

$$\underset{\text{Sodic amalgam.}}{Hg_n Na_m} + \underset{\text{Ammonic chloride.}}{mNH_4Cl} = \underset{\text{Ammonic amalgam.}}{Hg_n(N^vH_4)_m} + \underset{\text{Sodic chloride.}}{mNaCl.}$$

Ammonic amalgam rapidly decomposes into mercury, ammonia, and hydrogen, the ammonia and hydrogen being liberated in the proportions of $2NH_3$ to H_2 :—

$$\underset{\text{Ammonic amalgam.}}{2Hg_n(N^vH_4)_m} = \underset{\text{Mercury.}}{2nHg} + \underset{\text{Ammonia.}}{2mNH_3} + mH_2.$$

Ammonium plays the part of a compound monad radical, and its salts are isomorphous with those of potassium; they are all volatile, unless the acid from which they are derived be fixed.

COMPOUND OF NITROGEN WITH CHLORINE.

NITROUS CHLORIDE.

NCl_3 ?

Preparation. — By the action of chlorine upon ammonic chloride :—

$$N^vH_4Cl \ + \ 3Cl_2 \ = \ N'''Cl_3 \ + \ 4HCl.$$

Ammonic chloride.   Nitrous chloride.   Hydrochloric acid.

The formula of this compound is not fixed with certainty ; it may contain hydrogen, and it is possible that the two compounds intermediate between ammonia and nitrous chloride may exist:—

$$NH_3, \quad NH_2Cl, \quad NHCl_2, \quad NCl_3.$$

COMPOUND OF NITROGEN WITH IODINE AND HYDROGEN.

NITROUS HYDRODINIODIDE.

Preparation.—By the action of ammonia on iodine a brown substance is obtained, which has the composition NHI_2. It is formed according to the following equation :—

$$3NH_3 \ + \ 2I_2 \ = \ NHI_2 \ + \ 2NH_4I.$$

Ammonia.   Nitrous   Ammonic
     hydrodiniodide.   iodide.

CHAPTER XIII.

HEXAD ELEMENTS.

SECTION I.

SULPHUR, S_2.

Atomic weight =32. *Molecular weight* =64. *Molecular volume* ☐ *at* 1000° C., *but only one-third of this at its boiling-*

point. 1 *litre of sulphur vapour weighs* 32 *criths. Atomicity"* ^{iv} *and* ^{vi}. *Evidence of atomicity* :—

Hydrosulphuric acid	$S''H_2$.
Triethylsulphine iodide	$S^{iv}Et_3I$.
Sulphuric dioxydichloride	$S^{vi}O_2Cl_2$.
Sodic nitrosulphate	$S^{vi}O(NO)_2Nao_2$.

Occurrence.—Found in the free state in volcanic districts, and widely diffused in combination with metals and oxygen, as sulphides and sulphates.

Manufactured from native sulphur, and from

Iron pyrites	FeS''_2.
Copper pyrites....................	$(FeCu)S_2''$.

Character.—Sulphur is capable of existing in several allotropic forms, of which the following are the most important :—

Condition.	Specific gravity.	Behaviour with bisulphide of carbon.
α. Octohedral	2·05	Soluble.
β. Prismatic	1·98	Transformed into α.
γ. Plastic	1·95	Insoluble.
δ. Powder	1·95	Insoluble.

When united exclusively with basylous elements or radicals, sulphur is almost invariably a dyad; and it is then the analogue of oxygen, as will be seen from the following formulæ :—

Oxygen compounds ... OK_2, OKH, CO_2, $COKo_2$.
Sulphur compounds ... SK_2, SKH, CS_2'', $CSKs_2$.

COMPOUNDS OF SULPHUR WITH BASYLOUS OR POSITIVE ELEMENTS.

Sulphuretted hydrogen	SH_2.
Hydrosulphyl	$'S'_2H_2$, or Hs_2.
Carbonic disulphide	CS_2''.

SULPHURETTED HYDROGEN, *Hydrosulphuric Acid,* *Sulphhydric Acid.*

SH_2. (H)—(S)—(H)

Molecular weight $=34$. *Molecular volume* ☐☐. 1 *litre weighs* 17 *criths. Solid at* $-85°{\cdot}5$ C. *Liquid under a pressure of* 17 *atmospheres at* $10°$ C.

Occurrence.—Evolved with other gases from volcanoes and fumaroles. Found also in hepatic mineral waters, and frequently in waters which contain both organic matters and sulphates.

Preparation.—1. By direct union of its elements :—

$$H_2 + S = SH_2.$$

2. By the action of hydrochloric or dilute sulphuric acid on ferrous sulphide :—

$$FeS'' + 2HCl = SH_2 + FeCl_2.$$

| Ferrous sulphide. | Hydrochloric acid. | Sulphuretted hydrogen. | Ferrous chloride. |

$$FeS'' + SO_2Ho_2 = SH_2 + SO_2Feo''.$$

| Ferrous sulphide. | Sulphuric acid. | Sulphuretted hydrogen. | Ferrous sulphate. |

3. By the action of hydrochloric acid on antimonious sulphide with the aid of a gentle heat :—

$$Sb_2S_3'' + 6HCl = 3SH_2 + 2SbCl_3.$$

| Antimonious sulphide. | Hydrochloric acid. | Sulphuretted hydrogen. | Antimonious chloride. |

Reactions.—1. It is immediately decomposed by chlorine, thus :—

$$SH_2 + Cl_2 = 2HCl + S.$$

2. It is also rapidly decomposed by many metallic compounds rich in oxygen, such as ferric oxide :—

$$'Fe_2'''O_3 + 3SH_2 = 2FeS'' + S + 3OH_2.$$

| Ferric oxide. | Sulphuretted hydrogen. | Ferrous sulphide. | | Water. |

3. The sulphhydrates and sulphides of the metals are produced

by the action of hydrosulphuric acid on the hydrates and oxides, thus :—

$$OKH \quad + \quad SH_2 \quad = \quad SKH \quad + \quad OH_2.$$

Potassic hydrate.   Sulphuretted hydrogen.   Potassic sulphhydrate.   Water.

$$BaHo_2 \quad + \quad 2SH_2 \quad = \quad BaHs_2 \quad + \quad 2OH_2.$$

Baric hydrate.   Sulphuretted hydrogen.   Baric sulphhydrate.   Water.

$$OAg_2 \quad + \quad SH_2 \quad = \quad SAg_2 \quad + \quad OH_2.$$

Argentic oxide.   Sulphuretted hydrogen.   Argentic sulphide.   Water.

$$CuO \quad + \quad SH_2 \quad = \quad CuS'' \quad + \quad OH_2.$$

Cupric oxide.   Sulphuretted hydrogen.   Cupric sulphide.   Water.

HYDROSULPHYL, *Hydric Persulphide.*
$'S_2'H_2$, or Hs_2.

(H)—(S)—(S)—(H)

Probable molecular weight $=66$.   *Sp. gr.* 1·769.

Preparation.—By pouring a solution of calcic disulphide into hydrochloric acid :—

$$'S_2'Ca'' \quad + \quad 2HCl \quad = \quad 'S_2'H_2 \quad + \quad CaCl_2.$$

Calcic disulphide.   Hydrochloric acid.   Hydrosulphyl.   Calcic chloride.

Character.—It is the analogue of hydroxyl in composition and functions.

CARBONIC DISULPHIDE, *Bisulphide of Carbon.*
CS_2.

Molecular weight $=76$.   *Molecular volume* ☐.   1 *litre of carbonic disulphide vapour weighs* 38 *criths. Specific gravity* 1·26.   *Boils at* 45°.

Preparation.—1. By passing sulphur over strongly ignited charcoal :—

$$C \ + \ S_2 \ = \ CS''_2.$$

Carbonic
disulphide.

2. By heating together charcoal and iron- or copper-pyrites :—

$$C \ + \ 2FeS''_2 \ = \ CS''_2 \ + \ 2FeS''.$$

Iron pyrites.   Carbonic   Ferrous
Ferric disulphide.   disulphide.   sulphide.

Decompositions.—1. Heated potassium burns in the vapour of carbonic disulphide, with formation of potassic sulphide and liberation of carbon :—

$$CS''_2 \ + \ 2K_2 \ = \ 2SK_2 \ + \ C.$$

Carbonic disulphide.   Potassic sulphide.

2. When brought into contact with a solution of an alkaline hydrate, carbonic disulphide is decomposed, a carbonate and a sulpho-carbonate being formed :—

$$6KH \ + \ 3CS''_2 \ = \ 2CS''Ks_2 \ + \ COKo_2 \ + \ 3OH_2.$$

Potassic   Carbonic   Potassic   Potassic   Water.
hydrate.   disulphide.   sulpho-carbonate. carbonate.

3. In contact with solutions of alkaline sulphides, carbonic disulphide also forms alkaline sulpho-carbonates :—

$$SK_2 \ + \ CS''_2 \ = \ CS''Ks_2.$$

Potassic sulphide.   Carbonic disulphide.   Potassic sulpho-carbonate.

SULPHO-CARBONIC ACID.
$CS''Hs_2.$

Preparation.—By the action of hydrochloric acid on ammonic sulpho-carbonate:—

$$CS''(NH_4S)_2 \ + \ 2HCl \ = \ CS''Hs_2 \ + \ 2NH_4Cl.$$

Ammonic   Hydrochloric   Sulpho-carbonic   Ammonic
sulpho-carbonate.   acid.   acid.   chloride.

E

COMPOUNDS OF SULPHUR WITH OXYGEN AND HYDROXYL.

In these compounds the sulphur is either a dyad, a tetrad, or a hexad.

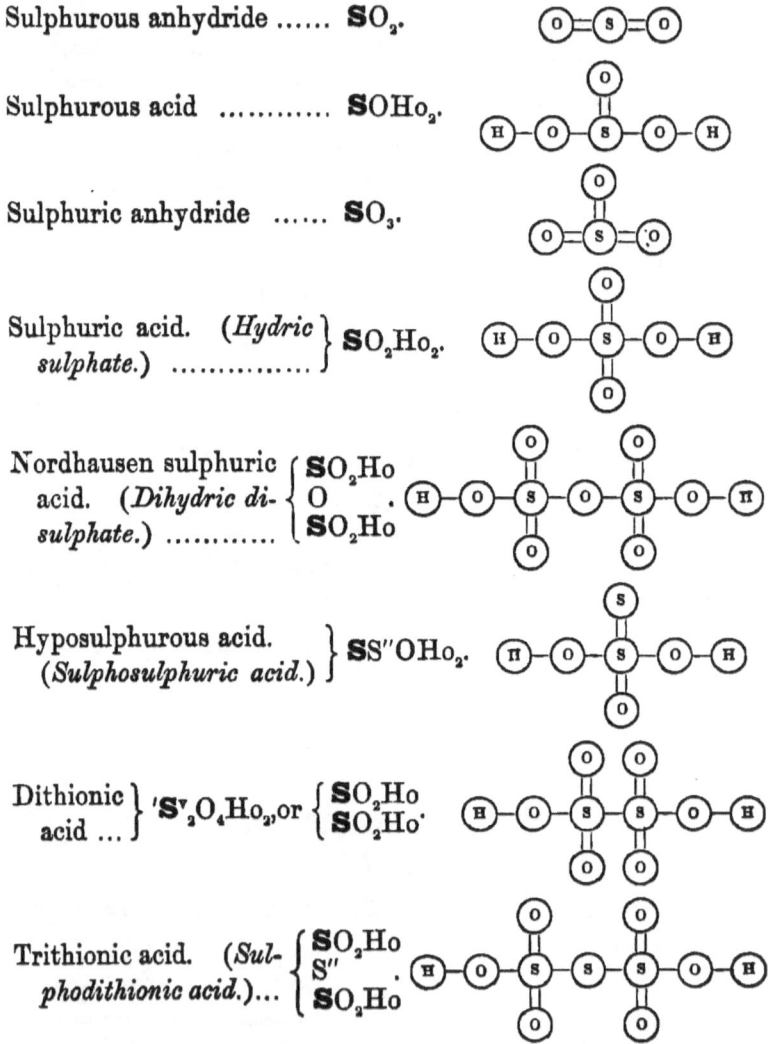

Sulphurous anhydride ...... **SO₂**.

Sulphurous acid .... **SOHo₂**.

Sulphuric anhydride ...... **SO₃**.

Sulphuric acid. *(Hydric sulphate.)* } **SO₂Ho₂**.

Nordhausen sulphuric acid. *(Dihydric di-sulphate.)* { **SO₂Ho** **O** **SO₂Ho**.

Hyposulphurous acid. *(Sulphosulphuric acid.)* } **SS″OHo₂**.

Dithionic acid ... } **ʹSʹ₂O₄Ho₂**, or { **SO₂Ho** **SO₂Ho**.

Trithionic acid. *(Sulphodithionic acid.)*... { **SO₂Ho** **S″** **SO₂Ho**.

Tetrathionic acid. (*Di-sulphodithionic acid.*)...
$$\begin{cases} \mathbf{SO_2Ho} \\ S'' \\ S'' \\ \mathbf{SO_2Ho} \end{cases}$$

.

Pentathionic acid. (*Tri-sulphodithionic acid.*) ......
$$\begin{cases} \mathbf{SO_2Ho} \\ S'' \\ S'' \\ S'' \\ \mathbf{SO_2Ho} \end{cases}$$

.

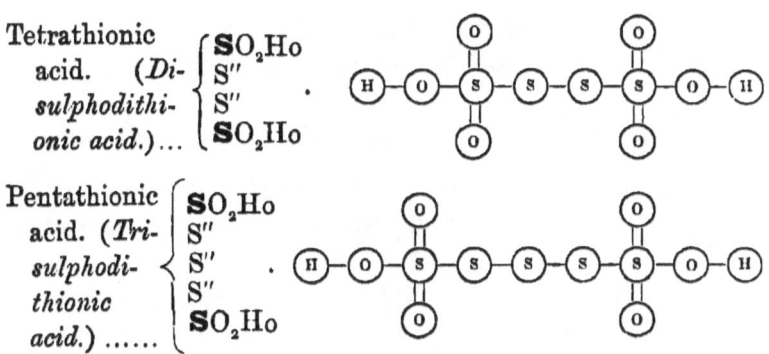

SULPHUROUS ANHYDRIDE.

SO_2.

Molecular weight $=64$. *Molecular volume* □. *1 litre weighs 32 criths. Solid at* $-76°$. *Liquid under the pressure of two atmospheres at* $7°$ C.

Occurrence.—1. As a volcanic product.

2. In the air of towns.

3. Evolved in the roasting of copper pyrites and other sulphureous ores.

Preparation.—1. By the combustion of sulphur in air or in oxygen :—

$$S \quad + \quad O_2 \quad = \quad SO_2.$$

2. By heating sulphuric acid with copper or mercury :—

$$2SO_2Ho_2 \quad + \quad Cu \quad = \quad SO_2 \quad + \quad SO_2Cuo'' \quad + \quad 2OH_2.$$
Sulphuric acid.　　　　　　　Sulphurous　Cupric sulphate.　　Water.
　　　　　　　　　　　　　anhydride.

$$2SO_2Ho_2 \quad + \quad Hg \quad = \quad SO_2 \quad + \quad SO_2Hgo'' \quad + \quad 2OH_2.$$
Sulphuric acid.　　　　　　　Sulphurous　Mercuric sulphate.　　Water.
　　　　　　　　　　　　　anhydride.

3. By heating charcoal with sulphuric acid :—

$$2SO_2Ho_2 \quad + \quad C \quad = \quad 2SO_2 \quad + \quad CO_2 \quad + \quad 2OH_2.$$
Sulphuric acid.　　　　　Sulphurous　　Carbonic　　　Water.
　　　　　　　　　　anhydride.　　anhydride.

E 2

4. By heating a mixture of about three parts by weight of sulphur (two atoms) with four of manganic oxide (one atom) :—

$$S_2 \quad + \quad MnO_2 \quad = \quad SO_2 \quad + \quad MnS''.$$
Manganic oxide.   Sulphurous anhydride.   Manganous sulphide.

Reactions.—1. Dissolved by water, producing an acid liquid which, when cooled to 0°, deposits white cubical crystals of sulphurous acid :—

$$SO_2 \quad + \quad OH_2 \quad = \quad SOHo_2.$$
Sulphurous anhydride.   Water.   Sulphurous acid.

2. Sulphurous anhydride, when passed into solutions of the metallic hydrates, produces sulphites. If the sulphurous anhydride be in excess, an acid sulphite is obtained :—

$$OKH \quad + \quad SO_2 \quad = \quad SOHoKo.$$
Potassic hydrate.   Sulphurous anhydride.   Hydric potassic sulphite.

3. If the metallic hydrate be in excess, the normal sulphite is formed, thus :—

$$2OKH \quad + \quad SO_2 \quad = \quad SOKo_2 \quad + \quad OH_2.$$
Potassic hydrate.   Sulphurous anhydride.   Normal potassic sulphite.   Water.

4. Sulphurous acid, when acted upon by metallic hydrates, produces the same salts :—

$$OKH \quad + \quad SOHo_2 \quad = \quad SOHoKo \quad + \quad OH_2:$$
$$2OKH \quad + \quad SOHo_2 \quad = \quad SOKo_2 \quad + \quad 2OH_2.$$

5. Sulphurous anhydride, when passed over metallic peroxides, produces sulphates :—

$$PbO_2 \quad + \quad SO_2 \quad = \quad SO_2Pbo''.$$
Perplumbic oxide.   Sulphurous anhydride.   Plumbic sulphate.

Detection.—Sulphites are recognized by the pungent odour of sulphurous anhydride which they evolve on the addition of a strong acid, such as sulphuric acid :—

$$SOKo_2 \quad + \quad SO_2Ho_2 \quad = \quad SO_2Ko_2 \quad + \quad SO_2 \quad + \quad OH_2.$$
Potassic sulphite.   Sulphuric acid.   Potassic sulphate.   Sulphurous anhydride.   Water.

When solutions of sulphites are mixed with solutions of argentic nitrate, a white precipitate of argentic sulphite is formed :—

$$\mathbf{SO}Ko_2 \;+\; \mathbf{2NO_2}Ago \;=\; \mathbf{SO}Ago_2 \;+\; \mathbf{2NO_2}Ko.$$

Potassic sulphite.　Argentic nitrate.　Argentic sulphite.　Potassic nitrate.

When this argentic sulphite is boiled with water, it becomes black, owing to the separation of metallic silver :—

$$\mathbf{SO}Ago_2 \;+\; \mathbf{OH_2} \;=\; \mathbf{SO_2}Ho_2 \;+\; Ag_2.$$

Argentic sulphite.　Water.　Sulphuric acid.

SULPHURIC ANHYDRIDE.

SO_3.

Molecular weight = 80. *Molecular volume* ☐☐. 1 *litre of sulphuric anhydride vapour weighs* 40 *criths. Fuses at* 24°·5. *Boils at* 52°·6.

Preparation.—1. By passing a mixture of sulphurous anhydride and oxygen over ignited spongy platinum :—

$$\mathbf{SO_2} \;+\; O \;=\; \mathbf{SO_3}.$$

Sulphurous anhydride.　Sulphuric anhydride.

2. By heating Nordhausen sulphuric acid :—

$$\left\{ \begin{array}{l} \mathbf{SO_2}Ho \\ O \\ \mathbf{SO_2}Ho \end{array} \right. \;=\; \mathbf{SO_2}Ho_2 \;+\; \mathbf{SO_3}.$$

Nordhausen sulphuric acid.　Sulphuric acid.　Sulphuric anhydride.

3. By heating the so-called anhydrous sodic bisulphate (disodic disulphate) :—

$$\left\{ \begin{array}{l} \mathbf{SO_2}Nao \\ O \\ \mathbf{SO_2}Nao \end{array} \right. \;=\; \mathbf{SO_2}Nao_2 \;+\; \mathbf{SO_3}.$$

Anhydrous sodic bisulphate (Disodic disulphate).　Sodic sulphate.　Sulphuric anhydride.

4. By heating sulphuric acid with phosphoric anhydride :—

$$SO_2Ho_2 \ + \ P_2O_5 \ = \ SO_3 \ + \ 2PO_2Ho.$$
Sulphuric acid.   Phosphoric anhydride.   Sulphuric anhydride.   'Meta-phosphoric acid.

SULPHURIC ACID.
SO_2Ho_2.

Molecular weight = 98.   *Molecular volume* ⊞.   *Dissociation.*
1 *litre of sulphuric acid vapour weighs* 24·5 *criths.   Sp.*
gr. 1·85.   *Boils at* 325°.

Preparation.—1. By the action of hydroxyl upon sulphurous anhydride :—

$$SO_2 \ + \ Ho_2 \ = \ SO_2Ho_2.$$
Sulphurous anhydride.   Hydroxyl.   Sulphuric acid.

2. By the exposure of sulphurous acid or metallic sulphites to air or oxygen :—

$$SOHo_2 \ + \ O \ = \ SO_2Ho_2.$$
Sulphurous acid.   Sulphuric acid.

$$SONao_2 \ + \ O \ = \ SO_2Nao_2.$$
Sodic sulphite.   Sodic sulphate.

3. By the addition of water to sulphuric anhydride :—

$$SO_3 \ + \ OH_2 \ = \ SO_2Ho_2.$$
Sulphuric anhydride.   Water.   Sulphuric acid.

4. By the action of nitric peroxide and oxygen on sulphurous anhydride and subsequent decomposition by water of the white crystalline compound thus produced (Brüning and De la Provostaye) :—

$$2SO_2 \ + \ 'N^{iv}{}_2O_4 \ + \ O = \begin{cases} SO_2(N'O_2) \\ O \\ SO_2(N'O_2) \end{cases}.$$
Sulphurous anhydride   Nitric peroxide.   White crystalline compound.

$$\left\{\begin{array}{l} \mathbf{SO_2(N'O_2)} \\ \mathbf{O} \\ \mathbf{SO_2(N'O_2)} \end{array}\right. + \mathbf{2OH_2} = \mathbf{2SO_2Ho_2} + \mathbf{N_2O_3}.$$

White crystalline compound *.   Water.   Sulphuric acid.   Nitrous anhydride.

In the manufacture of sulphuric acid on the large scale, the nitrous anhydride is again acted on by water and transformed into nitric acid and nitric oxide :—

$$\mathbf{3N_2O_3} + \mathbf{OH_2} = \mathbf{2NO_2Ho} + \mathbf{2'N''_2O_2}.$$

Nitrous anhydride.   Water.   Nitric acid.   Nitric oxide.

The nitric oxide by the action of oxygen reproduces nitric peroxide, which is then ready to undergo the same processes a second time. The nitric acid is at the same time reduced to nitric peroxide by the action of sulphurous anhydride :—

$$\mathbf{SO_2} + \mathbf{2NO_2Ho} = \mathbf{SO_2Ho_2} + \mathbf{'N^{iv}_2O_4}.$$

Sulphurous anhydride.   Nitric acid.   Sulphuric acid.   Nitric peroxide.

The crude sulphuric acid may be freed from traces of nitrous anhydride (which it always contains) by the addition of some ammonic sulphate :—

$$\mathbf{SO_2(NH_4O)_2} + \mathbf{N_2O_3} = \mathbf{SO_2Ho_2} + \mathbf{3OH_2} + \mathbf{N_4}.$$

Ammonic sulphate.   Nitrous anhydride.   Sulphuric acid.   Water.

Character.—Sulphuric acid forms several classes of salts :—

Hydric potassic sulphate ....} $\mathbf{SO_2KoHo}$.

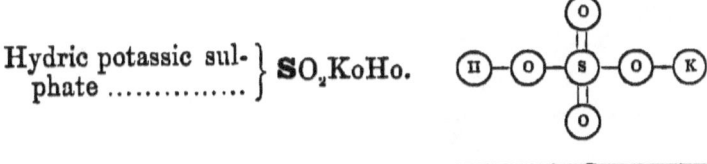

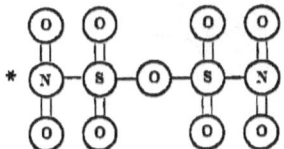

Potassic sulphate ... $SO_2Ko_2.$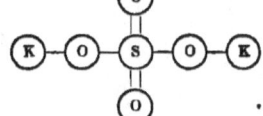

Anhydrous sodic bi- ⎧ SO_2Nao
sulphate. (*Disodic* ⎨ O .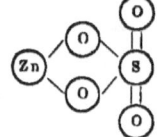
disulphate.) ...... ⎩ SO_2Nao

Zincic sulphate ...... $SO_2Zno''.$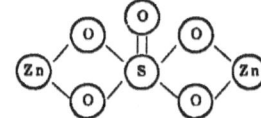

Tetrabasic zincic sul- ⎫
phate.   (*Dizincic* ⎬ $SOZno''_2.$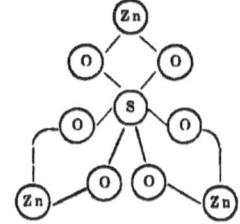
sulphate.) .... ⎭

Hexabasic zincic sul- ⎫
phate. (*Trizincic* ⎬ $SZno''_3.$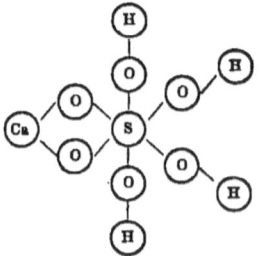
sulphate.) .... ⎭

Crystallized   gyp- ⎫
sum. (*Tetrahydric* ⎬ $SHo_4Cao''.$
calcic sulphate.)... ⎭

Gypsum dried at 100°. (*Dihydric calcic sulphate.*) ... $\rbrace$ $SOHo_2Cao''$.

Gypsum dried at 260°. (*Calcic sulphate.*) .... $\rbrace$ SO_2Cao''.

HYPOSULPHUROUS ACID, *Sulphosulphuric Acid.*

$$SS''OHo_2 \text{ (hypothetical).}$$

Preparation of Hyposulphites (Sulphosulphates).—1. By boiling a solution of sodic sulphite with sulphur:—

$$SONao_2 \;+\; S \;=\; SS''ONao_2.$$
Sodic sulphite.   Sodic hyposulphite
(Sodic sulphosulphate).

2. By exposure of an alkaline persulphide to the air:—

$$'S'_2Ca'' \;+\; O_3 \;=\; SS''OCao''.$$
Calcic   Calcic
persulphide.   hyposulphite.

Reaction.—The hyposulphites, when acted upon by acids, evolve sulphurous anhydride, whilst sulphur is precipitated:—

$$SS''ONao_2 \;+\; 2HCl \;=\; 2NaCl \;+\; OH_2 \;+\; S \;+\; SO_2.$$
Sodic hyposulphite   Hydrochloric   Sodic   Water.   Sulphurous
(Sodic sulpho-   acid.   chloride.   anhydride.
sulphate).

E 5

DITHIONIC ACID, *Hyposulphuric Acid.*

$$'S^v_2O_4Ho_2.$$

Preparation.—Powdered manganic oxide is suspended in water and a current of sulphurous anhydride passed through the liquid, when the manganic oxide gradually dissolves. The solution contains manganous dithionate or hyposulphate:—

$$\mathbf{MnO_2} \quad + \quad \mathbf{2SO_2} \quad = \quad 'S^v_2O_4Mno''.$$

Manganic oxide.   Sulphurous anhydride.   Manganous dithionate.

This solution is next treated with baric sulphide, which precipitates manganous sulphide, baric dithionate existing in the solution:—

$$'S^v_2O_4Mno'' \quad + \quad \mathbf{BaS''} \quad = \quad \mathbf{MnS''} \quad + \quad 'S^v_2O_4Bao''.$$

Manganous dithionate.   Baric sulphide.   Manganous sulphide.   Baric dithionate.

By adding sulphuric acid to a solution of the baric dithionate, baric sulphate is precipitated and dithionic acid remains in solution:—

$$'S^v_2O_4Bao'' \quad + \quad \mathbf{SO_2Ho_2} \quad = \quad \mathbf{SO_2Bao''} \quad + \quad 'S^v_2O_4Ho_2.$$

Baric dithionate.   Sulphuric acid.   Baric sulphate.   Dithionic acid.

TRITHIONIC ACID, *Sulphodithionic Acid,*
Sulphuretted Hyposulphuric Acid.

$$\left\{ \begin{array}{l} \mathbf{SO_2Ho} \\ \mathbf{S''} \\ \mathbf{SO_2Ho} \end{array} \right.$$

Preparation.—By digesting hydric potassic sulphite with sulphur, potassic trithionate and potassic hyposulphite (*sulphosulphate*) are formed:—

$$6\mathbf{SOKoHo} + 2\mathbf{S} = 2 \left\{ \begin{array}{l} \mathbf{SO_2Ko} \\ \mathbf{S''} \\ \mathbf{SO_2Ko} \end{array} \right. + \mathbf{SS''OKo_2} + 3\mathbf{OH_2}.$$

Hydric potassic sulphite.   Potassic trithionate.   Potassic Hyposulphite.   Water.

The two salts so produced, when decomposed by hydrofluo-silicic acid, yield trithionic acid, sulphurous acid, and sulphur:—

$$\mathbf{SS''OKo_2} + 2\left\{\begin{array}{l}\mathbf{SO_2Ko}\\ \mathbf{S''}\\ \mathbf{SO_2Ko}\end{array}\right._2 + 3H_2Si^{iv}F_6 = 3K_2Si^{iv}F_6$$

Potassic hyposulphite.	Potassic trithionate.	Hydrofluosilicic acid.	Potassic silicofluoride.

$$+ 2\left\{\begin{array}{l}\mathbf{SO_2Ho}\\ \mathbf{S''}\\ \mathbf{SO_2Ho}\end{array}\right. + \mathbf{SOHo_2} + S.$$

Trithionic acid.   Sulphurous acid.

TETRATHIONIC ACID, *Disulphodithionic Acid,*
Bisulphuretted Hyposulphuric Acid.

$$\left\{\begin{array}{l}\mathbf{SO_2Ho}\\ \mathbf{S''}\\ \mathbf{S''}\\ \mathbf{SO_2Ho}\end{array}\right.$$

Preparation.—When iodine is added to baric hyposulphite (*sulphosulphate*), baric iodide and baric tetrathionate are produced:—

$$2\mathbf{SS''OBao''} + I_2 = \mathbf{BaI_2} + \left\{\begin{array}{l}\mathbf{SO_2}\\ \mathbf{S''}\\ \mathbf{S''}\\ \mathbf{SO_2}\end{array}\right\}Bao''.$$

Baric hyposulphite.   Baric iodide.   Baric tetrathionate.

This salt, when decomposed by sulphuric acid, yields tetra-thionic acid.

PENTATHIONIC ACID, *Trisulphodithionic Acid,*
Trisulphuretted Hyposulphuric Acid.

$$\left\{\begin{array}{l}\mathbf{SO_2Ho}\\ \mathbf{S''}\\ \mathbf{S''}\\ \mathbf{S''}\\ \mathbf{SO_2Ho}\end{array}\right.$$

Preparation.—This acid is obtained by the action of hydrosulphuric acid on sulphurous anhydride :—

$$5SH_2 \;+\; 5SO_2 \;=\; \begin{cases} SO_2Ho \\ S'' \\ S'' \\ S'' \\ SO_2Ho \end{cases} \;+\; 4OH_2 \;+\; S_5.$$

Sulphuretted　　　　Sulphurous　　　　Pentathionic acid.　　　Water.
hydrogen.　　　　anhydride.

SELENIUM, Se₂.

Atomic weight $=79$. *Molecular weight* $=158$. *Molecular volume* ☐. 1 *litre of selenium vapour weighs* 79 *criths.* *Sp. gr.* 4·8. *Fuses a little above* 100°. *Boils at about* 700°. *Atomicity* ", ⁱᵛ, *and* ᵛⁱ. *Evidence of atomicity* :—

Hydroselenic acid......... $Se''H_2$.
Selenious chloride......... $Se^{iv}Cl_4$.
Selenic acid .... $Se^{vi}O_2Ho_2$.

Occurrence.—In small quantities in some mineral sulphides.

COMPOUNDS OF SELENIUM WITH HYDROGEN.

SELENIURETTED HYDROGEN, *Hydroselenic Acid.*
SeH_2.

Molecular weight $=81$. *Molecular volume* ☐. 1 *litre weighs* 40·5 *criths.*

Preparation.—By the action of hydrochloric acid upon ferrous selenide :—

$$FeSe'' \;+\; 2HCl \;=\; SeH_2 \;+\; FeCl_2.$$

Ferrous selenide.　Hydrochloric　　Hydroselenic　Ferrous chloride.
　　　　　　　　acid.　　　　acid.

Character.—Like hydrosulphuric acid, it produces precipitates in solutions of most of the heavy metals.

There are two chlorides of selenium : $'Se'_2Cl_2$ and $SeCl_4$.

COMPOUNDS OF SELENIUM WITH OXYGEN AND HYDROXYL.

Selenious anhydride .... SeO_2.
Selenious acid .... $SeOHo_2$.
Selenic acid .... SeO_2Ho_2.

These bodies closely resemble the corresponding sulphur compounds.

Selenious anhydride is formed by burning selenium in oxygen :—

$$Se_2 \quad + \quad 2O_2 \quad = \quad 2SeO_2.$$
Selenious
anhydride.

Selenious acid is formed by dissolving the anhydride in boiling water and crystallizing.

Potassic seleniate is prepared by fusing selenium or metallic selenides with nitre. The acid is obtained by transforming the potassic salt into a plumbic salt, and subsequently decomposing the latter with hydrosulphuric acid.

TELLURIUM, Te_2.

Atomic weight = 128. *Probable molecular weight* = 256. *Sp. gr.* 6·2. *Fuses at* 490°–500°. *Atomicity* ″, ⁱᵛ, *and* ᵛⁱ. *Evidence of atomicity* :—

Hydrotelluric acid.............. $Te''H_2$.
Tellurous chloride.............. $Te^{iv}Cl_4$.
Telluric acid .... $Te^{vi}O_2Ho_2$.

This element is of even less importance than selenium, which it closely resembles.

The following compounds are known :—

Hydrotelluric acid (*telluretted hydrogen*)	**Te**H$_2$.
Hypotellurous chloride	**Te**Cl$_2$.
Tellurous chloride	**Te**Cl$_4$.
Tellurous anhydride	**Te**O$_2$.
Telluric anhydride	**Te**O$_3$.
Tellurous acid...................................	**Te**OHo$_2$?
Telluric acid	**Te**O$_2$Ho$_2$.

CHAPTER XIV.

MONAD ELEMENTS.

SECTION II. (*continued from* Chap. VIII.).

BROMINE, Br$_2$.

Atomic weight = 80.   *Molecular weight* = 160.   *Molecular volume* ☐☐.   1 *litre of bromine vapour weighs* 80 *criths*. *Sp. gr.* 3·18.   *Fuses at* − 20°.   *Boils at* 63°.   *Atomicity* '. *Evidence of atomicity* :—

Hydrobromic acid	HBr.
Potassic bromide..................	KBr.
Argentic bromide	AgBr.

Occurrence.—In small quantities in some saline mineral waters. In sea-water, and the waters of the Dead Sea.

Preparation.—1. By the treatment, with chlorine, of the mother-liquors of saline waters containing bromides, and extracting the liberated bromine by ether :—

$$2\text{KBr} + \text{Cl}_2 = 2\text{KCl} + \text{Br}_2.$$
<small>Potassic bromide.   Potassic chloride.</small>

2. By heating together sulphuric acid, sodic bromide, and manganic oxide:—

$$2NaBr + \mathbf{MnO_2} + 2SO_2Ho_2 = Br_2$$
Sodic bromide.   Manganic oxide.   Sulphuric acid.

$$+ \mathbf{SO_2Nao_2} + \mathbf{SO_2Mno''} + 2OH_2.$$
Sodic sulphate.   Manganous sulphate.   Water.

Character.—Bromine unites with several metals directly, and with great energy.   Antimony and arsenic burn in it with briliancy.

At 0° bromine combines with water, forming a crystalline compound, $Br_2, 10OH_2$.

HYDROBROMIC ACID.

HBr.

Molecular weight =81. *Molecular volume* ☐. 1 *litre of hydrobromic acid weighs* 40·5 *criths. Fuses at* −73°. *Boils at* −69°.

Preparation.—1. By passing a mixture of hydrogen and bromine vapour through a red-hot tube, or by burning hydrogen in a mixture of bromine vapour and air:—

$$H_2 + Br_2 = 2HBr.$$
Hydrobromic acid.

2. By heating potassic bromide with phosphoric acid:—

$$3KBr + \mathbf{POHo_3} = \mathbf{POKo_3} + 3HBr.$$
Potassic bromide.   Phosphoric acid.   Potassic phosphate.   Hydrobromic acid.

Sulphuric acid cannot be employed for this operation, as a portion of the hydrobromic acid is then decomposed, bromine being liberated:—

$$\mathbf{SO_2Ho_2} + 2HBr = Br_2 + 2OH_2 + SO_2.$$
Sulphuric acid.   Hydrobromic acid.   Water.   Sulphurous anhydride.

3. By the action of water upon phosphorous tribromide:—

$$\mathbf{P'''Br_3} + 3OH_2 = \mathbf{P^vOHHo_2} + 3HBr.$$
Phosphorous tribromide.   Water.   Phosphorous acid.   Hydrobromic acid.

4. By gradually dropping bromine into water containing amorphous phosphorus :—

$$P_2 \ + \ 3Br_2 \ + \ 6OH_2 \ = \ 2P^vOHHo_2 \ + \ 6HBr.$$

Water.   Phosphorous acid.   Hydrobromic acid.

5. By passing sulphuretted hydrogen through water containing bromine :—

$$2SH_2 \ + \ 2Br_2 \ = \ 4HBr \ + \ S_2.$$

Sulphuretted hydrogen.   Hydrobromic acid.

Reactions.—1. Decomposed by chlorine with liberation of bromine :—

$$2HBr \ + \ Cl_2 \ = \ 2HCl \ + \ Br_2.$$

Hydrobromic acid.   Hydrochloric acid.

2. By the action of atmospheric oxygen a small quantity of bromine is liberated, but the decomposition is soon arrested :—

$$4HBr \ + \ O_2 \ = \ 2OH_2 \ + \ 2Br_2.$$

Hydrobromic acid.   Water.   Bromine.

3. In contact with metallic oxides, hydrates, and salts, bromides are formed.

COMPOUNDS OF BROMINE WITH OXYGEN AND HYDROXYL.

Hypobromous anhydride OBr_2.

Hypobromous acid ...... $OBrH$.

Bromic acid .... $\begin{cases} OBr \\ O \\ OH \end{cases}$.

The graphic formulæ of these compounds are analogous to those of the corresponding chlorine compounds, given at page 45.

HYPOBROMOUS ANHYDRIDE.

OBr_2.

Preparation.—By passing bromine vapour over dry mercuric oxide :—

$$\mathbf{Hg}O \quad + \quad 2Br_2 \quad = \quad \mathbf{Hg}Br_2 \quad + \quad OBr_2.$$

Mercuric   Mercuric   Hypobromous
oxide.   bromide.   anhydride.

HYPOBROMOUS ACID.

$OBrH$.

Preparation.—1. By passing hypobromous anhydride into water :—

$$OBr_2 \quad + \quad OH_2 \quad = \quad 2OBrH.$$

Hypobromous   Water.   Hypobromous
anhydride   acid.

2. By agitating mercuric oxide with bromine-water :—

$$2\mathbf{Hg}O \quad + \quad OH_2 \quad + \quad 2Br_2 \quad = \quad 2OBrH \quad + \left\{ \begin{array}{l} \mathbf{Hg}Br \\ O \\ \mathbf{Hg}Br \end{array} \right. .$$

Mercuric   Water.   Hypobromous   Mercuric
oxide.   acid.   oxybromide.

BROMIC ACID.

$$\left\{ \begin{array}{l} OBr \\ OHo \end{array} \right.$$

Preparation.—By acting upon a solution of baric bromate with sulphuric acid :—

$$\left\{ \begin{array}{l} OBr \\ O \\ Bao'' \\ O \\ OBr \end{array} \right. \quad + \quad \mathbf{S}O_2Ho_2 \quad = \quad 2 \left\{ \begin{array}{l} OBr \\ OHo \end{array} \right. \quad + \quad \mathbf{S}O_2Bao''.$$

Baric bromate.   Sulphuric   Bromic   Baric
acid.   acid.   ·sulphate.

Reaction.—By boiling, bromic acid decomposes into water, bromine, and oxygen :—

$$4\left\{\begin{array}{l} \mathbf{O}Br \\ \mathbf{O}Ho \end{array}\right. = 2Br_2 \ + \ 2\mathbf{O}H_2 \ + \ 5O_2.$$

Bromic acid.   Water.

Preparation of bromates.—1. By adding bromine to a solution of a metallic hydrate, and separating the bromate by crystallization: —

$$6KHo \ + \ 3Br_2 \ = \ 5KBr \ + \ \left\{\begin{array}{l} \mathbf{O}Br \\ \mathbf{O}Ko \end{array}\right. + \ 3\mathbf{O}H_2.$$

Potassic   Potassic   Potassic   Water.
hydrate.   bromide.   bromate.

2. By the action of potassic hydrate on bromine pentachloride :—

$$6KHo \ + \ BrCl_5 \ = \ 5KCl \ + \left\{\begin{array}{l} \mathbf{O}Br \\ \mathbf{O}Ko \end{array}\right. + \ 3\mathbf{O}H_2.$$

Potassic   Bromine   Potassic   Potassic   Water.
hydrate.   pentachloride.   chloride.   bromate.

Character of bromates.—Some of the bromates when heated lose oxygen, being transformed into bromides :—

$$2\left\{\begin{array}{l} \mathbf{O}Br \\ \mathbf{O}Ko \end{array}\right. = \ 2KBr \ + \ 3O_2.$$

Potassic   Potassic
bromate.   bromide.

Others evolve bromine and a portion of their oxygen, leaving metallic oxides :—

$$2\left\{\begin{array}{l} \mathbf{O}Br \\ O \\ Mgo'' \\ O \\ \mathbf{O}Br \end{array}\right. = \ 2\mathbf{Mg}O \ + \ 2Br_2 \ + \ 5O_2.$$

Magnesic   Magnesic
bromate.   oxide.

IODINE, I_2.

Atomic weight $=127$.   *Molecular weight* $=254$.   *Molecular volume* ☐.   *1 litre of iodine vapour weighs 127 criths.* *Sp. gr.* 4·95.   *Fuses at* 107°.   *Boils at* 180°.   *Atomicity* '. *Evidence of atomicity* :—

Hydriodic acid..................... HI.

Potassic iodide .... KI.

Argentic iodide .... AgI.

Occurrence.—In mineral springs, in sea-water, and in considerable quantities in sea-plants.

Manufacture.—Sea-weeds are burnt and the ash is extracted with water. The liquid is evaporated, and, after a considerable quantity of sodic carbonate and chloride has crystallized out, the mother-liquor, which contains potassic iodide, is distilled with sulphuric acid and manganic oxide :—

$$2KI + \mathbf{Mn}O_2 + 2SO_2Ho_2 = SO_2Ko_2 + \mathbf{SO}_2Mno'' + I_2 + 2\mathbf{O}H_2.$$

| Potassic iodide. | Manganic oxide. | Sulphuric acid. | Potassic sulphate. | Manganous sulphate. | Water. |

Reactions.—1. Iodine is precipitated from its solutions by chlorine and bromine :—

$$2KI \quad + \quad Cl_2 \quad = \quad 2KCl \quad + \quad I_2.$$

Potassic iodide.   Potassic chloride.

$$2KI \quad + \quad Br_2 \quad = \quad 2KBr \quad + \quad I_2.$$

Potassic iodide.   Potassic bromide.

2. Iodine unites directly with many metals.

HYDRIODIC ACID.

HI.

Molecular weight =128. *Molecular volume* ☐☐. 1 *litre of hydriodic acid weighs* 64 *criths. Fuses at* —55°.

Preparation.—1. By passing iodine vapour and hydrogen through a red-hot tube or over spongy platinum gently heated:—

$$H_2 \quad + \quad I_2 \quad = \quad 2HI.$$

2. By the action of dilute sulphuric acid on baric iodide, or of phosphoric acid on any iodide :—

$$\mathbf{BaI_2} \quad + \quad \mathbf{SO_2Ho_2} \quad = \quad 2HI \quad + \quad \mathbf{SO_2Bao''}.$$

| Baric iodide. | Sulphuric acid. | Hydriodic acid. | Baric sulphate. |

3. By decomposing phosphorous triiodide by water :—

$$\mathbf{PI_3} \quad + \quad 3OH_2 \quad = \quad \mathbf{POHHo_2} \quad + \quad 3HI.$$

| Phosphorous triiodide. | Water. | Phosphorous acid. | Hydriodic acid. |

4. By heating together water, potassic iodide, iodine, and phosphorus :—

$$4KI \quad + \quad P_2 \quad + \quad 5I_2 \quad + \quad 8OH_2 \quad = \quad 14HI \quad + \quad 2\mathbf{POHoKo_2}$$

| Potassic iodide. | | | Water. | Hydriodic acid. | Hydric dipotassic phosphate. |

5. A solution of hydriodic acid is obtained by passing sulphuretted hydrogen through water in which iodine is suspended:—

$$2\mathbf{SH_2} \quad + \quad 2I_2 \quad = \quad 4HI \quad + \quad S_2.$$

| Sulphuretted hydrogen. | | Hydriodic acid. | |

Reactions.—1. Decomposed by chlorine and bromine, with liberation of iodine :—

$$2HI \quad + \quad Cl_2 \quad = \quad 2HCl \quad + \quad I_2.$$

| Hydriodic acid. | | Hydrochloric acid. | |

$$2HI \quad + \quad Br_2 \quad = \quad 2HBr \quad + \quad I_2.$$

| Hydriodic acid. | | Hydrobromic acid. | |

2. It is gradually but completely decomposed by atmospheric oxygen; the iodine, which at first remains dissolved in the hydriodic acid, is after a time deposited in crystals :—

$$4HI \quad + \quad O_2 \quad = \quad 2\mathbf{OH_2} \quad + \quad 2I_2.$$

| Hydriodic acid. | | Water. | |

3. With metallic oxides, hydrates, and some salts it forms iodides. Even argentic chloride is transformed by hydriodic acid into argentic iodide :—

$$AgCl \quad + \quad HI \quad = \quad AgI \quad + \quad HCl.$$

| Argentic chloride. | Hydriodic acid. | Argentic iodide. | Hydrochloric acid. |

4. Hydriodic acid is rapidly decomposed by mercury, with liberation of hydrogen :—

$$2HI \quad + \quad 2Hg \quad = \quad 'Hg'_2I_2 \quad + \quad H_2.$$

Hydriodic acid.   Mercurous iodide.

COMPOUNDS OF IODINE WITH OXYGEN AND HYDROXYL.

Iodic anhydride .... $\left\{\begin{array}{l} \textbf{O}I \\ O \\ O \quad . \\ O \\ \textbf{O}I \end{array}\right.$

Iodic acid .... $\left\{\begin{array}{l} \textbf{O}I \\ \textbf{O}Ho \cdot \end{array}\right.$

Periodic anhydride......... $\left\{\begin{array}{l} \textbf{O}I \\ O \\ O \\ O \quad . \\ O \\ O \\ \textbf{O}I \end{array}\right.$

Periodic acid .... $\left\{\begin{array}{l} \textbf{O}I \\ O \quad . \\ \textbf{O}Ho \end{array}\right.$

The graphic formulæ of these compounds are analogous to those of the corresponding chlorine compounds given at p. 45.

IODIC ANHYDRIDE.

I_2O_5,   or $\left\{\begin{array}{l} \textbf{O}I \\ O \\ O \quad . \\ O \\ \textbf{O}I \end{array}\right.$

Preparation.—By heating iodic acid to 170°, when it separates into iodic anhydride and water:—

$$2\left\{\begin{matrix}\mathbf{OI}\\\mathbf{OHo}\end{matrix}\right. = \mathbf{OH_2} + \left\{\begin{matrix}\mathbf{OI}\\\mathbf{O}\\\mathbf{O}\\\mathbf{O}\\\mathbf{OI}\end{matrix}\right. .$$

Iodic acid.   Iodic anhydride.

Reaction.—When strongly heated, it decomposes into iodine and oxygen.

IODIC ACID.

$$\left\{\begin{matrix}\mathbf{OI}\\\mathbf{OHo}\end{matrix}\right.$$

Preparation.—1. By the action of sulphuric acid upon baric iodate:—

$$\left\{\begin{matrix}\mathbf{OI}\\\mathbf{O}\\\mathbf{Bao''}\\\mathbf{O}\\\mathbf{OI}\end{matrix}\right. + \mathbf{SO_2Ho_2} = 2\left\{\begin{matrix}\mathbf{OI}\\\mathbf{OHo}\end{matrix}\right. + \mathbf{SO_2Bao''}.$$

Baric iodate.   Sulphuric acid.   Iodic acid.   Baric sulphate.

2. By oxidizing iodine with strong boiling nitric acid:—

$$6\mathbf{NO_2Ho} + \mathbf{I_2} = 2\left\{\begin{matrix}\mathbf{OI}\\\mathbf{OHo}\end{matrix}\right. + 2\mathbf{OH_2} + 2\mathbf{N_2O_3} + \mathbf{N^{iv}_2O_4}.$$

Nitric acid.   Iodic acid.   Water.   Nitrous anhydride.   Nitric peroxide.

3. By acting upon iodine and water with chlorine:—

$$\mathbf{I_2} + 6\mathbf{OH_2} + 5\mathbf{Cl_2} = 2\left\{\begin{matrix}\mathbf{OI}\\\mathbf{OHo}\end{matrix}\right. + 10\mathbf{HCl}.$$

Water.   Iodic acid.   Hydrochloric acid.

Reactions.—1. In contact with hydriodic acid it forms water and iodine:—

$$\left\{\begin{matrix}\mathbf{OI}\\\mathbf{OHo}\end{matrix}\right. + 5\mathbf{HI} = 3\mathbf{OH_2} + 3\mathbf{I_2}.$$

Iodic acid.   Hydriodic acid.   Water.

2. It is reduced by many other deoxidizing agents.

Preparation of Iodates.—1. By treating solutions of metallic hydrates with iodine, and separating the iodate by crystallization :—

$$6KHo \ + \ 3I_2 \ = \ 5KI \ + \ \left\{ \begin{matrix} OI \\ OKo \end{matrix} \right. \ + \ 3OH_2.$$

Potassic   Potassic   Potassic   Water.
hydrate.   iodide.   iodate.

2. By dissolving iodine in potassic hydrate and treating the mixture with chlorine :—

$$12KHo \ + \ I_2 \ + \ 5Cl_2 \ = \ 10KCl \ + \ 2 \left\{ \begin{matrix} OI \\ OKo \end{matrix} \right. \ + \ 6OH_2.$$

Potassic   Potassic   Potassic   Water.
hydrate.   chloride.   iodate.

3. By heating together potassic chlorate and iodine :—

$$I_2 \ + \ \left\{ \begin{matrix} OCl \\ OKo \end{matrix} \right. \ = \ ICl \ + \ \left\{ \begin{matrix} OI \\ OKo \end{matrix} \right. .$$

Potassic   Iodine   Potassic
chlorate.   monochloride.   iodate.

Character of iodates.—Some of the iodates when heated split into iodides and oxygen, others into metallic oxides, iodine, and oxygen.

Iodic acid gives several well-defined anhydro-salts.

PERIODIC ANHYDRIDE.

$$I_2O_7, \ \text{or} \ \left\{ \begin{matrix} OI \\ O \\ O \\ O \\ O \\ O \\ OI \end{matrix} \right. .$$

Preparation.—By heating periodic acid to 160° :—

$$2 \left\{ \begin{matrix} OI \\ O \\ OHo \end{matrix} \right. \ = \ I_2O_7 \ + \ OH_2.$$

Periodic   Periodic   Water.
acid.   anhydride.

Reaction.—When heated it is decomposed into oxygen and iodic anhydride, and ultimately into iodine and oxygen.

PERIODIC ACID.

$$\left\{ \begin{array}{l} \mathbf{O}\text{I} \\ \text{O} \\ \mathbf{O}\text{Ho} \end{array} \right.$$

Preparation.—By decomposing plumbic periodate with sulphuric acid :—

$$\left\{ \begin{array}{l} \mathbf{O}\text{I} \\ \text{O} \\ \text{O} \\ \text{Pbo}'' \\ \text{O} \\ \text{O} \\ \mathbf{O}\text{I} \end{array} \right. + \ \mathbf{S}\text{O}_2\text{Ho}_2 \ = \ 2 \left\{ \begin{array}{l} \mathbf{O}\text{I} \\ \text{O} \\ \mathbf{O}\text{Ho} \end{array} \right. + \ \mathbf{S}\text{O}_2\text{Pbo}''.$$

| Plumbic periodate. | Sulphuric acid. | Periodic acid. | Plumbic sulphate. |

Preparation of Periodates.—Sodic periodate may be prepared by passing chlorine through mixed solutions of sodic hydrate and sodic iodate :—

$$\left\{ \begin{array}{l} \mathbf{O}\text{I} \\ \mathbf{O}\text{Nao} \end{array} \right. + \ 2\text{NaHo} \ + \ \text{Cl}_2 \ = \ \left\{ \begin{array}{l} \mathbf{O}\text{I} \\ \text{O} \\ \mathbf{O}\text{Nao} \end{array} \right. + \ \mathbf{O}\text{H}_2 \ + \ 2\text{NaCl}.$$

| Sodic iodate. | Sodic hydrate. | Sodic periodate. | Water. | Sodic chloride. |

FLUORINE, F_2.

Atomic weight $=19$. *Molecular weight* $=38$ (?). *Molecular volume* ☐. 1 *litre weighs* 19 *criths* (?). *Atomicity* '. *Evidence of atomicity* :—

Hydrofluoric acid.............. HF.

Occurrence.—In combination with metals in fluorspar, cryolite, apatite, and other minerals.

Little is known of fluorine in the uncombined condition.

COMPOUND OF FLUORINE WITH HYDROGEN.

HYDROFLUORIC ACID.

HF.

Molecular weight $=20$. *Molecular volume* ⬛. 1 *litre weighs* 10 *criths.*

Preparation.—By heating calcic fluoride with sulphuric acid in a leaden or platinum vessel :—

$$\mathbf{CaF_2} \;+\; \mathbf{SO_2Ho_2} \;=\; 2HF \;=\; \mathbf{SO_2Cao''}.$$

<table>
<tr><td>Calcic
fluoride.</td><td>Sulphuric
acid.</td><td>Hydrofluoric
acid.</td><td>Calcic
sulphate.</td></tr>
</table>

CHAPTER XV.

TETRAD ELEMENTS.

SECTION I.   (*Continued from* Chapter XI.)

SILICON, *Silicium*, Si.

Atomic weight $=28\cdot5$. *Sp. gr.* (*graphitoidal*) $=2\cdot49$. *Atomicity* ^{iv}.
Evidence of atomicity :—

Silicic chloride $\mathbf{SiCl_4}$.
Silicic fluoride.............. $\mathbf{SiF_4}$.

Occurrence.—Silicon is one of the most widely diffused elements.   It is found, in combination with oxygen and metals, in a very large number of minerals.

α. *Amorphous Silicon.*

Preparation.—1. By heating potassic silicofluoride with potassium :—

$$SiK_2F_c \;+\; 2K_2 \;=\; Si \;+\; 6KF.$$

<table>
<tr><td>Potassic
silicofluoride</td><td></td><td></td><td>Potassic
fluoride.</td></tr>
</table>

F

2. By heating sodium in a current of the vapour of silicic chloride :—

$$\mathbf{SiCl_4} \; + \; 2Na_2 \; = \; Si \; + \; 4NaCl.$$

Silicic chloride.   Sodi chloride.

Reactions.—1. Silicon is dissolved by aqueous hydrofluoric acid, and converted into hydrofluosilicic acid :—

$$Si \; + \; 6HF \; = \; SiH_2F_6 \; + \; 2H_2.$$

Hydrofluoric acid.   Hydrofluosilicic acid.

2. When fused with potassic hydrate, or boiled in its solution, it yields potassic silicate :—

$$Si \; + \; 4KHo \; = \; \mathbf{SiKo_4} \; + \; 2H_2.$$

Potassic hydrate.   Potassic silicate.

3. Heated in the air, it burns, producing silicic anhydride.

β. Graphitoidal Silicon.

Preparation. — By fusing amorphous silicon with aluminium, and boiling the compound in hydrochloric or hydrofluoric acid, which dissolves the aluminium, leaving the silicon in the form of hexagonal plates with a metallic lustre.

Character.—May be heated to whiteness in oxygen without burning.

Is gradually oxidized by a mixture of nitric and hydrofluoric acids.

Is slowly attacked by fused potassic hydrate.

γ. Adamantine Silicon.

Preparation.—By heating aluminium very strongly in a current of the vapour of silicic chloride. The aluminic chloride which is formed volatilizes, leaving the adamantine silicon behind :—

$$\mathbf{3SiCl_4} \; + \; 2Al_2 \; = \; 2'\mathbf{Al'''_2Cl_6} \; + \; Si_3.$$

Silicic chloride.   Aluminic chloride.

SILICIC HYDRIDE.
SiH_4.

Molecular weight $=32\cdot5$.

Has not been obtained free from hydrogen.

Preparation.—1. By decomposing dilute sulphuric acid by a feeble electric current passing from electrodes of aluminium containing silicon, when the silicic hydride is evolved at the negative pole.

2. By decomposing magnesic silicide with hydrochloric acid :—

$$SiMg''_2 + 4HCl = 2MgCl_2 + SiH_4.$$

Magnesic silicide.   Hydrochloric acid.   Magnesic chloride.   Silicic hydride.

Reaction.—Inflames spontaneously in air, producing water and silicic anhydride :—

$$SiH_4 + 2O_2 = SiO_2 + 2OH_2.$$

Silicic hydride.   Silicic anhydride.   Water.

SILICIC CHLORIDE.
$SiCl_4$.

Molecular weight $=170\cdot5$. *Molecular volume* ☐. 1 *litre weighs* $85\cdot25$ *criths. Sp. gr. of liquid* $1\cdot52$. *Boils at* $59°$.

Preparation.—1. By burning silicon in chlorine.

2. By heating a mixture of carbon and silicic anhydride in a stream of chlorine :—

$$SiO_2 + 2C + 2Cl_2 = SiCl_4 + 2CO.$$

Silicic anhydride.   Silicic chloride.   Carbonic oxide.

Reaction.—By contact with water it produces silicic and hydrochloric acids :—

$$SiCl_4 + 4OH_2 = SiHo_4 + 4HCl.$$

Silicic Chloride.   Water.   Silicic acid.   Hydrochloric acid.

F 2

SILICIC BROMIDE.

$SiBr_4$.

Molecular weight $=348\cdot5$.   *Sp. gr.* $2\cdot813$ at $0°$.   *Boils at* $153°$.

Preparation.—By the same method as that employed for making the chloride, bromine vapour being substituted for chlorine.

Reaction.—Decomposed by water in the same manner as the chloride.

SILICIC FLUORIDE.

SiF_4.

Molecular weight $=104\cdot5$.   *Molecular volume* ☐.   1 *litre weighs* $52\cdot25$ *criths.*   *Fuses at* $-140°$ C.   *Condensable gas.*

Preparation.—By heating together silicic anhydride, calcic fluoride, and sulphuric acid:—

$$SiO_2 \;+\; 2CaF_2 \;+\; 2SO_2Ho_2 \;=\; SiF_4$$

Silicic anhydride.   Calcic fluoride.   Sulphuric acid.   Silicic fluoride.

$$+\; 2SOHo_2Cao''.$$

Dihydric calcic sulphate.

Reaction.—By contact with water it produces silicic and hydrofluosilicic acids:—

$$3SiF_4 \;+\; 4OH_2 \;=\; SiHo_4 \;+\; 2SiH_2F_6.$$

Silicic fluoride.   Water.   Silicic acid.   Hydrofluosilicic acid.

By contact with metallic oxides, hydrates, and salts, hydrofluosilicic acid produces silicofluorides, some of which, as the potassic and baric compounds, are insoluble in water:—

$$SiH_2F_6 \;+\; 2KHo \;=\; SiK_2F_6 \;+\; 2OH_2.$$

Hydrofluosilicic acid.   Potassic hydrate.   Potassic silicofluoride.   Water.

COMPOUNDS OF SILICON WITH OXYGEN AND HYDROXYL.

Silicic anhydride........... SiO_2.
Silicic acid .... $SiHo_4$ and $SiOHo_2$.
Chryson .... $Si_6H_6O_4$.
Leukon .... $Si_3H_4O_5$.

SILICIC ANHYDRIDE.

SiO_2.

Molecular weight $=60.5$. *Sp. gr.* 2.69.

Occurrence.—In the pure state in many minerals, as quartz, agate, &c.

Preparation.—By heating silicic acid to 100°.

SILICIC ACID.

Tetrabasic... $SiHo_4$. Dibasic... $SiOHo_2$.

Preparation.—1. By treating a solution of a soluble silicate with hydrochloric acid :—

$$SiNao_4 \quad + \quad 4HCl \quad = \quad SiHo_4 \quad + \quad 4NaCl.$$
Sodic   Hydrochloric   Silicic   Sodic
silicate.   acid.   acid.   chloride.

2. By passing a stream of carbonic anhydride through a solution of a soluble silicate :—

$$SiNao_4 \quad + \quad 4OH_2 \quad + \quad 4CO_2 \quad = \quad SiHo_4 \quad + \quad 4COHoNao.$$
Sodic   Water.   Carbonic   Silicic   Hydric sodic
silicate.   anhydride.   acid.   carbonate.

A reaction similar to this is the cause of the disintegration of granitic rocks.

3. By passing silicic fluoride through water. (See p. 100.)

4. The dibasic silicic acid is said to be produced by the evaporation *in vacuo* at 16° of a solution of the tetrabasic acid in water.

The acid prepared by the first three of the above processes has probably the formula $SiHo_4$; by drying in the air a compound remains containing

$$Si_6H_8O_{16} \;=\; Si_6O_8Ho_8.$$

This last acid, heated to 100°, loses more water, being transformed into

$$Si_6H_4O_{14} \;=\; Si_6O_{10}Ho_4.$$

5. By the action of water on tetrethylic silicate, a compound is produced containing

$$Si_4H_8O_{12} \;=\; Si_4O_4Ho_8.$$

SILICATES.

The soluble alkaline silicates may be prepared by fusing silicic anhydride, in the form of sand or flints, or insoluble natural silicates, with alkaline hydrates or carbonates.

The silicates form a very important class of minerals. The following list contains a few examples :—

Sand.		
Flint.		
Rock crystal.	*Silicic anhydride*	SiO_2.
Quartz.		
Opal.		
Chalcedony.		
Peridote.	*Dimagnesic silicate*	$SiMgo''_2$.
Phenacite.	*Diglucinic silicate*	$SiGlo''_2$.
Willemite.	*Dizincic silicate*	$SiZno''_2$.
Zircon.	*Dizirconic silicate*	$SiZro''_2$.
Enstatite.	*Monomagnesic silicate* ...	$SiOMgo''$.

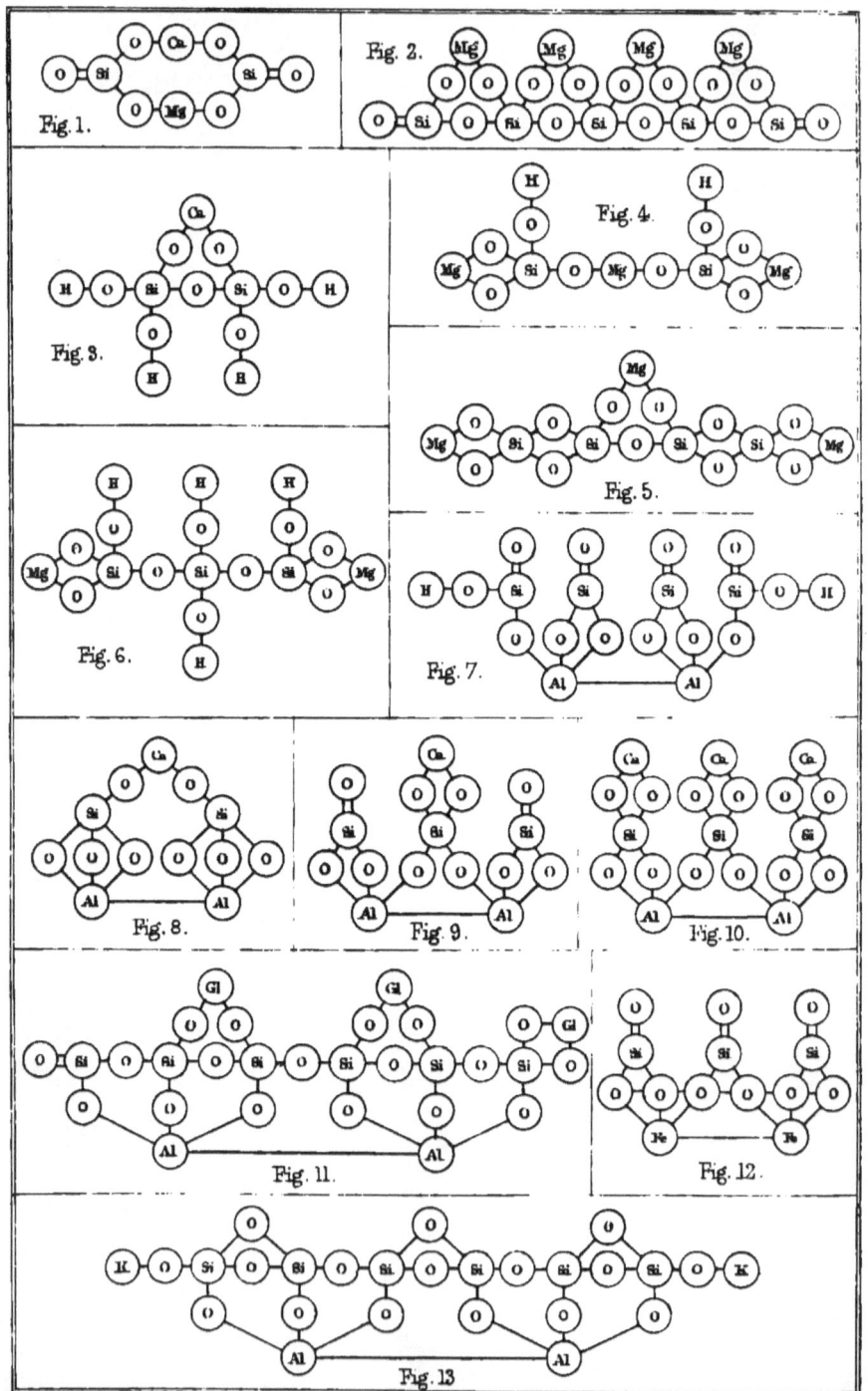

Fig.1.

Fig. 2.

Fig.3.

Fig. 4.

Fig. 5.

Fig.6.

Fig. 7.

Fig. 8.

Fig. 9.

Fig. 10.

Fig. 11.

Fig. 12.

Fig. 13

Diopside. *Calcic magnesic disilicate.* See fig. 1. ....⎱ SiO $Cao''Mgo''.$ SiO

Talc. *Tetramagnesic pentasilicate.* See fig. 2 .... $Si_5O_6Mgo''_4.$

Okenite. *Tetrahydric calcic disilicate.* See fig 3 ....
$$\begin{cases} SiHo_2 \\ O \quad Cao''. \\ SiHo_2 \end{cases}$$

Serpentine. *Dihydric trimagnesic disilicate.* See fig. 4 ....
$$\begin{cases} SiHoMgo'' \\ Mgo'' \\ SiHoMgo'' \end{cases}$$

Steatite. *Trimagnesic tetrasilicate.* See fig. 5 .... $Si_4O_5Mgo''_3.$

Meerschaum. *Tetrahydric dimagnesic trisilicate.* See fig. 6 ....
$$\begin{cases} SiHoMgo'' \\ O \\ SiHo_2 \\ O \\ SiHoMgo'' \end{cases}$$

Pyrophyllite. *Dihydric aluminic tetrasilicate.* See fig. 7 ....
$$\begin{matrix} SiOHo \\ SiO \\ SiO \quad Al_2o^{vi}. \\ SiOHo \end{matrix}$$

Anorthite. *Aluminic calcic disilicate.* See fig. 8............................⎱ $Si_2('Al'''_2O_6)^{vi}Cao'.$

Labradorite. *Aluminic calcic trisilicate.* See fig. 9 ....
$$\begin{matrix} SiO \\ SiCao''-Al_2o^{vi}. \\ SiO \end{matrix}$$

Grossularia. *Aluminic tricalcic trisilicate.* See fig. 10. ....
$$\begin{matrix} SiCao'' \\ SiCao''-Al_2o^{vi}. \\ SiCao'' \end{matrix}$$

Emerald. *Triglucinic aluminic hexasilicate.* See fig. 11 ....⎱ $Si_6O_6Al_2o^{vi}Glo''_3.$

Chloropal. *Ferric trisilicate.* See fig. 12 ....
$$\begin{cases} SiO \\ SiO-Fe_2o^{vi},3OH_2. \\ SiO \end{cases}$$

Felspar. Orthose. *Dipotassic aluminic hexasilicate.* See fig. 13 ...⎱ $Si_6O_8Ko_2Al_2o^{vi}.$

SILICIC SULPHIDE.

SiS''_2.

Preparation.—By passing the vapour of carbonic disulphide over silicic anhydride heated to redness :—

$$SiO_2 \;+\; CS''_2 \;=\; SiS''_2 \;+\; CO_2.$$

Silicic anhydride.   Carbonic disulphide.   Silicic sulphide.   Carbonic anhydride.

Reaction.—By the action of water, hydrosulphuric acid is evolved, and the solution contains silicic acid :—

$$SiS_2 \;+\; 4OH_2 \;=\; SiHo_4 \;+\; 2SH_2.$$

Silicic sulphide.   Water.   Silicic acid.   Sulphuretted hydrogen.

TIN, Sn.

Atomic weight = 118. *Molecular weight unknown. Sp. gr.* 7·28. *Fuses at* 228°. *Atomicity* " *and* ⁱᵛ, *also pseudo-triatomic.*

The following are the names and formulæ of the principal compounds of this metal :—

Stannous chloride .... $SnCl_2$.

Stannic chloride .... $SnCl_4$.

Stannous oxide .... SnO.

Stannic oxide or anhydride. SnO_2.

Distannous oxydichloride.. $\begin{cases} SnCl \\ O \\ SnCl \end{cases}$ (Cl)—(Sn)—(O)—(Sn)—(Cl)

Stannous hydrate $SnHo_2$.   (H)—(O)—(Sn)—(O)—(H)

- Stannic acid............... $SnOHo_2$.   (H)—(O)—(Sn)—(O)—(H) with (O) above Sn

Dipotassic stannite......... $SnKo_2$.

Dipotassic stannate......... $SnOKo_2, 4OH_2$.

Distannic trioxide .... $\begin{cases} \mathbf{Sn}^O_{\ \ }O, \\ \mathbf{Sn}_O^{\ \ }O, \end{cases}$

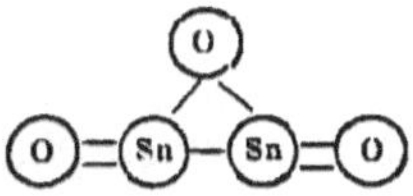

or

Stannous stannate $\mathbf{Sn}OSno''.$

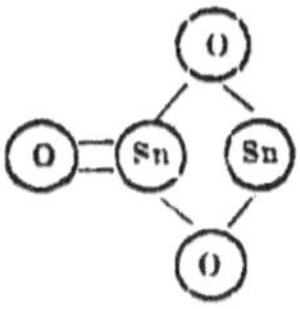

Metastannic acid (dried at 100°) .... $\begin{cases} \mathbf{Sn}Ho_3 \\ O \\ \mathbf{Sn}Ho_2 \\ O \\ \mathbf{Sn}O \quad . \\ O \\ \mathbf{Sn}Ho_2 \\ O \\ \mathbf{Sn}Ho_3 \end{cases}$

Dipotassic metastannate... $\begin{cases} \mathbf{Sn}Ho_2Ko \\ O \\ \mathbf{Sn}Ho_2 \\ O \\ \mathbf{Sn}O \\ O \\ \mathbf{Sn}Ho_2 \\ O \\ \mathbf{Sn}Ho_2Ko \end{cases}$

Stannous sulphide .... $\mathbf{Sn}S''.$

Stannic sulphide .... $\mathbf{Sn}S''_2.$

Distannic trisulphide ...... $\begin{cases} \mathbf{Sn}S''S'', \\ \mathbf{Sn}S''S'', \end{cases}$

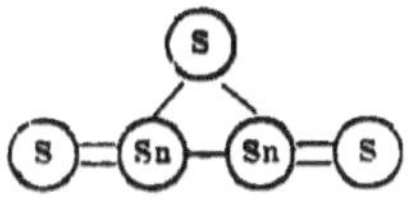

or

Stannous sulphostannate.. $\mathbf{Sn}SSns''.$

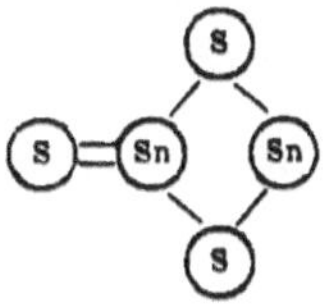

F 5

Stannous sulphate .... SO_2Sno''.

TITANIUM, Ti.

Atomic weight $=50$. *Molecular weight unknown.* *Sp. gr.* 5·3. *Atomicity* " *and* ^{iv}, *also pseudo-triatomic.*

The following are the names and formulæ of the chief compounds of titanium :—

Titanic tetrachloride ...... $TiCl_4$.

Dititanic hexachloride ... $\begin{cases} TiCl_3 \\ TiCl_3 \end{cases}$.

Titanous oxide.............. TiO.

Titanic oxide or anhydride (Rutile, Anatase, Brookite) .... $\Big\}$ TiO_2.

Titanic acid .... $TiOHo_2$.

Titanic sulphide .... TiS''_2.

Dititanic dinitride .... $\begin{cases} TiN''' \\ TiN''' \end{cases}$

Trititanic tetranitride...... $Ti_3N'''_4$

CHAPTER XVI.

PENTAD ELEMENTS.

SECTION I.   (*Continued from* Chapter XII.)

PHOSPHORUS, P₄.

Atomic weight =31. *Molecular weight* =124. *Molecular
volume* ☐☐. 1 *litre of phosphorus vapour weighs* 62
criths. *Sp. gr.* 1·83. *Fuses at* 44–45°. *Boils at* 290°.
Atomicity ''' *and* ˅. *Evidence of atomicity* :—

Phosphorous trihydride............... $P'''H_3$
Phosphorous trichloride .... $P'''Cl_3$.
Phosphoric chloride .... P^vCl_5.
Phosphonic iodide .... P^vH_4I.

Occurrence.—In combination as a constituent of several
minerals, and in small quantities in most rocks and soils.

In plants, and in the brain, nerves, urine, and bones of
animals.

Manufacture.—Calcined bones or Sombrerite, both of which
consist chiefly of calcic phosphate, are digested with sulphuric
acid, by which the tricalcic diphosphate is converted into tetrahydric calcic diphosphate :—

$$P_2O_2Cao''_3 + 2SO_2Ho_2 = P_2O_2Ho_4Cao'' + 2SO_2Cao''.$$
Tricalcic diphos-   Sulphuric   Tetrahydric calcic   Calcic
phate (Bone-ash).   acid.   diphosphate.   sulphate.

The tetrahydric calcic phosphate is extracted with water
from the calcic sulphate, evaporated, mixed with charcoal, dried
and distilled, when phosphorus, carbonic oxide and tricalcic
diphosphate are produced :—

$$3P_2O_4Cao'' + C_{10} = P_2O_2Cao''_3 + 10CO + P_4.$$
Calcic meta-   Tricalcic   Carbonic
phosphate.   diphosphate.   oxide.

AMORPHOUS PHOSPHORUS.   *Allotropic Phosphorus.   Red Phosphorus.*

Obtained by heating common phosphorus to 230°–250° in close vessels.

Neither the number nor the arrangement of the atoms in the molecule of this variety of phosphorus is known.

COMPOUNDS OF PHOSPHORUS WITH HYDROGEN.

Phosphorus forms three compounds with hydrogen, which cannot be obtained by the direct combination of their elements.

Solid phosphoretted hydrogen... $\begin{cases} \mathbf{P}(\mathbf{P}'''\mathbf{H})'' \\ \mathbf{P}(\mathbf{P}'''\mathbf{H})'' \end{cases}$?

Liquid ditto .... $'\mathbf{P}''_2\mathbf{H}_4$.

Gaseous ditto .... $\mathbf{PH}_3$.

GASEOUS PHOSPHORETTED HYDROGEN.

$\mathbf{PH}_3$.

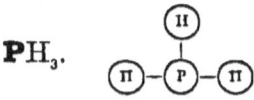

Molecular weight $=34$.   *Molecular volume* ▢.   1 *litre weighs* 17 *criths*.

Preparation.—1. By heating hypophosphorous acid :—

$$2\mathbf{P}OH_2Ho \quad = \quad \mathbf{P}H_3 \quad + \quad \mathbf{P}OHo_3.$$

Hypophosphorous acid.   Phosphoretted hydrogen.   Phosphoric acid.

2. By heating phosphorous acid :—

$$4\mathbf{P}OHHo_2 \quad = \quad \mathbf{P}H_3 \quad + \quad 3\mathbf{P}OHo_3.$$

Phosphorous acid.   Phosphoretted hydrogen.   Phosphoric acid.

3. By heating phosphorus with solution of sodic or potassic hydrate :—

$$3ONaH + P_4 + 3OH_2 = 3POH_2Nao + PH_3.$$

Sodic hydrate.   Water.   Sodic hypophosphite.   Phosphoretted hydrogen.

The gas prepared by this process contains free hydrogen and the vapour of liquid phosphoretted hydrogen.

Reactions.—1. By combustion in oxygen it yields phosphoric acid :—

$$PH_3 + 2O_2 = POHo_3.$$

Phosphoretted hydrogen.   Phosphoric acid.

2. When passed through a solution of cupric sulphate, it causes a black precipitate of cupric phosphide :—

$$2PH_3 + 3SO_2Cuo'' = P_2Cu''_3 + 3SO_2Ho_2.$$

Phosphoretted hydrogen.   Cupric sulphate.   Cupric phosphide.   Sulphuric acid.

3. When passed through a solution of argentic nitrate, metallic silver and nitric and phosphoric acids are formed :—

$$PH_3 + 8NO_2Ago + 4OH_2 = POHo_3 + 4Ag_2 + 8NO_2Ho.$$

Phosphoretted hydrogen.   Argentic nitrate.   Water.   Phosphoric acid.   Nitric acid.

4. It unites directly with hydriodic and hydrobromic acids when they are presented to it in the nascent state, forming compounds isomorphous with the corresponding substances in the nitrogen series :—

$$PH_3 + 3I_2 = PI_3 + 3HI ;$$

Phosphoretted hydrogen.   Phosphorous triiodide.   Hydriodic acid.

$$3PH_3 + 3HI = 3PH_4I.$$

Phosphoretted hydrogen.   Hydriodic acid.   Phosphonic iodide.

In this behaviour phosphoretted hydrogen bears a striking analogy to ammonia, although, unlike the latter compound, it does not unite with other acids.

LIQUID PHOSPHORETTED HYDROGEN.

$$'\mathbf{P}''_2\mathbf{H}_4, \text{ or } \begin{cases} \mathbf{PH}_2 \\ \mathbf{PH}_2 \end{cases}$$

Molecular weight $= 66$.

Preparation.—By the action of water or very dilute hydrochloric acid upon calcic phosphide, $'\mathbf{P}''_2\mathbf{Ca}''_2$, the gas evolved being transmitted through a freezing-mixture :—

$$'\mathbf{P}''_2\mathbf{Ca}''_2 \quad + \quad 4\mathbf{OH}'_2 \quad = \quad '\mathbf{P}''_2\mathbf{H}_4 \quad + \quad 2\mathbf{Ca}''\mathbf{Ho}_2.$$

| Calcic phosphide. | Water. | Liquid phosphoretted hydrogen. | Calcic hydrate. |

The calcic phosphide is prepared by passing the vapour of phosphorus over lime heated to redness :—

$$14\mathrm{P} \quad + \quad 14\mathbf{CaO} \quad = \quad 2\mathbf{P}_2\mathbf{O}_3\mathbf{Cao}''_2 \quad + \quad 5'\mathbf{P}''_2\mathbf{Ca}''_2.$$

| | Lime. | Calcic pyrophosphate. | Calcic phosphide. |

Reaction.—Decomposed by sunlight into solid and gaseous phosphoretted hydrogen .—

$$5'\mathbf{P}''_2\mathbf{H}_4 \quad = \quad 6\mathbf{PH}_3 \quad + \quad \begin{cases} \mathbf{P}(\mathrm{P}'''\mathrm{H})'' \\ \mathbf{P}(\mathrm{P}'''\mathrm{H})'' \end{cases} ?$$

| Liquid phosphoretted hydrogen. | Gaseous phosphoretted hydrogen. | Solid phosphoretted hydrogen. |

SOLID PHOSPHORETTED HYDROGEN.

$$\begin{cases} \mathbf{P}(\mathrm{P}'''\mathrm{H})'' \\ \mathbf{P}(\mathrm{P}'''\mathrm{H})'' \end{cases} ?$$

Molecular weight $= 126$?

Preparation.—By dissolving calcic phosphide in concentrated hydrochloric acid, or by the action of light upon the liquid phosphoretted hydrogen.

COMPOUNDS OF PHOSPHORUS WITH CHLORINE.

Phosphorus forms two compounds with chlorine :—

Phosphorous trichloride........ **PCl_3**.
Phosphoric chloride .... **PCl_5**.

PHOSPHOROUS TRICHLORIDE.

PCl_3.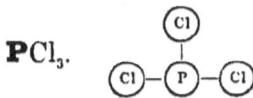

Molecular weight $= 137\cdot5$. *Molecular volume* ☐. 1 *litre of phosphorous trichloride vapour weighs* $68\cdot75$ *criths*. *Sp. gr.* $1\cdot45$. *Boils at* $74°$.

Preparation.—By the action of chlorine upon phosphorus :—

$$P_2 \quad + \quad 3Cl_2 \quad = \quad 2PCl_3.$$

Reaction.—By the action of water it yields hydrochloric and phosphorous acids :

$$\underset{\substack{\text{Phosphorous}\\\text{chloride.}}}{PCl_3} + \underset{\text{Water.}}{3OH_2} = \underset{\substack{\text{Hydrochloric}\\\text{acid.}}}{3HCl} + \underset{\substack{\text{Phosphorous}\\\text{acid.}}}{POHHo_2}.$$

PHOSPHORIC CHLORIDE.

PCl_5.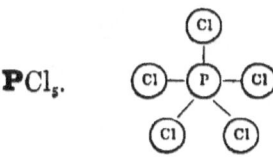

Molecular weight $= 208\cdot5$. *Molecular volume* ☐ *to* ⊞.

1 litre of phosphoric chloride weighs 52·1 *to* 104·25 *criths. Volatilizes below* 100°.

Preparation.—By the action of chlorine upon phosphorous chloride :—

$$PCl_3 + Cl_2 = PCl_5.$$

Phosphorous   Phosphoric
chloride.   chloride.

Reactions.—1. By the action of an excess of water it produces hydrochloric acid and phosphoric acid :—

$$PCl_5 + 4OH_2 = 5HCl + POHo_3.$$

Phosphoric   Water.   Hydrochloric   Phosphoric
chloride.   acid.   acid.

2. When submitted to the action of alcohols and acids, the chlorides of the radicals of the alcohols and acids are obtained, thus :—

$$\left\{ \begin{matrix} CH_3 \\ CH_2Ho \end{matrix} \right. + PCl_5 = \left\{ \begin{matrix} CH_3 \\ CH_2Cl \end{matrix} \right. + HCl + POCl_3.$$

Ethylic   Phosphoric   Ethylic   Hydrochloric   Phosphoric
alcohol.   chloride.   chloride.   acid.   oxytrichloride.

$$\left\{ \begin{matrix} CH_3 \\ COHo \end{matrix} \right. + PCl_5 = \left\{ \begin{matrix} CH_3 \\ COCl \end{matrix} \right. + HCl + POCl_3.$$

Acetic acid.   Phosphoric   Acetylic   Hydrochloric   Phosphoric
              chloride.   chloride.   acid.   oxytrichloride.

COMPOUND OF PHOSPHORUS WITH CHLORINE AND OXYGEN.

PHOSPHORIC OXYTRICHLORIDE.

$$POCl_3.$$

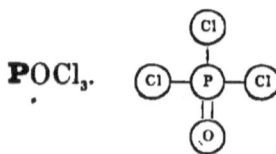

Molecular weight = 153·5. *Molecular volume* ☐☐. *1 litre of*

phosphoric oxytrichloride vapour weighs 76·75 *criths. Sp. gr.* 1·7. *Boiling-point* 110°.

Preparation.—1. By the action of a limited quantity of water on phosphoric chloride:—

$$PCl_5 + OH_2 = POCl_3 + 2HCl.$$

Phosphoric chloride.   Water.   Phosphoric oxytrichloride.   Hydrochloric acid.

2. By passing oxygen through boiling phosphorous trichloride:—

$$PCl_3 + O = POCl_3.$$

Phosphorous trichloride.   Phosphoric oxytrichloride.

3. By heating phosphoric chloride with phosphoric anhydride:—

$$P_2O_5 + 3PCl_5 = 5POCl_3.$$

Phosphoric anhydride.   Phosphoric chloride.   Phosphoric oxytrichloride.

4. It is formed as a secondary product in the preparation of the chlorides of alcohol and acid radicals as above described (p. 112).

Reactions.—1. By contact with water it is transformed into hydrochloric and phosphoric acids :—

$$POCl_3 + 3OH_2 = POHo_3 + 3HCl.$$

Phosphoric oxytrichloride.   Water.   Phosphoric acid.   Hydrochloric acid.

2. By distillation with the salts of organic acids it yields the chloracids:—

$$3\left\{\begin{matrix} CH_3 \\ CONao \end{matrix}\right. + POCl_3 = 3\left\{\begin{matrix} CH_3 \\ COCl \end{matrix}\right. + PONao_3.$$

Sodic acetate.   Phosphoric oxytrichloride.   Acetylic chloride.   Sodic phosphate.

COMPOUND OF PHOSPHORUS WITH CHLORINE AND SULPHUR.

PHOSPHORIC SULPHOTRICHLORIDE.

$$PS''Cl_3.$$

Molecular weight = 169·5. *Boils at* 128°.

Preparation.—By the action of sulphuretted hydrogen upon phosphoric chloride:—

$$PCl_5 \quad + \quad SH_2 \quad = \quad PS''Cl_3 \quad + \quad 2HCl.$$

Phosphoric   Sulphuretted   Phosphoric   Hydrochloric
chloride.   hydrogen.   sulphotrichloride.   acid.

Reaction.—When boiled with sodic hydrate, it yields sodic chloride and trisodic sulphophosphate:—

$$6ONaH \quad + \quad PS''Cl_3 \quad = \quad 3NaCl \quad + \quad PS''Nao_3 \quad + \quad 3OH_2.$$

Sodic   Phosphoric   Sodic   Trisodic   Water.
hydrate.   sulphotrichloride.   chloride.   sulpho-
                                               phosphate.

COMPOUNDS OF PHOSPHORUS WITH OXYGEN AND HYDROXYL.

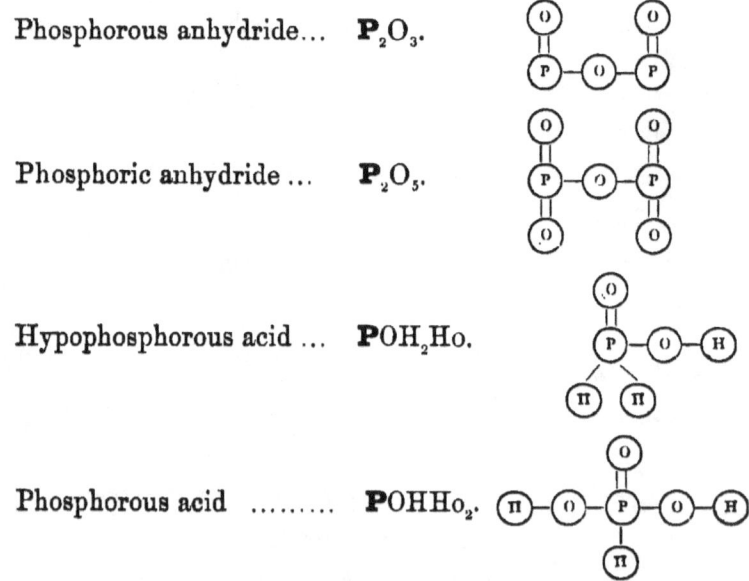

Phosphorous anhydride... P_2O_3.

Phosphoric anhydride ... P_2O_5.

Hypophosphorous acid ... POH_2Ho.

Phosphorous acid .... $POHHo_2$.

Phosphoric acid } $\mathbf{P}OHo_3.$
(tribasic) ...

Metaphosphoric } $\mathbf{P}O_2Ho.$
acid (monobasic).

Pyrophosphoric } $\mathbf{P}_2O_3Ho_4.$
acid (tetra-
basic) ....

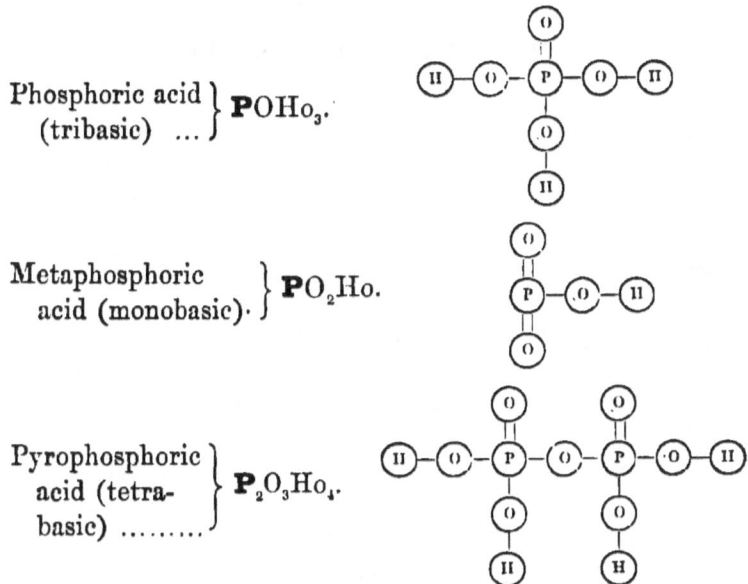

Hexabasic phosphoric acid......... $\mathbf{P}_4O_7Nao_6.$
Sodium salt (Fleitmann and Henneberg) (*Hexasodic tetra-
phosphate*):—

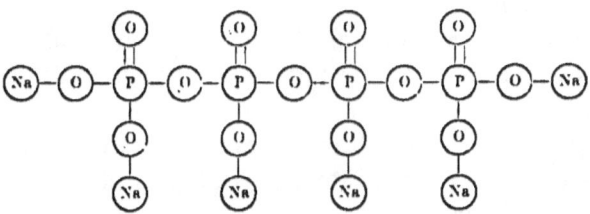

Dodecabasic phosphoric acid .... $\mathbf{P}_{10}O_{19}Nao_{12}.$
Sodium salt (Fleitmann and Henneberg) (*Dodecasodic deca-
phosphate*):—

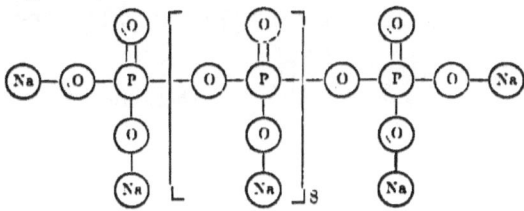

PHOSPHOROUS ANHYDRIDE.

$$P_2O_3.$$

Molecular weight $=110.$

Preparation.—By the slow oxidation of phosphorus in a gentle current of dry air.

Reaction.—In contact with water it produces phosphorous acid :—

$$P_2O_3 \quad + \quad 3OH_2 \quad = \quad 2POHHo_2.$$
Phosphorous anhydride.   Water.   Phosphorous acid.

PHOSPHOROUS ACID.

$$POHHo_2.$$

Molecular weight $=82.$

Preparation.—1. By the action of water on phosphorous anhydride as above.

2. By the slow oxidation of phosphorus in moist air.

3. By the action of water upon phosphorous chloride (see p. 111).

4. By passing chlorine through phosphorus under hot water.

Reactions.—1. When heated, it yields phosphoric acid and phosphoretted hydrogen :—

$$4POHHo_2 \quad = \quad 3POHo_3 \quad + \quad PH_3.$$
Phosphorous acid.   Phosphoric acid.   Phosphoretted hydrogen.

2. It absorbs oxygen from the air, yielding phosphoric acid :—

$$2POHHo_2 \quad + \quad O_2 \quad = \quad 2POHo_3.$$
Phosphorous acid.   Phosphoric acid.

PHOSPHORIC ANHYDRIDE.

$$\mathbf{P_2O_5}.$$

Molecular weight $=142$.

Preparation.—By burning phosphorus in excess of dry air or oxygen.

Reaction.—By contact with water it forms metaphosphoric acid :—

$$\underset{\substack{\text{Phosphoric}\\\text{anhydride.}}}{\mathbf{P_2O_5}} + \underset{\text{Water.}}{\mathbf{OH_2}} = \underset{\substack{\text{Metaphosphoric}\\\text{acid.}}}{\mathbf{2PO_2Ho}}.$$

METAPHOSPHORIC ACID.

$$\mathbf{PO_2Ho}.$$

Molecular weight $=80$.

Preparation.—1. By dissolving phosphoric anhydride in water (see above).

2. By heating phosphoric acid to redness—

$$\underset{\substack{\text{Phosphoric}\\\text{acid.}}}{\mathbf{POHo_3}} = \underset{\substack{\text{Metaphos-}\\\text{phoric acid.}}}{\mathbf{PO_2Ho}} + \underset{\text{Water.}}{\mathbf{OH_2}}.$$

Preparation of metaphosphates.—The metaphosphates may be produced—

1. By igniting a dihydric phosphate with a fixed base :—

$$\underset{\substack{\text{Dihydric sodic}\\\text{phosphate.}}}{\mathbf{POHo_2Nao}} = \underset{\substack{\text{Sodic}\\\text{metaphosphate.}}}{\mathbf{PO_2Nao}} + \underset{\text{Water.}}{\mathbf{OH_2}}.$$

2. By igniting a monohydric phosphate which contains one atom of a volatile base :—

$$\underset{\substack{\text{Hydric sodic ammonic}\\\text{phosphate.}}}{\mathbf{POHoNao(N^vH_4O)}} = \underset{\substack{\text{Sodic}\\\text{metaphosphate.}}}{\mathbf{PO_2Nao}} + \underset{\text{Ammonia.}}{\mathbf{NH_3}} + \underset{\text{Water.}}{\mathbf{OH_2}}.$$

3. By igniting a dihydric pyrophosphate :—

$$\underset{\substack{\text{Dihydric disodic}\\\text{pyrophosphate.}}}{\mathbf{P_2O_3Ho_2Nao_2}} = \underset{\substack{\text{Sodic}\\\text{metaphosphate.}}}{\mathbf{2PO_2Nao}} + \underset{\text{Water.}}{\mathbf{OH_2}}.$$

PYROPHOSPHORIC ACID.

$$P_2O_3Ho_4.$$

Molecular weight $= 178.$

Preparation.—By decomposing plumbic pyrophosphate by hydrosulphuric acid :—

$$P_2O_3Pbo''_2 \quad + \quad 2SH_2 \quad = \quad 2PbS'' \quad + \quad P_2O_3Ho_4.$$

| Plumbic pyrophosphate. | Sulphuretted hydrogen. | Plumbic sulphide. | Pyrophosphoric acid. |

Pyrophosphates are prepared by heating monohydric phosphates containing two atoms of a fixed base :—

$$2POHoNao_2 \quad = \quad P_2O_3Nao_4 \quad + \quad OH_2.$$

| Hydric disodic phosphate. | Sodic pyrophosphate. | Water. |

PHOSPHORIC ACID, *Orthophosphoric Acid.*

$$POHo_3.$$

Molecular weight $= 98.$

Preparation.—1. By boiling a solution of phosphoric anhydride or of metaphosphoric acid in water :—

$$P_2O_5 \quad + \quad 3OH_2 \quad = \quad 2POHo_3.$$

| Phosphoric anhydride. | Water. | Phosphoric acid. |

2. By the oxidation of amorphous phosphorus with nitric acid, and then boiling the product with water.

3. By the action of water upon phosphoric chloride and phosphoric oxytrichloride (see pp. 112 and 113).

4. By the combustion of phosphoretted hydrogen in air or oxygen :—

$$PH_3 \quad + \quad 2O_2 \quad = \quad POHo_3.$$

| Phosphoretted hydrogen. | Phosphoric acid. |

5. By decomposing tricalcic diphosphate (bone-ash) with a large excess of sulphuric acid:—

$$\mathbf{P_2O_2Cao''_3} + \mathbf{3SO_2Ho_2} + \mathbf{6OH_2} = \mathbf{2POHo_3}$$

Tricalcic   Sulphuric   Water.   Phosphoric
diphosphate.   acid.   acid.

$$+ \mathbf{3SHo_4Cao''}.$$

Gypsum.   Tetrahydric
calcic sulphate.

Reaction.—When heated to 213°, it produces pyrophosphoric acid:—

$$\mathbf{2POHo_3} = \mathbf{P_2O_3Ho_4} + \mathbf{OH_2}.$$

Phosphoric   Pyrophosphoric   Water.
acid.   acid.

The phosphates are a numerous and important class of salts. The following list contains some of the most interesting:—

Common sodic phosphate (*Hydric disodic phosphate*)	$\mathbf{POHoNao_2}, 12\mathbf{OH_2}.$
Trisodic phosphate	$\mathbf{PONao_3}, 12\mathbf{OH_2}.$
Hydric sodic potassic phosphate	$\mathbf{POHoNaoKo}, \mathbf{OH_2}.$
Apatite (*Francolite*)	$\mathbf{P_3O_3Cao''_4}\left(\begin{smallmatrix}O\\F\end{smallmatrix}Ca''\right).$
Triple phosphate (*Diammonic* *dimagnesic diphosphate*) ...	$\mathbf{P_2O_2Amo_2Mgo''_2}, 12\mathbf{OH_2}.$
Vivianite	$\mathbf{P_2O_2Feo''_3}, 8\mathbf{OH_2}.$
Wavellite	$\mathbf{P_4O('Al'''_2O_6)^{vi}_3}, 12\mathbf{OH_2}.$
Pyromorphite	$\mathbf{P_3O_3Pbo''_4}\left(\begin{smallmatrix}O\\Cl\end{smallmatrix}Pb''\right).$

ARSENIC, As₄.

Atomic weight $=75$. *Molecular weight* $=300$. *Molecular volume* ☐☐. *1 litre of arsenic vapour weighs* 150 *criths. Sp. gr.* 5·6 *to* 5·9. *Volatile at* 180°. *Atomicity''' and* ˙. *Evidence of atomicity:—*

Arseniuretted hydrogen .... $As'''H_3$.

Arsenious chloride................. $As'''Cl_3$.

Tetrethylarsonic chloride .... As^vEt_4Cl.

Occurrence.—In nature, in various ores, and sometimes in the free state. In some mineral waters, and in the mud of rivers.

Preparation.—By reducing, with charcoal, arsenious anhydride, which is produced in the roasting of many ores :—

$$As_2O_3 \;+\; 3C \;=\; As_2 \;+\; 3CO.$$

Arsenious
anhydride.   Carbonic
oxide.

COMPOUND OF ARSENIC WITH HYDROGEN.

ARSENIURETTED HYDROGEN, *Arsenious Hydride.*
AsH_3.

Molecular weight $=78$. *Molecular volume* ☐☐. 1 *litre weighs*
39 *criths. Boils at* $-40°$.

Preparation.—1. In the pure state by the action of sulphuric acid on an alloy of arsenic and zinc :—

$$As_2Zn''_3 \;+\; 3SO_2Ho_2 \;=\; 3SO_2Zno'' \;+\; 2AsH_3.$$

Arsenious
zincide.   Sulphuric
acid.   Zincic
sulphate.   Arseniuretted
hydrogen.

2. By the action of nascent hydrogen upon soluble arsenic compounds, as by the introduction of arsenious acid into an apparatus evolving hydrogen :—

$$AsHo_3 \;+\; 3H_2 \;=\; AsH_3 \;+\; 3OH_2.$$

Arsenious
acid.   Arseniuretted
hydrogen.   Water.

Reactions.—1. When burnt with free access of air, it gives water and arsenious anhydride :—

$$2AsH_3 \;+\; 3O_2 \;=\; As_2O_3 \;+\; 3OH_2.$$

Arseniuretted
hydrogen.   Arsenious
anhydride.   Water.

2. When burnt with a limited supply of air, it yields water and free arsenic :—

$$4\mathbf{AsH_3} \ + \ 3O_2 \ = \ As_4 \ + \ 6OH_2.$$

Arseniuretted
hydrogen.   Water.

3. When exposed to a red heat, it is decomposed into arsenic and hydrogen.

4. Passed through a solution of argentic nitrate, it yields a precipitate of metallic silver, arsenious and nitric acids remaining in solution :—

$$6\mathbf{NO_2Ago} \ + \ 3OH_2 \ + \ \mathbf{AsH_3} \ = \ 6NO_2Ho$$

Argentic nitrate.   Water.   Arseniuretted
hydrogen.   Nitric acid.

$$+ \ \mathbf{AsHo_3} \ + \ 3Ag_2.$$

Arsenious acid.

COMPOUND OF ARSENIC WITH CHLORINE.
ARSENIOUS CHLORIDE.
$\mathbf{AsCl_3}$.

Molecular weight $=181\cdot5$. *Molecular volume* ☐☐. 1 *litre of arsenious chloride vapour weighs* 90·75 *criths. Sp. gr.* 2·205. *Boils at* 132°.

Preparation.—1. By the action of dry chlorine upon arsenic :—

$$As_2 \ + \ 3Cl_2 \ = \ 2\mathbf{AsCl_3}.$$

Arsenious
chloride.

2. By distilling arsenic with mercuric chloride (corrosive sublimate) :—

$$As_2 \ + \ 6\mathbf{HgCl_2} \ = \ 3'\mathbf{Hg'_2Cl_2} \ + \ 2\mathbf{AsCl_3}.$$

Mercuric
chloride.   Mercurous
chloride.   Arsenious
chloride.

3. By distilling sodic chloride, arsenious anhydride, and sulphuric acid :—

$$\mathbf{As_2O_3} \ + \ 6NaCl \ + \ 6\mathbf{SO_2Ho_2} \ = \ 2\mathbf{AsCl_3}$$

Arsenious
anhydride.   Sodic
chloride.   Sulphuric
acid.   Arsenious
chloride.

$$+ \ 6\mathbf{SO_2HoNao} \ + \ 3OH_2.$$

Hydric sodic
sulphate.   Water.

G

Reaction.—With excess of water it forms arsenious and hydrochloric acids :—

$$\mathbf{As}Cl_3 \;+\; 3\mathbf{OH}_2 \;=\; 3HCl \;+\; \mathbf{As}Ho_3.$$

Arsenious chloride.   Water.   Hydrochloric acid.   Arsenious acid.

COMPOUNDS OF ARSENIC WITH OXYGEN AND HYDROXYL.

Arsenious anhydride .... $\mathbf{As}_2O_3$.
Arsenic anhydride .... $\mathbf{As}_2O_5$.
Arsenious acid .... $\mathbf{As}Ho_3$.
Arsenic acid .... $\mathbf{As}OHo_3$.

ARSENIOUS ANHYDRIDE.
$\mathbf{As}_2O_3$.

Molecular weight $=198$.   *Molecular volume* □. 1 *litre of arsenious anhydride vapour weighs* 198 *criths* (*anomalous*). *Sp. gr.* 3·7.

Occurrence.—Very rare in nature.
Preparation.—1. By burning arsenic in air or oxygen.
2. By roasting arsenical ores in certain metallurgical operations.

ARSENIOUS ACID.
$\mathbf{As}Ho_3$.

Molecular weight $=126$.

Only known in solution.

Arsenious acid forms many salts, of which the following are examples :—

Dihydric potassic arsenite (*Fowler's solution*) .... $\mathbf{As}Ho_2Ko$.
Hydric cupric arsenite (*Scheele's green*) .... $\mathbf{As}HoCuo''$.
Triargentic arsenite .... $\mathbf{As}Ago_3$.

A monobasic arsenious acid, $\mathbf{As}OHo$, corresponding to nitrous acid, appears to exist, one of its compounds, $\mathbf{As}OAmo$, being known. Arsenious acid when boiled with cupric acetate yields Schweinfurt green, $3\mathbf{As_2O_2Cuo''}$, $\mathbf{Cu}(C_2H_3O_2)_2$.

ARSENIC ANHYDRIDE.
$\mathbf{As_2O_5}$.

Molecular weight $=230$.

Preparation.—By heating arsenic acid nearly to redness :—

$$2\mathbf{As}OHo_3 \;=\; 3OH_2 \;+\; \mathbf{As_2O_5}.$$
$$\text{Arsenic acid.} \qquad \text{Water.} \qquad \substack{\text{Arsenic}\\\text{anhydride.}}$$

ARSENIC ACID.
$\mathbf{As}OHo_3$.

Molecular weight $=142$.

Preparation.—By treating arsenious anhydride with nitric acid :—

$$\mathbf{As_2O_3} \;+\; 2\mathbf{NO_2Ho} \;+\; 2OH_2 \;=\; 2\mathbf{As}OHo_3 \;+\; \mathbf{N_2O_3}.$$
$$\substack{\text{Arsenious}\\\text{anhydride.}} \quad \text{Nitric acid.} \quad \text{Water.} \quad \text{Arsenic acid.} \quad \substack{\text{Nitrous}\\\text{anhydride.}}$$

Salts are known derived from acids of the three following formulæ :—

$\mathbf{As}O_2Ho,$	$\mathbf{As}OHo_3,$	$\mathbf{As_2}O_3Ho_4,$
Metarsenic acid.	Arsenic acid.	Pyrarsenic acid.
corresponding to	corresponding to	corresponding to
$\mathbf{P}O_2Ho$	$\mathbf{P}OHo_3.$	$\mathbf{P_2}O_3Ho_4.$
Metaphosphoric acid.	Phosphoric acid.	Pyrophosphoric acid.

and

$\mathbf{N}O_2Ho.$
Nitric acid.

G 2

COMPOUNDS OF ARSENIC WITH SULPHUR AND HYDROSULPHYL.

Realgar .... $\left\{\begin{array}{l} \mathbf{As}S'' \\ \mathbf{As}S'' \end{array}\right. = \mathbf{'As''_2S''_2}.$

Sulpharsenious anhydride (*Arsenious sulphide*) .... $\left.\right\}$ $\mathbf{As_2S''_3}.$

Sulpharsenic anhydride (*Arsenic sulphide*) .... $\left.\right\}$ $\mathbf{As_2S''_5}.$

Sulpharsenious acid..................... $\mathbf{As}Hs_3.$

Sulpharsenic acid...................... $\mathbf{As}S''Hs_3.$

REALGAR, *Diarsenious Disulphide.*

$$\mathbf{'As''_2S''_2}, \quad \text{or} \quad \left\{\begin{array}{l} \mathbf{As}S'' \\ \mathbf{As}S'' \end{array}\right.$$

Molecular weight $=214$.   *Sp. gr.* $3\cdot5$.

Occurrence.—Found native.

Preparation.—By heating sulphur with arsenious anhydride :—

$$S_7 \quad + \quad 2\mathbf{As_2O_3} \quad = \quad 3SO_2 \quad + \quad 2\mathbf{'As''_2S''_2}.$$
Arsenious   Sulphurous   Realgar.
anhydride.   anhydride.

SULPHARSENIOUS ANHYDRIDE, *Arsenious Sulphide, Orpiment.*

$$\mathbf{As_2S''_3}.$$

Molecular weight $=246$.   *Sp. gr.* $3\cdot5$.

Occurrence.—Found native.

Preparation.—By passing sulphuretted hydrogen through a solution of arsenious anhydride in hydrochloric acid :—

$$2\mathbf{As}Cl_3 \quad + \quad 3\mathbf{S}H_2 \quad = \quad 6HCl \quad + \quad \mathbf{As_2S''_3}.$$
Arsenious   Sulphuretted   Hydrochloric   Arsenious
chloride.   hydrogen.   acid.   sulphide.

Reaction.—Arsenious sulphide dissolves in caustic alkali, producing an arsenite and a sulpharsenite :—

$$As_2S''_3 + 4OKH = AsHoKo_2 + AsHsKs_2 + OH_2.$$

Arsenious sulphide.   Potassic hydrate.   Hydric dipotassic arsenite.   Sulphhydric disulphopotassic sulpharsenite.   Water.

By the addition of an acid, the arsenious sulphide is reprecipitated :—

$$AsHoKo_2 + AsHsKs_2 + 4HCl = 4KCl$$

Hydric dipotassic arsenite.   Sulphhydric disulphopotassic sulpharsenite.   Hydrochloric acid.   Potassic chloride.

$$+ As_2S''_3 + 3OH_2.$$

Arsenious sulphide.   Water.

Proustite is a native sulphargentic sulpharsenite, $AsAgs_3$.

SULPHARSENIC ANHYDRIDE, *Arsenic Sulphide.*
As_2S_5.

Molecular weight $=310$.

Preparation.—By fusing together arsenious sulphide and sulphur. Sulpharseniates may be obtained by passing sulphuretted hydrogen through solutions of arseniates :—

$$AsOHoNao_2 + 4SH_2 = AsSHsNas_2 + 4OH_2.$$

Hydric disodic arseniate.   Sulphuretted hydrogen.   Sulphhydric disulphosodic sulpharseniate.   Water.

ANTIMONY, Sb_4?

Atomic weight $=122$. *Probable molecular weight* $=488$. *Sp. gr.* $6·7$. *Fuses at* $430°$. *Atomicity* ''' *and* ^{v}. *Evidence of atomicity* :—

Antimonious chloride .... $Sb'''Cl_3$.
Antimonic tetretho-chloride
  (*Tetrethyl-stibonic chloride*). } Sb^vEt_4Cl.
Antimonic chloride .... Sb^vCl_5.

Occurrence.—To a small extent in the native state. Alloyed with metals in a few minerals. Sometimes in the form of oxide, but principally in the form of grey antimony ore or *stibnite*, which consists of antimonious sulphide.

Preparation.—1. By fusing the native sulphide and introducing metallic iron, which removes the sulphur:—

$$Sb_2S''_3 \ + \ 3Fe \ = \ 3FeS'' \ + \ Sb_2.$$

Antimonious   Ferrous
sulphide.   sulphide.

2. The native sulphide is roasted in contact with the air, when it is partially converted into antimonious oxide:—

$$2Sb_2S''_3 \ + \ 9O_2 \ = \ 2Sb_2O_3 \ + \ 6SO_2.$$

Antimonious   Antimonious   Sulphurous
sulphide.   oxide.   anhydride.

The roasted mineral is then fused with charcoal and sodic carbonate. The reaction takes place in two stages: first, the remaining sulphide is converted into oxide by the sodic carbonate, and subsequently the oxide is reduced by the carbon:—

1. $$Sb_2S''_3 \ + \ 3CONao_2 \ = \ 3CO_2 \ + \ 3SNa_2 \ + \ Sb_2O_3;$$

Antimonious   Sodic   Carbonic   Sodic   Antimonious
sulphide.   carbonate.   anhydride.   sulphide.   oxide.

2. $$Sb_2O_3 \ + \ 3C \ = \ 3CO \ + \ Sb_2.$$

Antimonious   Carbonic
oxide.   oxide.

3. Antimony may be obtained in the pure condition by reducing, with charcoal, the oxide formed by the action of nitric acid upon crude antimony.

COMPOUND OF ANTIMONY WITH HYDROGEN.

ANTIMONIURETTED HYDROGEN, *Antimonious Hydride.*

$$SbH_3.$$

Molecular weight $= 125$?

This compound is unknown in the pure condition.

Preparation.—1. By the action of hydrochloric acid upon an alloy of zinc and antimony :—

$$\mathbf{Sb_2Zn''_3} \ + \ 6HCl \ = \ 2\mathbf{SbH_3} \ + \ 3\mathbf{ZnCl_2}.$$

Antimonious   Hydrochloric   Antimonious   Zincic
zincide.   acid.   hydride.   chloride.

2. By the action of nascent hydrogen evolved from zinc and sulphuric acid upon soluble antimony compounds. In both these reactions the antimonious hydride is always mixed with much hydrogen :—

$$\mathbf{SbCl_3} \ + \ 3H_2 \ = \ \mathbf{SbH_3} \ + \ 3HCl.$$

Antimonious   Antimonious   Hydrochloric
chloride.   hydride.   acid.

Reactions.—1. When burnt in air or in oxygen, it yields water and antimonious oxide :—

$$2\mathbf{SbH_3} \ + \ 3O_2 \ = \ \mathbf{Sb_2O_3} \ + \ 3OH_2.$$

Antimonious   Antimonious   Water.
hydride.   oxide.

2. When burnt with a limited supply of air the hydrogen only is oxidized, the antimony being deposited :—

$$4\mathbf{SbH_3} \ + \ 3O_2 \ = \ Sb_4 \ + \ 6OH_2.$$

Antimonious   Water.
hydride.

3. Decomposed into its elements, like arsenious hydride, when passed through a red-hot tube.

4. When transmitted through a solution of argentic nitrate, it produces a precipitate of antimonious argentide, thus differing from arsenious hydride (see p. 121) :—

$$3\mathbf{NO_2}Ago \ + \ \mathbf{SbH_3} \ = \ 3\mathbf{NO_2}Ho \ + \ \mathbf{SbAg_3}.$$

Argentic   Antimonious   Nitric   Antimonious
nitrate.   hydride.   acid.   argentide.

From the composition of this compound, and from that of some of its analogues, the composition of antimonious hydride is inferred.

Antimonious hydride........................ $\mathbf{SbH_3}$.

Antimonious bromide .... $\mathbf{SbBr_3}$.

Antimonious argentide .... $\mathbf{SbAg_3}$.

Antimonious zincide .... $\mathbf{Sb_2Zn''_3}$.

Antimonious ethide. (*Triethylstibine.*) $\mathbf{SbEt_3}$.

Antimonious amylide. (*Triamylstibine.*) $\mathbf{SbAy_3}$.

COMPOUNDS OF ANTIMONY WITH CHLORINE.

Antimonious chloride.................. $\mathbf{SbCl_3}$.

Antimonic chloride.................... $\mathbf{SbCl_5}$.

ANTIMONIOUS CHLORIDE.

$\mathbf{SbCl_3}$.

Molecular weight $= 228{\cdot}5$. *Molecular volume* ☐. 1 *litre of
antimonious chloride vapour weighs* 114·25 *criths. Fuses
at* 72°. *Boils at* 223°.

Preparation.—1. By passing chlorine over excess of metallic
antimony or antimonious sulphide, and purifying by distillation :—

$$Sb_2 \; + \; 3Cl_2 \; = \; 2\mathbf{SbCl_3}.$$
Antimonious
chloride.

$$2\mathbf{Sb_2S''_3} \; + \; 9Cl_2 \; = \; 4\mathbf{SbCl_3} \; + \; 3'\mathbf{S'_2Cl_2}.$$
Antimonious   Antimonious   Disulphur
sulphide.   chloride.   dichloride.

2. By dissolving antimonious sulphide in hydrochloric acid,
or antimony in hydrochloric acid containing a little nitric acid,
evaporating and distilling the product :—

$$\mathbf{Sb_2S''_3} \; + \; 6HCl \; = \; 3\mathbf{SH_2} \; + \; 2\mathbf{SbCl_3}.$$
Antimonious   Hydrochloric   Sulphuretted   Antimonious
sulphide.   acid.   hydrogen.   chloride.

$$Sb_2 + 6HCl \; + \; 6\mathbf{NO_2Ho} \; = \; 2\mathbf{SbCl_3} \; + \; 6\mathbf{OH_2} \; + \; 3'\mathbf{N^{iv}_2O_4}.$$
Hydrochloric   Nitric   Antimonious   Water.   Nitric
acid.   acid.   chloride.   peroxide.

3. By distilling antimony or antimonious sulphide with mercuric chloride:—

$$\text{Sb}_4 + 4\textbf{HgCl}_2 = 2\textbf{SbCl}_3 + \textbf{'Sb''}_2\text{Hg''}_2 + \textbf{'Hg'}_2\text{Cl}_2.$$

Mercuric   Antimonious   Dimercuric   Mercurous
chloride.   chloride.   diantimonide.   chloride.

$$\textbf{Sb}_2\textbf{S''}_3 + 3\textbf{HgCl}_2 = 2\textbf{SbCl}_3 + 3\textbf{HgS''}.$$

Antimonious   Mercuric   Antimonious   Mercuric
sulphide.   chloride.   chloride.   sulphide.

4. By distilling antimonious sulphate with sodic chloride:—

$$\textbf{S}_3\textbf{O}_6\text{Sbo'''}_2 + 6\text{NaCl} = 2\textbf{SbCl}_3 + 3\textbf{SO}_2\text{Nao}_2.$$

Antimonious   Sodic   Antimonious   Sodic
sulphate.   chloride.   chloride.   sulphate.

Reaction.—With water it produces antimonious oxychloride:—

$$\textbf{SbCl}_3 + \textbf{OH}_2 = 2\text{HCl} + \textbf{SbOCl}.$$

Antimonious   Water.   Hydrochloric   Antimonious
chloride.   acid.   oxychloride.

Long-continued action of water transforms this compound into antimonious oxide:—

$$2\textbf{SbOCl} + \textbf{OH}_2 = 2\text{HCl} + \textbf{Sb}_2\textbf{O}_3.$$

Antimonious   Water.   Hydrochloric   Antimonious
oxychloride.   acid.   oxide.

ANTIMONIC CHLORIDE.

$\textbf{SbCl}_5.$

Molecular weight $= 299\cdot5$.   *Fuses at* $0°$.

Preparation.—1. By acting upon antimony with excess of chlorine:—

$$\text{Sb}_2 + 5\text{Cl}_2 = 2\textbf{SbCl}_5.$$

Antimonic
chloride.

2. By passing chlorine over antimonious chloride, the latter liquefies, producing antimonic chloride:—

$$\textbf{SbCl}_3 + \text{Cl}_2 = \textbf{SbCl}_5.$$

Antimonious   Antimonic
chloride.   chloride.

Reactions.—1. With a small quantity of water it forms antimonic oxytrichloride, analogous to phosphoric oxytrichloride:—

$$\underset{\substack{\text{Antimonic} \\ \text{chloride.}}}{\textbf{SbCl}_5} + \underset{\substack{\text{Water.}}}{\textbf{OH}_2} = \underset{\substack{\text{Antimonic} \\ \text{oxytrichloride.}}}{\textbf{SbOCl}_3} + \underset{\substack{\text{Hydrochloric} \\ \text{acid.}}}{\text{2HCl.}}$$

2. An excess of water transforms antimonic chloride into orthantimonic acid or pyrantimonic acid, corresponding to pyrophosphoric acid:—

$$\underset{\substack{\text{Antimonic} \\ \text{chloride.}}}{\textbf{SbCl}_5} + \underset{\substack{\text{Water.}}}{\textbf{4OH}_2} = \underset{\substack{\text{Orthantimonic} \\ \text{acid.}}}{\textbf{SbOHo}_3} + \underset{\substack{\text{Hydrochloric} \\ \text{acid.}}}{\text{5HCl}:}$$

$$\text{or } \underset{\substack{\text{Antimonic} \\ \text{chloride.}}}{\textbf{2SbCl}_5} + \underset{\substack{\text{Water.}}}{\textbf{7OH}_2} = \underset{\substack{\text{Pyrantimonic} \\ \text{acid.}}}{\textbf{Sb}_2\textbf{O}_3\textbf{Ho}_4} + \underset{\substack{\text{Hydrochloric} \\ \text{acid.}}}{\text{10HCl.}}$$

3. By the action of sulphuretted hydrogen antimonic sulphotrichloride is formed:—

$$\underset{\substack{\text{Antimonic} \\ \text{chloride.}}}{\textbf{SbCl}_5} + \underset{\substack{\text{Sulphuretted} \\ \text{hydrogen.}}}{\textbf{SH}_2} = \underset{\substack{\text{Antimonic} \\ \text{sulphotrichloride.}}}{\textbf{SbS''Cl}_3} + \underset{\substack{\text{Hydrochloric} \\ \text{acid.}}}{\text{2HCl.}}$$

Antimonious bromide, $\textbf{SbBr}_3$, resembles antimonious chloride; it fuses at 90°, boils at 270°, and by the action of water is converted into the *oxybromide,* $\textbf{SbOBr}$.

Antimonious iodide, $\textbf{SbI}_3$, when acted upon by water forms the *oxyiodide,* $\textbf{SbOI}$.

The corresponding *fluoride,* $\textbf{SbF}_3$, is said to exist and to be soluble in water without decomposition.

OXIDES AND ACIDS OF ANTIMONY.

Antimonious oxide or anhydride ...... $\textbf{Sb}_2\textbf{O}_3.$

Diantimonic tetroxide .... $\textbf{'Sb}^{iv}_2\textbf{O}_4.$

Antimonic anhydride .... Sb_2O_5.
Metantimonious acid .... $SbOHo$.
Orthantimonic acid .... $SbOHo_3$?
Metantimonic acid .... SbO_2Ho.
Pyrantimonic acid .... $Sb_2O_3Ho_4$.

ANTIMONIOUS OXIDE, OR ANHYDRIDE.

$$Sb_2O_3.$$

Molecular weight $= 292$.

Occurrence.—In nature in the rare minerals *valentinite* and *senarmontite*.

Preparation.—1. By burning antimony in air:—

$$2Sb_2 + 3O_2 + 2Sb_2O_3.$$
Antimonious oxide.

2. By pouring a solution of antimonious chloride in dilute hydrochloric acid into a boiling solution of sodic carbonate:—

$$2SbCl_3 + 3CONao_2 = Sb_2O_3 + 3CO_2 + 6NaCl.$$
Antimonious chloride. Sodic carbonate. Antimonious oxide. Carbonic anhydride. Sodic chloride.

3. By heating metantimonious acid to the temperature of boiling water:—

$$2SbOHo = OH_2 + Sb_2O_3.$$
Metantimonious acid. Water. Antimonious oxide.

Reactions.—1. When heated to redness in the air, it burns like tinder, forming diantimonic tetroxide:—

$$Sb_2O_3 + O = 'Sb^{iv}_2O.$$
Antimonious oxide. Diantimon tetroxide.

2. Readily reduced to the metallic state by ignition with charcoal, hydrogen, &c. :—

$$\underset{\substack{\text{Antimonious} \\ \text{oxide.}}}{Sb_2O_3} + C_3 = Sb_2 + \underset{\substack{\text{Carbonic} \\ \text{oxide.}}}{3CO.}$$

$$\underset{\substack{\text{Antimonious} \\ \text{oxide.}}}{Sb_2O_3} + 3H_2 = Sb_2 + \underset{\text{Water.}}{3OH_2.}$$

3. Readily dissolved by a hot solution of hydric potassic tartrate (cream of tartar), forming potassic antimonylic tartrate (tartar emetic) :—

$$2\begin{cases} COHo \\ CHHo \\ CHHo \\ COKo \end{cases} + \underset{\substack{\text{Antimonious} \\ \text{oxide.}}}{Sb_2O_3} = 2\begin{cases} CO(Sb'''O_2) \\ CHHo \\ CHHo \\ COKo \end{cases} + \underset{\text{Water.}}{OH_2.}$$

Hydric potassic tartrate (Cream of tartar).   Antimonious oxide.   Potassic antimonylic tartrate (Tartar emetic).

4. Dissolved by hydrochloric acid, forming antimonious chloride :—

$$\underset{\substack{\text{Antimonious} \\ \text{oxide.}}}{Sb_2O_3} + \underset{\substack{\text{Hydrochloric} \\ \text{acid.}}}{6HCl} = \underset{\substack{\text{Antimonious} \\ \text{chloride.}}}{2SbCl_3} + \underset{\text{Water.}}{3OH_2.}$$

METANTIMONIOUS ACID.

SbOHo.

Molecular weight $=155$.

Preparation.—By pouring a solution of antimonious chloride into a cold solution of sodic carbonate :—

$$\underset{\substack{\text{Antimonious} \\ \text{chloride.}}}{2SbCl_3} + \underset{\substack{\text{Sodic} \\ \text{carbonate.}}}{3CONao_2} + \underset{\text{Water.}}{OH_2} = \underset{\substack{\text{Metantimonious} \\ \text{acid.}}}{2SbOHo}$$

$$+ \underset{\substack{\text{Carbonic} \\ \text{anhydride.}}}{3CO_2} + \underset{\substack{\text{Sodic} \\ \text{chloride.}}}{6NaCl.}$$

Reactions.—1. Decomposed by heat (page 131).

2. Readily dissolved by alkaline hydrates, producing ill-defined antimonites.

DIANTIMONIC TETROXIDE.

$$\left\{ \begin{array}{l} SbO_2 \\ SbO_2 \end{array} \right. = \; 'Sb^{iv}_2O_4.$$

Molecular weight =308.

Occurrence.—Found native as *cervantite.*

Preparation.—1. By igniting antimonic oxide, or the white solid produced by the action of nitric acid upon metallic antimony :—

$$2Sb_2O_5 \; = \; 2'Sb^{iv}_2O_4 \; + \; O_2.$$

Antimonic oxide.   Diantimonic tetroxide.

2. By heating antimonious oxide in contact with the air :—

$$2Sb_2O_3 \; + \; O_2 \; = \; 2'Sb^{iv}_2O_4.$$

Antimonious oxide.   Diantimonic tetroxide.

ANTIMONIC ANHYDRIDE.

Sb_2O_5.

Molecular weight =324.   *Sp. gr.* 6·6.

Preparation.—By gently heating the corresponding acids :—

$$2SbOHo_3 \; = \; Sb_2O_5 \; + \; 3OH_2.$$

Orthantimonic acid.   Antimonic anhydride.   Water.

$$2SbO_2Ho \; = \; Sb_2O_5 \; + \; OH_2.$$

Metantimonic acid.   Antimonic anhydride.   Water.

$$Sb_2O_3Ho_4 \; = \; Sb_2O_5 \; + \; 2OH_2.$$

Pyrantimonic acid.   Antimonic anhydride.   Water.

Reactions.—1. When heated, it is decomposed into dianti-monic tetroxide and oxygen (see page 133).

2. Fused with potassic carbonate, it produces potassic met-antimoniate :—

$$\underset{\substack{\text{Antimonic}\\\text{anhydride.}}}{\textbf{Sb}_2\textbf{O}_5} + \underset{\substack{\text{Potassic}\\\text{carbonate.}}}{\textbf{COKo}_2} = \underset{\substack{\text{Potassic}\\\text{metantimoniate.}}}{\textbf{2SbO}_2\textbf{Ko}} + \underset{\substack{\text{Carbonic}\\\text{anhydride.}}}{\textbf{CO}_2.}$$

ORTHANTIMONIC ACID ?

$\textbf{SbOHo}_3$?

Preparation.—Said to be formed by the action of water upon antimonic chloride (see p. 130).

METANTIMONIC ACID.

$\textbf{SbO}_2\textbf{Ho.}$

Preparation.—1. By the action of nitric acid containing a little hydrochloric acid on metallic antimony :—

$$\text{Sb}_2 + \underset{\text{Nitric acid.}}{\textbf{4NO}_2\textbf{Ho}} = \underset{\substack{\text{Metantimonic}\\\text{acid.}}}{\textbf{2SbO}_2\textbf{Ho}} + \underset{\substack{\text{Nitrous}\\\text{anhydride.}}}{\textbf{N}_2\textbf{O}_3}$$

$$+ \underset{\text{Nitric oxide.}}{\textbf{N}''_2\textbf{O}_2} + \underset{\text{Water.}}{\textbf{OH}_2.}$$

2. By the spontaneous dehydration of orthantimonic acid, or of pyrantimonic acid :—

$$\underset{\substack{\text{Orthantimonic}\\\text{acid.}}}{\textbf{SbOHo}_3} = \underset{\text{Water.}}{\textbf{OH}_2} + \underset{\substack{\text{Metantimonic}\\\text{acid.}}}{\textbf{SbO}_2\textbf{Ho}} ;$$

$$\underset{\substack{\text{Pyrantimonic}\\\text{acid.}}}{\textbf{Sb}_2\textbf{O}_3\textbf{Ho}_4} = \underset{\text{Water.}}{\textbf{OH}_2} + \underset{\substack{\text{Metantimonic}\\\text{acid.}}}{\textbf{2SbO}_2\textbf{Ho.}}$$

Reaction.—By the action of alkaline hydrates it produces either metantimoniates or orthantimoniates :—

$$\underset{\substack{\text{Metantimonic}\\\text{acid.}}}{\mathbf{SbO_2Ho}} + \underset{\substack{\text{Potassic}\\\text{hydrate.}}}{\mathbf{OKH}} = \underset{\substack{\text{Potassic}\\\text{metantimoniate.}}}{\mathbf{SbO_2Ko}} + \underset{\substack{\text{Water.}}}{\mathbf{OH_2}};$$

$$\underset{\substack{\text{Metantimonic}\\\text{acid.}}}{\mathbf{SbO_2Ho}} + \underset{\substack{\text{Potassic}\\\text{hydrate.}}}{\mathbf{OKH}} = \underset{\substack{\text{Dihydric potassic}\\\text{orthantimoniate.}}}{\mathbf{SbOHo_2Ko}}.$$

PYRANTIMONIC ACID, *Parantimonic Acid.*
(*Metantimonic acid of Frémy.*)
$$\mathbf{Sb_2O_3Ho_4}.$$

Preparation.—By acidifying solutions of pyrantimoniates :—

$$\underset{\substack{\text{Dihydric dipotassic}\\\text{pyrantimoniate.}}}{\mathbf{Sb_2O_3Ko_2Ho_2}} + \underset{\substack{\text{Hydrochloric}\\\text{acid.}}}{\mathbf{2HCl}} = \underset{\substack{\text{Pyrantimonic}\\\text{acid.}}}{\mathbf{Sb_2O_3Ho_4}} + \underset{\substack{\text{Potassic}\\\text{chloride.}}}{\mathbf{2KCl}}.$$

Dihydric dipotassic pyrantimoniate is prepared by fusing antimonic anhydride with excess of potassic hydrate, and extracting the mass with water, when an alkaline solution containing dihydric dipotassic pyrantimoniate $\mathbf{Sb_2O_3Ho_2Ko_2}$ is formed. This solution produces precipitates in solutions of sodium salts, the sodic pyrantimoniate thus formed containing $\mathbf{Sb_2O_3Ho_2Nao_2, 6OH_2}$.

COMPOUND OF ANTIMONY WITH OXYGEN AND SULPHUR.

ANTIMONIOUS OXYDISULPHIDE.

$$\left\{ \begin{array}{l} \mathbf{SbS''} \\ \mathbf{O} \\ \mathbf{SbS''} \end{array} \right. = \mathbf{Sb_2S''_2O}.$$

Molecular weight $= 324$.

Occurs as a rare mineral known as *red antimony.*

COMPOUNDS OF ANTIMONY AND SULPHUR.

Antimonious sulphide .... $Sb_2S''_3$.
Antimonic sulphide .... $Sb_2S''_5$.

ANTIMONIOUS SULPHIDE, *Sulphantimonious Anhydride.*

$$Sb_2S''_3.$$

Molecular weight $= 340$.

Occurrence.—In nature as *stibnite* or *grey antimony ore*.

Preparation.—1. By heating together antimony and sulphur, or antimonious oxide and sulphur in the proper proportions:—

$$2Sb_2 + 3S_2 = 2Sb_2S_3.$$
Antimonious
sulphide.

$$2Sb_2O_3 + S_9 = 2Sb_2S''_3 + 3SO_2.$$
Antimonious   Antimonious   Sulphurous
oxide.   sulphide.   anhydride.

2. By passing sulphuretted hydrogen through a solution of antimonious chloride:—

$$2SbCl_3 + 3SH_2 = Sb_2S''_3 + 6HCl.$$
Antimonious   Sulphuretted   Antimonious   Hydrochloric
chloride.   hydrogen.   sulphide.   acid.

Reactions.—1. Decomposed by hot hydrochloric acid (see p. 128).

2. Soluble with decomposition in solutions of alkaline hydrates:—

$$Sb_2S''_3 + 6KHo = SbKs_3 + SbKo_3 + 3OH_2.$$
Antimonious   Potassic   Tripotassic   Tripotassic   Water.
sulphide.   hydrate.   sulphanti-   antimonite.
                            monite.

Addition of an acid reproduces and precipitates the antimonious sulphide:—

$$SbKs_3 + SbKo_3 + 6HCl = Sb_2S''_3 + 6KCl + 3OH_2.$$
Trisulpho-   Tripotassic   Hydro-   Antimo-   Potassic   Water.
potassic   antimonite.   chloric   nious   chloride.
sulphanti-   acid.   sulphide.
monite.

3. Soluble in alkaline sulphhydrates :—

$$\mathbf{Sb_2S''_3} + \mathbf{6KHs} = \mathbf{2SbKs_3} + \mathbf{3SH_2}.$$

Antimonious sulphide.   Potassic sulph-hydrate.   Trisulphopotassic sulphanti-monite.   Sulphuretted hydrogen.

SULPHANTIMONITES.

Many sulphantimonites occur in nature :—

Orthosulphantimonites.

General formulæ :—$\mathbf{Sb}Ms_3$ and $\mathbf{Sb_2}Ms''_3$.

Dark-red silver. *Trisulphargentic sulphan-timonite* .... $\mathbf{SbAgs_3}$.

Boulangerite. *Trisulphoplumbic sulphan-timonite* .... $\mathbf{Sb_2Pbs''_3}$.

Bournonite. *Disulphoplumbic sulphocu-prous sulphantimonite* .... $\mathbf{Sb_2Pbs''_2(Cu_2S''_2)''}$.

Metasulphantimonites.

General formulæ :—$\mathbf{SbS''}Ms$ and $\mathbf{Sb_2S''_2}Ms''$.

Miargyrite. *Sulphargentic metasulphan-timonite* .... $\mathbf{SbS''Ags}$.

Zinkenite. *Sulphoplumbic metasulphanti-monite* .... $\mathbf{Sb_2S''_2Pbs''}$.

Antimony copper glance. *Sulphocuprous metasulphantimonite* .... $\mathbf{Sb_2S''_2(Cu_2S''_2)''}$.

Berthierite. *Sulphoferrous metasulphan-timonite* .... $\mathbf{Sb_2S''_2Fes''}$.

Pyrosulphantimonites.

General formulæ :—$\mathbf{Sb_2S''}Ms_4$ and $\mathbf{Sb_2S''Ms''_2}$.

Feather ore. *Sulphoplumbic pyrosulphan-timonite* .... $\mathbf{Sb_2S''Pbs''_2}$.

Fahl ore. *Sulphocuprosoferrous pyrosulph-antimonite* .... $\mathbf{Sb_2S''(Cu_2FeS''_3)''_2}$.

ANTIMONIC SULPHIDE, *Sulphantimonic Anhydride.*

$$Sb_2S''_5.$$

Molecular weight $= 404.$

Preparation.—1. By passing sulphuretted hydrogen through a solution of antimonic chloride :—

$$2SbCl_5 \ + \ 5SH_2 \ = \ Sb_2S''_5 \ + \ 10HCl.$$

| Antimonic chloride. | Sulphuretted hydrogen. | Antimonic sulphide. | Hydrochloric acid. |

2. By the addition of an acid to a solution of a sulphantimoniate :—

$$2SbS''Nas_3 \ + \ 6HCl \ = \ Sb_2S''_5 \ + \ 6NaCl \ + \ 3SH_2.$$

| Trisulphosodic sulphantimoniate. | Hydrochloric acid. | Antimonic sulphide. | Sodic chloride. | Sulphuretted hydrogen. |

Reactions.—1. Decomposed by boiling hydrochloric acid, into antimonious chloride, sulphuretted hydrogen, and sulphur :—

$$Sb_2S''_5 \ + \ 6HCl \ = \ 2SbCl_3 \ + \ 3SH_2 \ + \ S_2.$$

| Antimonic sulphide. | Hydrochloric acid. | Antimonious chloride. | Sulphuretted hydrogen. |

2. Soluble in solutions of alkaline sulphides :—

$$Sb_2S''_5 \ + \ 3SK_2 \ = \ 2SbS''Ks_3.$$

| Antimonic sulphide. | Potassic sulphide. | Trisulphopotassic sulphantimoniate. |

3. Soluble in solutions of alkaline hydrates :—

$$4Sb_2S''_5 \ + \ 24OKH \ = \ 3SbOKo_3 \ + \ 5SbS''Ks_3 \ + \ 12OH_2.$$

| Antimonic sulphide. | Potassic hydrate. | Tripotassic antimoniate. | Trisulphopotassic sulphantimoniate. | Water. |

BISMUTH, Bi_4?

Atomic weight $= 208$. *Sp. gr.* $9\cdot83$. *Fuses at* $265°$. *Atomicity* $'''$ *and* ^{v}. *Evidence of atomicity* :—

Bismuthous chloride	$Bi'''Cl_3$.
Bismuthous oxide	Bi'''_2O_3.
Bismuthous ethide	$Bi'''Et_3$.
Bismuthous dichlorethide	$Bi'''EtCl_2$.
Bismuthic anhydride	$Bi^v_2O_5$.

Occurrence.—Principally in the metallic state in nature.

Preparation.—1. On a large scale by fusion and separation from earthy impurities.

2. It may be obtained in the pure state by dissolving commercial bismuth in nitric acid, precipitating the basic nitrate by addition of water, and reducing the precipitate by ignition with charcoal.

No compound of bismuth with hydrogen is known.

COMPOUND OF BISMUTH WITH CHLORINE.

BISMUTHOUS CHLORIDE.
$BiCl_3$.

Molecular weight $= 314\cdot5$. *Molecular volume* ☐☐. 1 *litre of bismuthous chloride vapour weighs* $157\cdot25$ *criths.*

Preparation.—1. By passing dry chlorine over metallic bismuth :—

$$Bi_2 \ + \ 3Cl_2 \ = \ 2BiCl_3.$$
Bismuthous chloride.

2. By evaporating a solution of bismuth in hydrochloric acid containing a little nitric acid, and distilling.

3. By distilling metallic bismuth with mercuric chloride:—

$$Bi_2 \;+\; 6HgCl_2 \;=\; 2BiCl_3 \;+\; 3'Hg'_2Cl_2.$$

Mercuric   Bismuthous   Mercurous
chloride.   chloride.   chloride.

Reaction.—By the addition of water it is decomposed, forming bismuthous oxychloride :—

$$BiCl_3 \;+\; OH_2 \;=\; BiOCl \;+\; 2HCl.$$

Bismuthous   Water.   Bismuthous   Hydrochlo-
chloride.   oxychloride.   ric acid.

The following compounds are also known :—

Bismuthous bromide	$BiBr_3$.
Bismuthous iodide	BiI_3.
Bismuthous fluoride	BiF_3.
Bismuthous oxybromide	$BiOBr$.
Bismuthous oxyiodide..............	$BiOI$.
Dibismuthous tetrachloride	$\left\{ \begin{array}{l} BiCl_2. \\ BiCl_2. \end{array} \right.$

COMPOUNDS OF BISMUTH WITH OXYGEN AND HYDROXYL.

Dibismuthous dioxide	$\left\{ \begin{array}{l} BiO. \\ BiO. \end{array} \right.$
Bismuthous oxide	Bi_2O_3.
Dibismuthic tetroxide	$'Bi^{iv}_2O_4$.
Bismuthic anhydride..............	Bi_2O_5.
Bismuthous oxyhydrate, or meta-bismuthous acid .... }	$BiOHo$.
Metabismuthic acid	BiO_2Ho.

BISMUTHOUS OXIDE.

$$\mathbf{Bi_2O_3}.$$

Molecular weight $=464$. *Sp. gr.* $8 \cdot 2$.

Occurrence.—As the rare mineral *bismuth ochre.*

Preparation.—1. By burning bismuth in air or oxygen.

2. By heating the nitrate, carbonate, or hydrate:—

$$2\mathbf{N_3O_6Bi}o''' \;=\; \mathbf{Bi_2O_3} \;+\; 3\mathbf{N_2O_3} \;+\; 3O_2.$$

Bismuthous nitrate *. Bismuthous oxide. Nitrous anhydride.

$$2\mathbf{NO_2(Bi'''Ho_2O)} \;=\; \mathbf{Bi_2O_3} \;+\; 2\mathbf{OH_2} \;+\; \mathbf{N_2O_3} \;+\; O_2.$$

Bismuthous nitrate dihydrate †. Bismuthous oxide. Water. Nitrous anhydrid

$$\mathbf{C}O(BiO_2)_2 \;=\; \mathbf{Bi_2O_3} \;+\; \mathbf{C}O_2.$$

Bismuthylic carbonate ‡. Bismuthous oxide. Carbonic anhydride.

$$2\mathbf{BiOHo} \;=\; \mathbf{Bi_2O_3} \;+\; \mathbf{OH_2}.$$

Bismuthous oxyhydrate. Bismuthous oxide. Water.

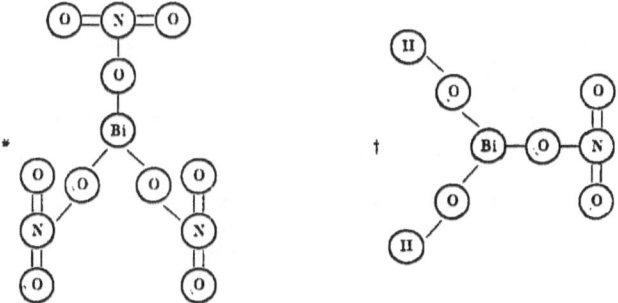

Bismuthous nitrate. Bismuthous nitrate dihydrate.

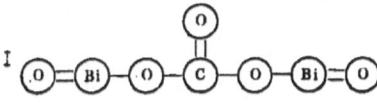

Bismuthylic carbonate.

3. By boiling bismuthous hydrate in solution of potassic hydrate, whereby it is converted into bismuthous oxide with loss of water.

Reactions.—Dissolved by hydrochloric, nitric, and sulphuric acids, forming the bismuthous chloride, nitrate, and sulphate:—

$$\mathbf{BiCl_3}. \qquad \mathbf{N_3O_6Bio'''}. \qquad \mathbf{S_3O_6Bio'''_2}.$$

| Bismuthous chloride. | Bismuthous nitrate. | Bismuthous sulphate. |

BISMUTHOUS OXYHYDRATE, *Metabismuthous Acid.*
BiOHo.

Preparation.—By pouring a solution of bismuthous nitrate in dilute nitric acid into dilute ammonia or potassic hydrate, and drying the precipitate, which, at first, probably contains orthobismuthous acid:—

$$\mathbf{N_3O_6Bio'''} \ + \ \mathbf{3OKH} \ = \ \mathbf{BiHo_3} \ + \ \mathbf{3NO_2Ko}.$$

| Bismuthous nitrate. | Potassic hydrate. | Orthobismuthous acid. | Potassic nitrate. |

$$\mathbf{BiHo_3} \ = \ \mathbf{BiOHo} \ + \ \mathbf{OH_2}.$$

| Orthobismuthous acid. | Metabismuthous acid. | Water. |

Reaction.—By heat or by boiling with caustic alkali, the water is expelled, and bismuthous oxide formed (see p. 141).

An unstable metabismuthite is produced by fusing bismuthous oxide with sodic carbonate:—

$$\mathbf{Bi_2O_3} \ + \ \mathbf{CONao_2} \ = \ \mathbf{2BiONao} \ + \ \mathbf{CO_2}.$$

| Bismuthous oxide. | Sodic carbonate. | Sodic metabismuthite. | Carbonic anhydride. |

BISMUTHIC OXIDE, OR ANHYDRIDE.
Bi_2O_5.

Preparation.—By heating bismuthic acid to 130°.

Reactions.—1. When heated to the boiling-point of mercury, it loses oxygen, being converted either into bismuthous oxide or dibismuthic tetroxide :—

$$\underset{\substack{\text{Bismuthic} \\ \text{oxide.}}}{\mathbf{Bi_2O_5}} = \underset{\substack{\text{Bismuthous} \\ \text{oxide.}}}{\mathbf{Bi_2O_3}} + O_2 :$$

$$\underset{\substack{\text{Bismuthic} \\ \text{oxide.}}}{2\mathbf{Bi_2O_5}} = \underset{\substack{\text{Dibismuthic} \\ \text{tetroxide.}}}{2'\mathbf{Bi}^{iv}_2O_4} + O_2.$$

2. When heated in a current of hydrogen, it is readily reduced to bismuthous oxide.

3. Heated with hydrochloric acid, it evolves chlorine, producing bismuthous chloride and water :—

$$\underset{\substack{\text{Bismuthic} \\ \text{oxide.}}}{\mathbf{Bi_2O_5}} + \underset{\substack{\text{Hydrochloric} \\ \text{acid.}}}{10\text{HCl}} = \underset{\substack{\text{Bismuthous} \\ \text{chloride.}}}{2\mathbf{BiCl_3}} + \underset{\substack{\text{Water.}}}{5\mathbf{OH_2}} + 2\text{Cl}_2.$$

4. Sulphurous acid converts it into bismuthous sulphate :—

$$\underset{\substack{\text{Bismuthic} \\ \text{oxide.}}}{3\mathbf{Bi_2O_5}} + \underset{\substack{\text{Sulphurous} \\ \text{acid.}}}{6\text{SOH}_2} = \underset{\substack{\text{Bismuthous} \\ \text{sulphate.}}}{2\mathbf{S_3O_6Bio}'''_2} + \underset{\substack{\text{Bismuthous} \\ \text{oxide.}}}{\mathbf{Bi_2O_3}} + \underset{\substack{\text{Water.}}}{6\mathbf{OH_2}}.$$

5. When heated with sulphuric or nitric acid, it evolves oxygen, producing bismuthous sulphate or nitrate :—

$$\underset{\substack{\text{Bismuthic} \\ \text{oxide.}}}{\mathbf{Bi_2O_5}} + \underset{\substack{\text{Sulphuric} \\ \text{acid.}}}{3\text{SO}_2\text{Ho}_2} = \underset{\substack{\text{Bismuthous} \\ \text{sulphate.}}}{\mathbf{S_3O_6Bio}'''_2} + \underset{\substack{\text{Water.}}}{3\mathbf{OH_2}} + O_2 ;$$

$$\underset{\substack{\text{Bismuthic} \\ \text{oxide.}}}{\mathbf{Bi_2O_5}} + \underset{\substack{\text{Nitric} \\ \text{acid.}}}{6\text{NO}_2\text{Ho}} = \underset{\substack{\text{Bismuthous} \\ \text{nitrate.}}}{2\mathbf{N_3O_6Bio}'''} + \underset{\substack{\text{Water.}}}{3\mathbf{OH_2}} + O_2.$$

METABISMUTHIC ACID.
BiO₂Ho.

$$\mathbf{BiO_2Ho.}$$

Preparation.—Obtained as a red deposit by passing chlorine through a solution of potassic hydrate containing bismuthous oxide in suspension :—

$$\underset{\substack{\text{Potassic} \\ \text{hydrate.}}}{4\mathbf{OKH}} + 2\text{Cl}_2 + \underset{\substack{\text{Bismuthous} \\ \text{oxide.}}}{\mathbf{Bi_2O_3}} = \underset{\substack{\text{Metabismuthic} \\ \text{acid.}}}{2\mathbf{BiO_2Ho}} + \underset{\substack{\text{Potassic} \\ \text{chloride.}}}{4\text{KCl}} + \underset{\substack{\text{Water.}}}{\mathbf{OH_2}}.$$

Reaction.—Dissolves in hot solution of potassic hydrate. By the addition of an acid to the liquid, a salt, said to have the composition

$$\mathbf{Bi_2O_4HoKo},$$

is precipitated.

COMPOUNDS OF BISMUTH WITH SULPHUR.

Dibismuthous disulphide $\mathbf{'Bi''_2S''_2}$.
Bismuthous sulphide .... $\mathbf{Bi_2S''_3}$.

DIBISMUTHOUS DISULPHIDE.

$$\left\{ \begin{array}{l} \mathbf{BiS''} \\ \mathbf{BiS'''} \end{array} \right. \text{ or } \mathbf{'Bi''_2S''_2}.$$

Molecular weight =480.   *Sp. gr.* 7·3.

Preparation.—By fusing bismuth and sulphur together in the proper proportions.

BISMUTHOUS SULPHIDE.

$$\mathbf{Bi_2S''_3}.$$

Molecular weight =512.   *Sp. gr.* 6·4.

Occurrence.—As the rare mineral *bismuth glance.*
Preparation.—1. By fusing sulphur and bismuth in the proper proportions.

2. By precipitating bismuth solutions by sulphuretted hydrogen :—

$$\mathbf{2BiCl_3} \quad + \quad \mathbf{3SH_2} \quad = \quad \mathbf{Bi_2S''_3} \quad + \quad \mathbf{6HCl}.$$

Bismuthous chloride.	Sulphuretted hydrogen.	Bismuthous sulphide.	Hydrochloric acid.

Reaction.—This compound is not dissolved by alkaline hydrates or sulphhydrates.

A few sulphobismuthites are found in nature:—

Kobellite. *Sulphoplumbic sulphobismuthite* **Bi**$_2$**Pbs**″$_3$.

Needle ore. *Disulphoplumbico-cuprous*
sulphobismuthite ....} **Bi**$_2$**Pbs**″$_2$(′**Cu**′$_2$**S**$_2$)″.

BISMUTHOUS DITELLURO-SULPHIDE.

Bi$_2$**Te**″$_2$**S**″.

Sp. gr. 7·5 *to* 7·8.

Occurrence.—In nature, as *telluric bismuth* or *tetradymite.*

CHAPTER XVII.

MONAD ELEMENTS.

SECTION III.

POTASSIUM, K$_2$.

Atomic weight =39. *Probable molecular weight* =78. *Sp. gr.*
0·865. *Fuses at* 55°. *Boils at a low red heat. Atomicity* ′.

Evidence of atomicity :—

Potassic chloride	KCl.
Potassic iodide	KI.
Potassic hydrate	KHo.
Potassic sulphide.................	**S**K$_2$.

Occurrence.—In rocks in the form of silicate, and in soils partly as carbonate. As chloride in solid saline deposits.

In the juices of almost all plants, generally in combination with organic acids.

In sea-water and in most mineral waters.

Preparation.—1. By the action of a powerful voltaic current upon potassic hydrate, when potassium and hydrogen are liberated at the negative pole :—

$$2OKH \;=\; K_2 \;+\; H_2 \;+\; O_2.$$

Potassic
hydrate.

2. By submitting potassic hydrate to the action of metallic iron at a strong white heat :—

$$4OKH \;+\; 3Fe \;=\; {}^{iv}(Fe_3)^{viii}O_4 \;+\; 2K_2 \;+\; 2H_2.$$

Potassic
hydrate.

Magnetic iron
oxide.

3. By igniting hydric potassic tartrate (cream of tartar) out of contact with air, and subsequently mixing the residue, consisting of potassic carbonate and carbon, with charcoal, and distilling at a very high temperature :—

First operation :—

$$2\begin{cases}COKo \\ CHHo \\ CHHo \\ COHo \end{cases} = COKo_2 \;+\; 5OH_2 \;+\; 4CO \;+\; 3C.$$

Hydric
potassic
tartrate.

Potassic
carbonate.

Water.

Carbonic
oxide.

Second operation :—

$$COKo_2 \;+\; 2C \;=\; 3CO \;+\; K_2.$$

Potassic
carbonate.

Carbonic
oxide.

Reactions.—1. Potassium decomposes water at the common temperature with great energy, the heat evolved being sufficient to cause the ignition of the liberated hydrogen :—

$$K_2 \;+\; 2OH_2 \;=\; 2KHo \;+\; H_2.$$

Water.

Potassic
hydrate.

2. When potassium is ignited in a stream of carbonic anhydride, a portion of the latter is decomposed :—

$$2K_2 \;+\; 3CO_2 \;=\; 2COKo_2 \;+\; C.$$

Carbonic
anhydride.

Potassic
carbonate.

COMPOUNDS OF POTASSIUM WITH CHLORINE, BROMINE, IODINE, AND FLUORINE.

Potassic chloride .... KCl.
Potassic bromide .... KBr.
Potassic iodide.............. KI.
Potassic fluoride .... KF.

Potassic iodide is prepared by digesting iron filings, water, and iodine together, filtering the colourless solution, and precipitating the iron by potassic carbonate:—

$$Fe + I_2 = \mathbf{Fe}I_2;$$
Ferrous
iodide.

$$\mathbf{Fe}I_2 + \mathbf{C}OKo_2 = 2KI + \mathbf{C}OFeo''.$$
Ferrous   Potassic   Potassic   Ferrous
iodide.   carbonate.   iodide.   carbonate.

COMPOUND OF POTASSIUM WITH HYDROXYL.

POTASSIC HYDRATE, *Caustic Potash, Potash.*
KHo, or OKH.

Preparation.—1. By boiling in an iron vessel a solution of potassic carbonate with calcic hydrate :—

$$\mathbf{C}OKo_2 + \mathbf{Ca}Ho_2 = 2KHo + \mathbf{C}OCao''.$$
Potassic   Calcic   Potassic   Calcic
carbonate.   hydrate.   hydrate.   carbonate.

2. By the action of potassium upon water (see p. 146).

Reactions.—By contact with acids potassic hydrate produces potassium salts:—

$$KHo + HCl = KCl + \mathbf{O}H_2:$$
Potassic   Hydrochloric   Potassic   Water.
hydrate.   acid.   chloride.

$$KHo + \mathbf{S}O_2Ho_2 = \mathbf{S}O_2HoKo + \mathbf{O}H_2:$$
Potassic   Sulphuric   Hydric potassic   Water.
hydrate.   acid.   sulphate.

$$2KHo + \mathbf{S}O_2Ho_2 = \mathbf{S}O_2Ko_2 + 2\mathbf{O}H_2.$$
Potassic   Sulphuric   Potassic   Water.
hydrate.   acid.   sulphate.

H 2

COMPOUNDS OF POTASSIUM WITH OXYGEN.

Potassic oxide............ OK_2.

Potassic dioxide .... $\begin{cases} OK \\ OK \end{cases}$

Potassic tetroxide ...... $\begin{cases} OK \\ O \\ O \\ OK \end{cases}$.

POTASSIC OXIDE.

OK_2.

Preparation.—1. By heating potassic hydrate with potassium :—

$$2KHo \;+\; K_2 \;=\; 2OK_2 \;+\; H_2.$$
Potassic hydrate.   Potassic oxide.

2. By fusing together, in a current of nitrogen, potassic peroxide and potassium :—

$$K_2O_4 \;+\; 3K_2 \;=\; 4OK_2.$$
Potassic peroxide.   Potassic oxide.

POTASSIC DIOXIDE.

K_2O_2.

Preparation.—Obtained by the action of water on potassic peroxide.

POTASSIC TETROXIDE, *Potassic Peroxide.*

$$K_2O_4.$$

Preparation.—By fusing potassium in a current of oxygen.

COMPOUND OF POTASSIUM WITH HYDROSULPHYL.

POTASSIC SULPHHYDRATE.

$$KHs.$$

Preparation.—By saturating potassic hydrate with sulphuretted hydrogen :—

$$KHo \quad + \quad SH_2 \quad = \quad KHs \quad + \quad OH_2.$$

| Potassic hydrate. | Sulphuretted hydrogen. | Potassic sulphhydrate. | Water. |

COMPOUNDS OF POTASSIUM WITH SULPHUR.

The following have been obtained :—

Dipotassic sulphide...... $SK_2.$

Dipotassic disulphide ... $K_2S_2.$

Dipotassic trisulphide... $K_2S_3.$

Dipotassic tetrasulphide. $K_2S_4.$

Dipotassic pentasulphide ......} ... $K_2S_5.$

Dipotassic heptasulphide ...} ... K_2S_7 ?

DIPOTASSIC SULPHIDE.

$$SK_2.$$

Preparation.—1. By the action of potassic hydrate on potassic sulphhydrate :—

$$KHo \quad + \quad KHs \quad = \quad SK_2 \quad + \quad OH_2.$$

| Potassic hydrate. | Potassic sulphhydrate. | Dipotassic sulphide. | Water. |

2. By igniting potassic sulphate with hydrogen or carbon:—

$$SO_2Ko_2 \;+\; 4H_2 \;=\; SK_2 \;+\; 4OH_2.$$

Potassic   Dipotassic   Water.
sulphate.   sulphide.

Reactions of dipotassic sulphide and the higher potassic sulphides :—1. By heating dipotassic sulphide with the necessary quantities of sulphur, it forms the higher potassic sulphides.

2. The potassic sulphhydrate and dipotassic sulphide, when acted upon by acids, yield sulphuretted hydrogen :—

$$KHs \;+\; HCl \;=\; KCl \qquad SH_2.$$

Potassic   Hydrochloric   Potassic   Sulphuretted
sulphhydrate.   acid.   chloride.   hydrogen.

$$SK_2 \;+\; 2HCl \;=\; 2KCl \;+\; SH_2.$$

Dipotassic   Hydrochloric   Potassic   Sulphuretted
sulphide.   acid.   chloride.   hydrogen.

3. The higher potassic sulphides, similarly treated, yield sulphuretted hydrogen and a precipitate of sulphur :—

$$K_2S_3 \;+\; 2HCl \;=\; 2KCl \;+\; SH_2 \;+\; S_2.$$

Dipotassic   Hydrochloric   Potassic   Sulphuretted
trisulphide.   acid.   chloride.   hydrogen.

4. A mixture of the higher potassic sulphides and potassic hyposulphite, known under the name of *hepar sulphuris* or *liver of sulphur*, may be prepared by heating potassic carbonate with sulphur :—

$$3COKo_2 \;+\; 4S_2 \;=\; 2K_2S_3 \;+\; SS''OKo_2 \;+\; 3CO_2;$$

Potassic   Dipotassic   Potassic   Carbonic
carbonate.   trisulphide.   hyposulphite.   anhydride.

$$3COKo_2 \;+\; 6S_2 \;=\; 2K_2S_5 \;+\; SS''OKo_2 \;+\; 3CO_2.$$

Potassic   Dipotassic   Potassic   Carbonic
carbonate.   pentasulphide.   hyposulphite.   anhydride.

5. The last mixture, when acted upon by acids, suffers successively the following decompositions : —

$$2K_2S_5 \;+\; SS''OKo_2 \;+\; 6HCl \;=\; 6KCl$$

Dipotassic   Potassic   Hydrochloric   Potassic
pentasulphide.   hyposulphite.   acid.   chloride.

$$\qquad +\; SS''OHo_2 \;+\; 2SH_2 \;+\; 4S_2;$$

Hyposulphurous   Sulphuretted
acid.   hydrogen.

then

$$SS''OHo_2 \;=\; SO_2 \;+\; S \;+\; OH_2.$$

Hyposulphurous   Sulphurous   Water.
acid.   anhydride.

POTASSIC CARBONATE.

$CO{K_o}_2$.

Preparation.—1. By lixiviating the ashes of land-plants.

2. By burning hydric potassic tartrate in a current of air.

SODIUM, Na_2.

Atomic weight $=23$.   *Probable molecular weight* $=46$.   *Sp. gr.*
0·97.   *Fuses at* $90°$.   *Boils at a red heat.   Atomicity '.*
Evidence of atomicity :—

Sodic chloride .... $NaCl$.
Sodic hydrate .... $ONaH$.
Sodic oxide .... ONa_2.

Occurrence.—In nature in the form of chloride. In sea-water
and most springs.   As silicate in several minerals.

Preparation.—1. By electrolyzing sodic hydrate.

2. By acting upon sodic hydrate with metallic iron at a
strong white heat.

Manufacture.—By distilling in an iron retort a mixture of
sodic carbonate and charcoal :—

$$CON{a_o}_2 \; + \; 2C \; = \; Na_2 \; + \; 3CO.$$
Sodic
carbonate.   Carbonic
oxide.

Reactions.—Similar to those of potassium, but less energetic.
The compounds of sodium very much resemble those of potassium.

SODIC CARBONATE.

$CON{a_o}_2$.

Manufacture.—1. Formerly by the lixiviation of the ashes
of marine plants.

2. By Leblanc's process, which consists in first transforming

sodic chloride into sodic sulphate by the action of sulphuric acid :—

$$2NaCl \ + \ SO_2Ho_2 \ = \ 2HCl \ + \ SO_2Nao_2.$$

| Sodic chloride. | Sulphuric acid. | Hydrochloric acid. | Sodic sulphate. |

The sodic sulphate (technically termed *salt cake*) is next heated with calcic carbonate and small coal. The carbon reduces the sodic sulphate to sulphide, and the calcic carbonate transforms the sodic sulphide into sodic carbonate, insoluble calcic oxysulphide being simultaneously produced :—

$$5SO_2Nao_2 \ + \ 20C \ = \ 5SNa_2 \ + \ 20CO;$$

| Sodic sulphate. | | Sodic sulphide. | Carbonic oxide. |

$$5SNa_2 \ + \ 7COCao'' \ = \ 5CONao_2$$

| Sodic sulphide. | Calcic carbonate. | Sodic carbonate. |

$$+ \ 5CaS'', 2CaO \ + \ 2CO_2.$$

| | Calcic oxysulphide. | Carbonic anhydride. |

The sodic carbonate, commonly called *soda ash,* is obtained by the extraction of the resulting mass with water.

LITHIUM, Li₂.

Atomic weight $=7$.   *Probable molecular weight* $=14$.   *Sp. gr.* $=0\cdot59$.   *Fuses at* $180°$.   *Atomicity* '.   *Evidence of atomicity* :—

     Lithic chloride .... LiCl.
     Lithic hydrate (Lithia) .... OLiH.

Occurrence.—In nature, in the minerals *petalite, spodumene, lepidolite,* and *triphylline,* and in small quantities in some mineral waters and ashes of plants.

The properties of lithium resemble those of potassium and sodium ; and the compounds of the three metals also exhibit considerable similarity.

CÆSIUM and RUBIDIUM.

The compounds of the two metals cæsium (Cs=133) and rubidium (Rb=85·5), which have been recently discovered, closely resemble those of potassium and sodium.

SECTION IV.

THALLIUM, Tl$_2$.

Atomic weight =204. *Probable molecular weight* =408. *Sp. gr.* 11·81 to 11·91. *Fuses at* 561°. *Atomicity' and perhaps'''.*

Evidence of atomicity :—

Thallic chloride .... TlCl.
Thallic oxide .... **OTl$_2$.**
Thallic perchloride .... **Tl'''Cl$_3$** ?

Occurrence.—In small quantities in certain varieties of pyrites, and in minute quantities in some mineral springs.

Preparation.—By extracting with water the deposit formed in the flues of sulphuric acid-chambers, and precipitating the thallium by hydrochloric acid. The chloride is converted into sulphate by the action of sulphuric acid; and when purified, a solution of the sulphate is decomposed by metallic zinc, which precipitates the metallic thallium.

The following list contains the principal compounds of this metal :—

Thallic chloride .... TlCl.
Thallic perchloride .... **TlCl$_3$** ?
Thallic oxide .... **OTl$_2$.**
Thallic peroxide .... $\left\{\begin{array}{l}\textbf{OTl}\\ \text{O } ?\\ \textbf{OTl}\end{array}\right.$
Thallic sulphide........................ **STl$_2$.**
Thallic nitrate .... **NO$_2$Tlo.**
Thallic sulphate........................ **SO$_2$Tlo$_2$.**
Thallic carbonate .... **COTlo$_2$.**

Two other compounds of thallium with chlorine have also been described, but they have not yet been completely investigated.

SILVER, Ag_2.

Atomic weight $=108$. *Probable molecular weight* $=216$. *Sp. gr.* $10\cdot4743$. *Fuses at about* $1000°$. *Atomicity* $'$. *Evidence of atomicity* :—

Argentic chloride........................ AgCl.
Argentic iodide .... AgI.
Argentic oxide .... OAg_2.

Occurrence.—In nature in the free state, and as sulphide in *silver glance*; as sulphantimonite in *dark-red silver-ore* (see p. 137), as chloride in *horn-silver*, as a compound of bromide and chloride ($2AgBr, 3AgCl$) in *embolite*, and also as carbonate.

Extraction.—1. The silver minerals are roasted with sodic chloride, by which the metal is converted into chloride; the mass is then mixed with water, scrap iron, and mercury, and agitated for some hours. The iron reduces the argentic chloride to the metallic state, and the silver is then dissolved by the mercury.

2. By crystallizing argentiferous lead. Nearly pure lead is first deposited, and the residue rich in silver is then cupelled.

Reactions.—1. Silver is blackened by sulphuretted hydrogen, argentic sulphide being formed.

2. Silver is acted upon by hot concentrated sulphuric acid :—

$$Ag_2 \quad + \quad 2SO_2Ho_2 \quad = \quad SO_2Ago_2 \quad + \quad 2OH_2 \quad + \quad SO_2.$$
Sulphuric   Argentic   Water.   Sulphurous
acid.   sulphate.   anhydride.

3. Nitric acid readily dissolves silver :—

$$3Ag_2 \quad + \quad 8NO_2Ho \quad = \quad 6NO_2Ago \quad + \quad 4OH_2 \quad + \quad 'N''_2O_2.$$
Nitric   Argentic   Water.   Nitric
acid.   nitrate.   oxide.

4. At a red heat silver decomposes hydrochloric acid :—

$$Ag_2 \quad + \quad 2HCl \quad = \quad H_2 \quad + \quad 2AgCl.$$
$$\underset{\substack{\text{Hydrochloric} \\ \text{acid.}}}{} \qquad\qquad \underset{\substack{\text{Argentic} \\ \text{chloride.}}}{}$$

There are three compounds of silver with oxygen :—

Argentous oxide .... OAg_4.

Argentic oxide........................... OAg_2.

Argentic peroxide .... $\begin{cases} OAg. \\ OAg. \end{cases}$

Argentous oxide is prepared by heating argentic citrate to 100° in a stream of hydrogen, dissolving the residue, which contains argentous citrate, in cold water, and precipitating the argentous oxide by potassic hydrate.

Argentic oxide is formed by precipitating argentic nitrate with a solution of baric hydrate, and drying the precipitate, which is probably argentic hydrate, AgHo. This is the salifiable oxide of silver.

Argentic peroxide is obtained by electrolyzing a solution of argentic nitrate, when it is deposited upon the positive pole.

Argentic chloride, bromide, and iodide are insoluble in water and nitric acid.

CHAPTER XVIII.

DYAD ELEMENTS.

SECTION II.

BARIUM, Ba.

Atomic weight =137. *Probable molecular weight* =137. *Sp. gr. between* 4·0 *and* 5·0. *Fuses below a red heat. Atomicity* ".

Evidence of atomicity :—

Baric chloride .... $Ba''Cl_2$.

Baric hydrate .... $Ba''Ho_2$.

Baric oxide.............................. $Ba''O$.

Occurrence. — In nature in the form of sulphate in the mineral *heavy spar*, and as carbonate ($CO Bao''$) in *witherite*.

Preparation.—1. By electrolyzing moistened baric hydrate, carbonate, nitrate, or chloride, the negative electrode being mercury. An amalgam of barium is thus formed, from which the mercury is removed by distillation.

2. By passing the vapour of potassium or sodium over baric oxide strongly heated in an iron tube, and extracting the metal by means of mercury.

3. By acting upon a solution of baric chloride with sodium amalgam, barium amalgam is produced.

Reaction.—Barium decomposes water at the common temperature :—

$$Ba \ + \ 2OH_2 \ = \ H_2 \ + \ BaHo_2.$$
$$\text{Water.} \qquad\qquad \text{Baric hydrate.}$$

COMPOUNDS OF BARIUM WITH OXYGEN.

Baric oxide..................... $BaO.$

Baric peroxide .... $Ba{}^O_O \Big\}$.

BARYTA, *Baric Oxide.*

BaO.

Preparation.—1. By converting the native carbonate into nitrate by the action of nitric acid, and then heating the nitrate to redness in an iron crucible :—

$$\text{C}O\text{Bao}'' \quad + \quad 2\text{NO}_2\text{Ho} \quad = \quad \begin{cases} \text{NO}_2 \\ \text{Bao}'' \\ \text{NO}_2 \end{cases} + \text{OH}_2 + \text{CO}_2 ;$$

Baric   Nitric acid.   Baric   Water.   Carbonic
carbonate *.   nitrate †.   anhydride.

$$2 \begin{cases} \text{NO}_2 \\ \text{Bao}'' \\ \text{NO}_2 \end{cases} = 2\text{BaO} \quad + \quad 2'\text{N}^{iv}{}_2\text{O}_4 \quad + \quad \text{O}_2.$$

Baric   Baric   Nitric
nitrate.   oxide.   peroxide.

2. The nitrate may be obtained from native baric sulphate by mixing the latter with charcoal and heating the mixture to a high red heat, by which the sulphate is converted into sulphide—

$$\text{SO}_2\text{Bao}'' \quad + \quad \text{C}_4 \quad = \quad \text{BaS}'' \quad + \quad 4\text{CO} ;$$

Baric   Baric   Carbonic
sulphate ‡.   sulphide.   oxide.

the residue is then treated with dilute nitric acid, when baric nitrate is formed.

Reaction.—In contact with water, baric oxide is converted, with great evolution of heat, into baric hydrate :—

$$\text{BaO} \quad + \quad \text{OH}_2 \quad = \quad \text{BaHo}_2.$$

Baric   Water.   Baric
oxide.   hydrate.

BARIC PEROXIDE.

$$\mathbf{Ba}{O \atop O}\Big\} .$$

Preparation.—1. By passing oxygen over baric oxide or baric hydrate heated to dull redness :—

$$2\mathbf{Ba}O \;+\; O_2 \;=\; 2\mathbf{Ba}{O \atop O}\Big\} .$$

Baric oxide.   Baric peroxide.

$$2\mathbf{Ba}Ho_2 \;+\; O_2 \;=\; 2\mathbf{Ba}{O \atop O}\Big\} \;+\; 2OH_2.$$

Baric hydrate.   Baric peroxide.   Water.

2. By heating baric oxide to redness in a crucible and gradually adding potassic chlorate :—

$$3\mathbf{Ba}O \;+\; ClO_2Ko \;=\; KCl \;+\; 3\mathbf{Ba}{O \atop O}\Big\} .$$

Baric oxide.   Potassic chlorate.   Potassic chloride.   Baric peroxide.

Reactions.—1. By the action of heat it splits into baric oxide and oxygen :—

$$2\mathbf{Ba}{O \atop O}\Big\} \;=\; 2\mathbf{Ba}O \;+\; O_2.$$

Baric peroxide.   Baric oxide.

2. By treatment with steam at the same temperature at which the peroxide was previously formed, it produces baric hydrate and oxygen :—

$$2\mathbf{Ba}{O \atop O}\Big\} \;+\; 2OH_2 \;=\; 2\mathbf{Ba}Ho_2 \;+\; O_2.$$

Baric peroxide.   Water.   Baric hydrate.

3. By the action of acids upon baric peroxide, hydroxyl is formed (p. 44).

COMPOUND OF BARIUM WITH HYDROXYL.

BARIC HYDRATE, *Caustic Baryta.*

$BaHo_2$.

Preparation.—1. By the action of water on baric oxide (p. 157).

2. By boiling in water, with cupric oxide, the mass containing baric sulphide, prepared by reducing baric sulphate with carbon :—

$$BaS'' \ + \ CuO \ + \ OH_2 \ = \ BaHo_2 \ + \ CuS''.$$

| Baric sulphide. | Cupric oxide. | Water. | Baric hydrate. | Cupric sulphide. |

Barium salts are formed by the action of acids upon baric hydrate, carbonate, or oxide.

DIHYDRIC BARIC DISULPHATE.

$$\begin{cases} SO_2Ho \\ Bao'' \\ SO_2Ho \end{cases}.$$

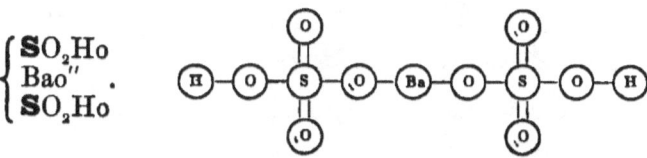

This compound is formed by boiling baric sulphate in concentrated sulphuric acid, when the salt crystallizes on cooling.

STRONTIUM, Sr.

Atomic weight $= 87.5$. *Probable molecular weight* $= 87.5$. *Sp. gr.* 2.5. *Fuses at a higher temperature than barium. Atomicity ". Evidence of atomicity :—*

Strontic chloride........................ $Sr''Cl_2$.
Strontic hydrate........................ $Sr''Ho_2$.
Strontic oxide........................ $Sr''O$.

Occurrence.—In the form of carbonate as the mineral *stron-tianite*, and as sulphate in *celestine*.

Preparation.—1. By the same methods as those employed in the preparation of barium.

2. By electrolyzing fused strontic chloride.

Character.—The compounds of strontium resemble those of barium in constitution, preparation, and properties.

The *strontic peroxide* can only be prepared by adding hydroxyl to a solution of strontic hydrate :—

$$\mathbf{Sr}\text{Ho}_2 \; + \; \text{Ho}_2 \; = \; \mathbf{Sr}{\overset{\text{O}}{\underset{\text{O}}{\Big\}}} \; + \; 2\text{OH}_2.$$

Strontic   Hydroxyl.   Strontic   Water.
hydrate.   peroxide.

Strontic carbonate is more easily decomposed by heat than baric carbonate.

CALCIUM, Ca.

Atomic weight $=40$.   *Probable molecular weight* $=40$.   *Sp. gr.* 1·6.   *Atomicity* ".   *Evidence of atomicity* :—

Calcic chloride...................... $\mathbf{Ca}''\text{Cl}_2$.
Calcic hydrate...................... $\mathbf{Ca}''\text{Ho}_2$.'
Calcic oxide .... $\mathbf{Ca}''\text{O}$.

Occurrence.—In nature as carbonate in the numerous *calc spars*, chalk, marble, &c. ; as tetrahydric calcic sulphate ($\mathbf{S}\text{Ho}_4\text{Cao}''$) in *gypsum*, *alabaster*, *selenite*, &c.; as phosphate in *apatite* and *phosphorite* (see p. 119); as fluoride in the *fluor spars* (see p. 96), and in combination with silicon, oxygen, and other metals in numerous minerals.

Preparation.—1. By processes similar to those employed for the preparation of barium and strontium.

2. By fusing together sodium, zinc, and calcic chloride, and subsequently heating the alloy of calcium and zinc so obtained, to a very high temperature in a crucible of gas-car-

bon, when the zinc volatilizes, leaving the calcium, which contains, however, a small quantity of iron.

Character.—The compounds of calcium resemble those of
barium and strontium.

Calcic oxide or *quicklime* (**CaO**) is manufactured on a large
scale by burning coal intermixed with chalk or limestone, when
carbonic anhydride is easily expelled from the chalk or limestone, leaving calcic oxide.

Calcic hydrate or *slaked lime* (**CaHo$_2$**) is formed by the action
of water upon calcic oxide; it is much less soluble in water
than the baric and strontic hydrates.

Calcic peroxide $\left(\mathbf{Ca}{\,}^{O}_{O} \right\}\ \right)$ is prepared like the corresponding

strontium compound.

By passing chlorine over calcic hydrate, a compound known
as *chloride of lime* or *bleaching-powder* is formed. This has been
supposed to consist of calcic chloride mixed with calcic hypochlorite, but it is more probably calcic chloro-hypochlorite, as
expressed by the following formulæ :—

Ca(OCl)Cl. ⓒˡ—⟨Ca⟩—ⓞ—ⓒˡ

The corresponding baric and strontic chloro-hypochlorites
Ba$^{OCl}_{Cl}$ and **Sr**$^{OCl}_{Cl}$, are known.

Barium, strontium, and calcium all form soluble dihydric
dicarbonates :—

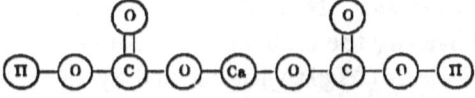

They are produced by passing an excess of carbonic anhydride through solutions of baric, strontic, and calcic hydrates.
The compounds are decomposed at 100°, carbonic anhydride
being evolved and carbonates precipitated :—

$$\left\{ \begin{array}{l} \mathbf{COHo} \\ \mathbf{Cao''} \\ \mathbf{COHo} \end{array} \right. = \mathbf{COCao''} + \mathbf{OH_2} + \mathbf{CO_2}.$$

Dihydric calcic   Calcic   Water.   Carbonic
dicarbonate.   carbonate.   anhydride.

1

MAGNESIUM, Mg.

Atomic weight $=24$.   *Probable molecular weight* $=24$.   *Sp. gr.*
    1·75. *Fuses at a red heat. Volatilizes at a bright-red heat.*
    Atomicity ".   *Evidence of atomicity* :—

Magnesic chloride.................. $\mathbf{Mg}''Cl_2$.
Magnesic oxide.................... $\mathbf{Mg}''O$.
Magnesic hydrate................ $\mathbf{Mg}''Ho_2$.

Occurrence.—In nature in *dolomite*, the calcic magnesic dicarbonate,

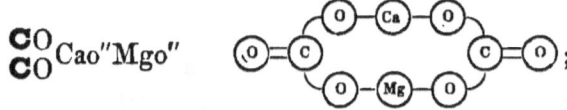

$$\begin{matrix}\mathbf{C}O \\ \mathbf{C}O\end{matrix} Cao''Mgo''$$

in *brucite* or magnesic hydrate, $\mathbf{Mg}Ho_2$, and in many minerals
containing silicon.

Preparation.—1. By electrolyzing fused magnesic chloride.

2. By fusing a mixture of magnesic chloride, potassic chloride, and sodium.

Reactions.—1. It very slowly decomposes water at the ordinary temperature, but more rapidly at a boiling heat.

2. It readily burns when heated to redness in the air.

Character.—Magnesium only forms one compound with oxygen, $\mathbf{Mg}O$, *magnesia*. It is obtained by burning magnesium
in air, or by heating the carbonate to redness.

Magnesic hydrate ($\mathbf{Mg}Ho_2$) is formed by the action of water
upon magnesic oxide, or by precipitating magnesic sulphate by
potassic hydrate :—

$$\underset{\substack{\text{Magnesic}\\\text{sulphate.}}}{SO_2Mgo''} \;+\; \underset{\substack{\text{Potassic}\\\text{hydrate.}}}{2OKH} \;=\; \underset{\substack{\text{Potassic}\\\text{sulphate.}}}{SO_2Ko_2} \;+\; \underset{\substack{\text{Magnesic}\\\text{hydrate.}}}{\mathbf{Mg}Ho_2}.$$

It scarcely dissolves in water.

Crystallized magnesic sulphate ($SOHo_2Mgo''$, $6OH_2$) is prepared by treating *dolomite*, the magnesic calcic dicarbonate, with

sulphuric acid, filtering from the nearly insoluble calcic sulphate, and crystallizing :—

$$\begin{matrix} CO \\ CO \end{matrix} Cao''Mgo'' + 2SO_2Ho_2 = SOHo_2Cao'' + SOHo_2Mgo''$$

<table>
<tr><td>Dolomite, calcic magnesic dicarbonate.</td><td>Sulphuric acid.</td><td>Dihydric calcic sulphate.</td><td>Dihydric magnesic sulphate.</td></tr>
</table>

$$+ \quad 2CO_2$$

Carbonic anhydride.

Magnesic sulphate is very soluble in water, thus differing from the baric, strontic, and calcic sulphates.

Magnesic sulphate, when mixed with potassic or ammonic sulphate, forms a disulphate, as, for instance,

$$\begin{matrix} SO_2Ko \\ SO_2Ko \end{matrix} Mgo''.$$

Dipotassic magnesic disulphate.

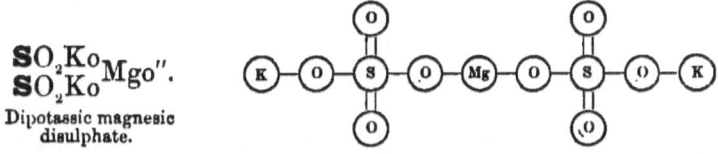

Many magnesic phosphates are known.

Diammonic dimagnesic diphosphate,

$$\begin{matrix} POAmo \\ POAmo \end{matrix} Mgo''_2,$$

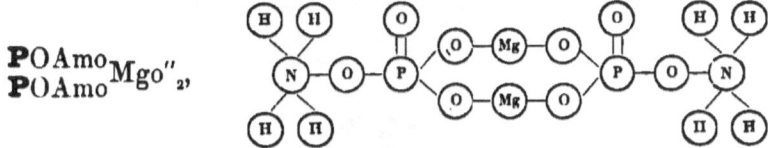

occurs in the seeds of some of the cereals, and sometimes in urine, and in the form of calculi: it is found in nature as *guanite* and *struvite*.

Magnesic carbonate ($COMgo''$) is found in nature as *magnesite*.

MAGNESIA ALBA, *Tetrahydric tetramagnesic tricarbonate.*

$$\mathbf{C}_3\mathrm{Mgo''}_4\mathrm{Ho}_4, \quad \text{or} \quad \begin{cases} \mathbf{C}\mathrm{Ho}_2 \\ \mathrm{Mgo''}_2 \\ \mathbf{C} \\ \mathrm{Mgo''}_2 \\ \mathbf{C}\mathrm{Ho}_2 \end{cases},$$

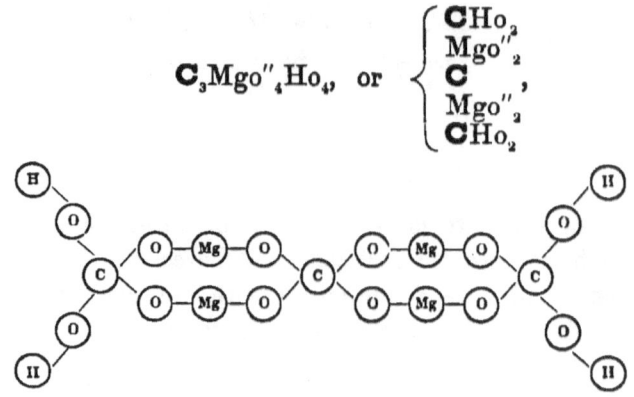

This compound is formed by boiling a solution of magnesic sulphate with sodic carbonate (Berzelius) :—

$$4\mathrm{SO}_2\mathrm{Mgo''} + 4\mathbf{C}\mathrm{ONao}_2 + 2\mathrm{OH}_2 = \mathbf{C}_3\mathrm{Mgo''}_4\mathrm{Ho}_4$$

Magnesic sulphate.　　Sodic carbonate.　　Water.　　Magnesia alba.

$$+ \ 4\mathrm{SO}_2\mathrm{Nao}_2 + \mathbf{C}\mathrm{O}_2.$$

Sodic sulphate.　　Carbonic anhydride.

ZINC, Zn.

Atomic weight $=65$. *Molecular weight* $=65$. *Molecular and atomic volume* ☐☐. *1 litre of zinc vapour weighs* 32·5 *criths. Sp. gr.* 6·8 *to* 7·2. *Fuses at* 500°. *Distils at a red heat. Atomicity* ". *Evidence of atomicity* :—

Zincic chloride	$\mathbf{Zn}''\mathrm{Cl}_2$.
Zincic oxide.........................	$\mathbf{Zn}''\mathrm{O}$.
Zincic hydrate.......................	$\mathbf{Zn}''\mathrm{Ho}_2$.

Occurrence.—In nature as oxide ($\mathbf{Zn}\mathrm{O}$) in *red zinc*, as sulphide ($\mathbf{Zn}\mathrm{S}''$) in the mineral *zinc blende*, carbonate ($\mathbf{C}\mathrm{OZno}''$)

in *calamine,* and as silicate in *electric calamine, williamite,* or *zinc glass.*

Manufacture.—Zinc blende or calamine is roasted in a current of air, whereby it is converted into zincic oxide :—

$$CO Zno'' = ZnO + CO_2.$$
Zincic carbo-   Zincic   Carbonic
nate (calamine).   oxide.   anhydride.

$$2ZnS'' + 3O_2 = 2ZnO + 2SO_2.$$
Zincic sulphide   Zincic oxide.   Sulphurous
(Zinc blende).   anhydride.

The roasted and powdered mineral is then heated with powdered coal, when the zinc is reduced and distils over :—

$$ZnO + C = Zn + CO.$$
Zincic   Carbonic
oxide.   oxide.

Reactions.—1. It slowly decomposes aqueous vapour at 100° :—

$$OH_2 + Zn = ZnO + H_2.$$
Water.   Zincic oxide.

2. Zinc is attacked by almost every acid at the common temperature.

3. When boiled in potassic, sodic, or even ammonic hydrate, hydrogen is evolved, and a mixed oxide formed :—

$$2OKH + Zn = ZnKo_2 + H_2.$$
Potassic   Dipotassic
hydrate.   zincic oxide.

COMPOUND OF ZINC WITH OXYGEN.

ZINCIC OXIDE.

ZnO.

Preparation.—1. Zincic oxide is obtained by burning zinc in air.

2. By passing steam over heated zinc.

3. By heating the precipitate formed by ammonic carbonate in solutions of zinc salts.

OTHER COMPOUNDS OF ZINC.

Zincic hydrate ($\mathbf{Zn}Ho_2$) is obtained as a white precipitate by the action of potassic hydrate on solutions of zinc salts :—

$$\underset{\text{Zincic sulphate.}}{SO_2Zno''} + \underset{\substack{\text{Potassic} \\ \text{hydrate.}}}{2OKH} = \underset{\substack{\text{Zincic} \\ \text{hydrate.}}}{\mathbf{Zn}Ho_2} + \underset{\substack{\text{Potassic} \\ \text{sulphate.}}}{SO_2Ko_2}.$$

The precipitate is dissolved by excess of potassic hydrate.

Crystallized zincic sulphate is isomorphous with crystallized magnesic sulphate, and contains seven molecules of water, six of which are easily expelled at a moderate heat, the last only being driven off at a somewhat high temperature. It also resembles magnesic sulphate in forming double salts with potassic and ammonic sulphates :—

Zincic sulphate (crystallized) .... $SOHo_2Zno''$, $6OH_2$.

Dipotassic zincic disulphate (crystallized) $\begin{cases} SO_2Ko \\ Zno'' \\ SO_2Ko \end{cases}$, $6OH_2$.

Zincic carbonate ($\mathbf{C}OZno''$) occurs in nature as *calamine*.

The precipitate obtained by adding a solution of sodic carbonate to a solution of a salt of zinc has a variable constitution. The reaction usually takes place thus :—

$$\underset{\text{Zincic sulphate.}}{5SO_2Zno''} + \underset{\text{Sodic carbonate.}}{5CONao_2} + \underset{\text{Water.}}{3OH_2} = \begin{cases} CHo(OZn''Ho)_2 \\ Zno'' \\ CHo(OZn''Ho)_2 \end{cases}$$
$$\underset{\text{Dihydric pentazincic dicarbonate tetrahydrate *.}}{}$$

$$+ \underset{\text{Sodic sulphate.}}{5SO_2Nao_2} + \underset{\text{Carbonic anhydride.}}{3CO_2}.$$

CHAPTER XIX.

DYAD ELEMENTS.

SECTION IV.

CADMIUM, Cd.

Atomic weight =112. *Molecular weight* =112. *Molecular and atomic volume* ☐☐. 1 *litre of cadmium vapour weighs* 56 *criths. Sp. gr.* 8·7. *Fuses below* 260°. *Easily volatile. Atomicity* ". *Evidence of atomicity* :—

Cadmic chloride .... $Cd''Cl_2$.
Cadmic oxide .... $Cd''O$.

Occurrence.—In nature in small quantities, associated with zinc; and in the form of sulphide as *greenockite*.

Preparation.—By distilling fractionally the more volatile part of the metal obtained in the manufacture of zinc, and then dissolving this more volatile product (which consists of zinc, cadmium, and a little copper) in hydrochloric or dilute sulphuric acid, precipitating the cadmium and copper with sulphuretted hydrogen, dissolving the mixed sulphides in dilute sulphuric acid, and adding an excess of solution of ammonic carbonate, which precipitates both cadmium and copper, but redissolves the latter. The cadmic carbonate is then ignited, and the resulting oxide reduced by charcoal.

Cadmic oxide (CdO) is prepared by heating the hydrate, carbonate, or nitrate.

Cadmic hydrate ($CdHo_2$) is obtained by precipitating a solution of a cadmic salt by sodic or potassic hydrate.

Cadmic sulphate ($SO_2Cdo'',4OH_2$) is obtained by dissolving cadmic oxide or carbonate in sulphuric acid. By heating this compound, or by partially decomposing it with alkaline hydrates, it is transformed into

Dicadmic sulphate dihydrate .... $SO_2(OCd''Ho)_2$

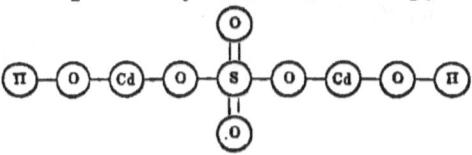

MERCURY, Hg.

Atomic weight =200. *Molecular weight* =200. *Molecular and atomic volume* ☐. 1 *litre of mercury vapour weighs* 100 *criths. Sp. gr.* 13·59. *Fuses at* −40°. *Boils at* 360°. *Atomicity* ", *also a pseudo-monad.*

The following list contains the principal compounds of this metal :—

Mercurous chloride (*horn-mercury*) ... } $'Hg'_2Cl_2$, or $\begin{cases} HgCl \\ HgCl \end{cases}$

Mercuric chloride ... $HgCl_2$.

Mercurous oxide...... $'Hg'_2O$, or $\begin{cases} Hg \\ Hg \end{cases}O$.

Mercuric oxide ...... HgO.

Mercurous sulphide.. $'Hg'_2S''$, or $\begin{cases} Hg \\ Hg \end{cases}S''$.

Mercuric sulphide (*vermilion, cinna-bar*) .... } HgS''.

Mercurous sulphate.. SO_2Hg_2o''.

Mercuric sulphate ... SO_2Hgo''.

Trimercuric sulphate (*Turpeth mineral*). } $SHgo''_3$.

Tetrahydric mercurous dinitrate... $N_2O_2Ho_4Hg_2o''$.

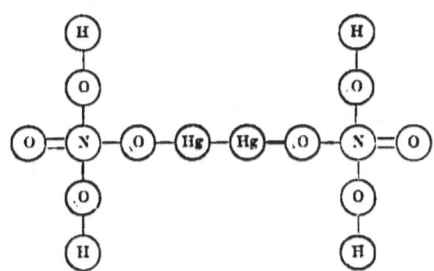

Dimercurous dinitrate... $\begin{cases} NOHg_2o'' \\ O \\ NOHg_2o'' \end{cases}$.

Hexahydric trimercu-rous tetranitrate ... } $N_4O_4Ho_6Hg_2o''_3$.

Mercurous dimercuric dinitrate .... } $N_2O_2Hg_2o''Hgo''_2$.

Tetrahydric mercuric dinitrate .... } $N_2O_2Ho_4Hgo''$.

Tetrahydric dimercu-ric dinitrate .... } $N_2OHo_4Hgo''_2$.

Dihydric trimercuric dinitrate .... } $N_2OHo_2Hgo''_3$.

Trimercuric carbonate... $CHgo''(Hg''_2O_3)''$.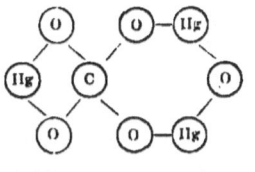

Tetramercuric carbonate. $CHgo''(Hg''_3O_4)''$.

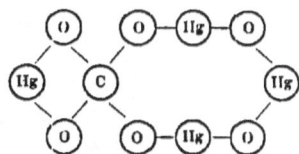

Mercurosodiammonic dichloride .... } $\begin{matrix} NH_3ClHg \\ NH_3ClHg \end{matrix}$ }.

I

Mercurosomercurodi- $\Big\}$ $\mathbf{N_2Hg''H_4'Hg'_2Cl_2}$.
   ammonic dichloride.

Mercurammonic chlo- $\left.\begin{array}{c}\\\\\\\end{array}\right\}$ $\mathbf{NH_2Hg''Cl}$.
   ride. (*White preci-*
   pitate.) ....

Trimercuric diamide $\mathbf{N_2Hg''_3}$.

COPPER, Cu.

Atomic weight $=63\cdot5$. *Probable molecular weight* $=63\cdot5$. *Sp. gr.*
    $8\cdot8$. *Fuses at about* 780°. *Atomicity*"; *also a pseudo-monad.*

The following are the principal compounds of this metal :—

Cuprous hydride......... $\left\{\begin{array}{l}\mathbf{CuH}\\ \mathbf{CuH}\cdot\end{array}\right.$

Cuprous chloride $\mathbf{'Cu'_2Cl_2}$ or $\left\{\begin{array}{l}\mathbf{CuCl}\\ \mathbf{CuCl}\cdot\end{array}\right.$

Cupric chloride .... $\mathbf{CuCl_2, 2OH_2}$.

Cuprous hydrate, $\mathbf{4'Cu'_2O, OH_2}$, or $\left\{\begin{array}{l}\mathbf{CuHo}\\ \mathbf{Cu}\\ \mathbf{Cu}\\ \mathbf{Cu}\\ \mathbf{Cu}\\ \mathbf{Cu}\\ \mathbf{Cu}\\ \mathbf{CuHo}\end{array}\right.$

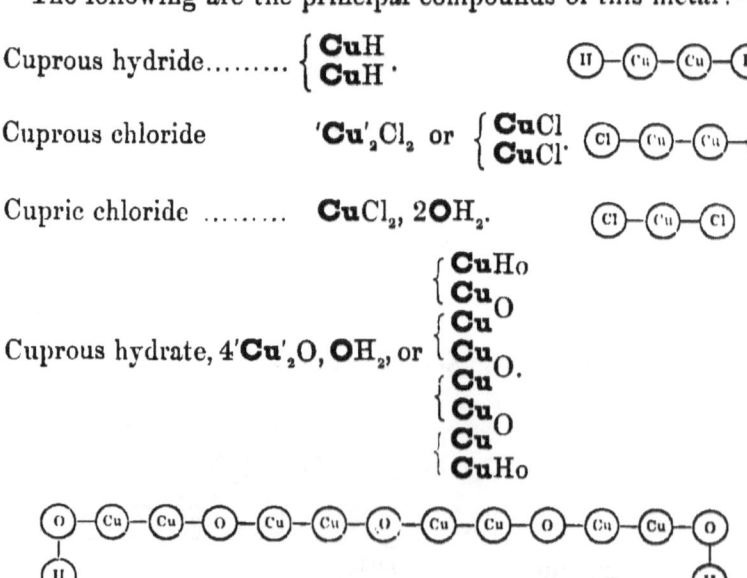

Cupric hydrate .... $\mathbf{Cu}Ho_2$.

Cuprous quadrantoxide. $\left\{\begin{array}{l}\mathbf{Cu}\\ Cu''\\ Cu''\\ \mathbf{Cu}\end{array}\right\}O.$

Cuprous oxide. (*Red copper ore*, or *ruby ore*.) .... $'\mathbf{Cu'}_2O$ or $\left\{\begin{array}{l}\mathbf{Cu}\\ \mathbf{Cu}\end{array}O.\right.$

Cupric oxide $\mathbf{Cu}O.$

Cuprous sulphide. (*Copper glance*.) ... $'\mathbf{Cu'}_2S''$ or $\left\{\begin{array}{l}\mathbf{Cu}\\ \mathbf{Cu}\end{array}S''.\right.$

Cupric sulphide. (*Indigo copper* or *blue copper*.) .... $\mathbf{Cu}S''.$

Cupric sulpho-hydrate.. $5\mathbf{Cu}S'', \mathbf{Cu}Ho_2,$ or $\left\{\begin{array}{l}\mathbf{Cu}Ho\\ S''\\ Cu''\\ S''\\ Cu''\\ S''\\ Cu''\\ S''\\ Cu''\\ S''\\ \mathbf{Cu}Ho\end{array}\right.$.

Cupric nitrate .... $\dfrac{\mathbf{N}O_2}{\mathbf{N}O_2}Cuo'', 4\mathbf{OH}_2,$ or $\dfrac{\mathbf{N}Ho_4}{\mathbf{N}Ho_4}Cuo''.$

Dihydric cupric sulphate $\mathbf{S}OHo_2Cuo'', 4\mathbf{OH}_2.$

Hydric tricupric sulphate trihydrate ... $\mathbf{S}OHo(OCu''Ho)_3.$

Dihydric tetracupric sulphate tetrahydrate.(*Brochantite.*) } $\mathbf{S}Ho_2(OCu''Ho)_4.$

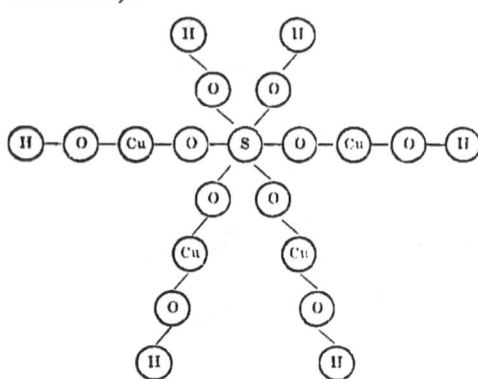

Hydric pentacupric sulphate pentahydrate............... } $\mathbf{S}Ho(OCu''Ho)_5,2\mathbf{O}H_2.$

Ammoniocupric sulphate. (*Dihydric diammonic cuprodiammonic sulphate.*) } $\mathbf{S}Ho_2Amo_2\begin{bmatrix}N^vH_3\\N^vH_3\end{bmatrix}Cu''O_2\end{bmatrix}''$

Dipotassic cupric disulphate .... $\begin{cases}\mathbf{S}O_2Ko\\Cuo''\\\mathbf{S}O_2Ko\end{cases}.$

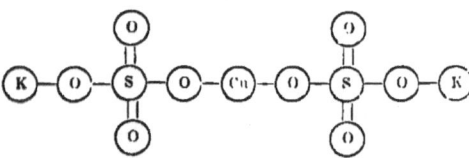

Dicupric carbonate. (*Mysorin.*) .... } $\mathbf{C}Cuo_2''.$

Dicupric carbonate dihydrate. (*Malachite.*) } $\mathbf{C}O(OCu''Ho)_2.$

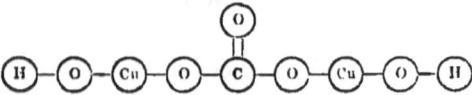

Dihydric tricupric dicarbonate. (*Blue malachite, azurite, mountain-blue* or *copper-azure*.) ...... } $\left\{\begin{array}{l}\mathbf{C}\text{HoCuo}'' \\ \text{Cuo}'' \\ \mathbf{C}\text{HoCuo}''\end{array}\right.$

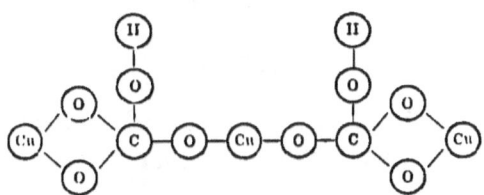

Cuprodiammonic carbonate. (*Ammoniocupric carbonate*.) .... } $\mathbf{C}\text{O}\left[\begin{array}{l}\text{N}^{\text{V}}\text{H}_3 \\ \text{N}^{\text{V}}\text{H}_3\end{array}\text{Cu}''\text{O}_2\right]''.$

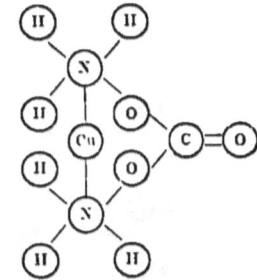

Hydric cupric silicate hydrate. (*Dioptase*.) } $\mathbf{Si}\text{OHo}(\text{OCu}''\text{Ho}).$

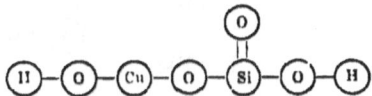

CHAPTER XX.

TRIAD ELEMENTS.

SECTION II.

GOLD, Au_2.

Atomic weight $= 196 \cdot 7$. *Probable molecular weight* $= 393 \cdot 4$. *Sp. gr.* 19·3 *to* 19·5. *Fuses at about* 1100°–1200°. *Atomicity* ′ *and* ‴.

The following are the names and formulæ of the chief compounds of gold :—

Aurous chloride............ **AuCl.**

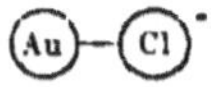

Auric chloride **$AuCl_3$.**

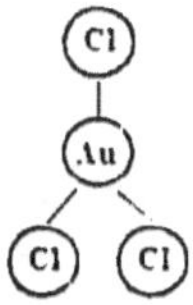

Aurous iodide **AuI.**

Auric iodide .... **AuI_3.**

Aurous oxide.............. **Au_2O.**

Auric oxide. (*Auric anhydride.*).................... $\begin{cases} \textbf{AuO} \\ \textbf{O} \\ \textbf{AuO} \end{cases}$.

Potassic aurate .... **$AuOKo, 3OH_2$.**

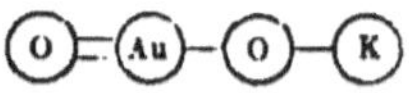

Aurous sulphide **Au_2S''.**

Auric sulphide .... $\begin{cases} \textbf{AuS''} \\ \textbf{S''} \\ \textbf{AuS''} \end{cases}$.

CHAPTER XXI.

TETRAD ELEMENTS.

SECTION II.

ALUMINIUM, Al.

Atomic weight =27·5. *Molecular weight unknown. Specific gravity* 2·6. *Fuses at about* 450°. *Atomicity* [iv], *but is always a pseudo-triad. Evidence of atomicity :—Analogy with iron and chromium.*

Annexed are the names and formulæ of the most important compounds of this metal :—

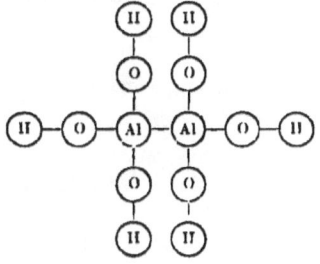

Aluminic chloride ... $\mathbf{Al'''_2Cl_6}$ or $\begin{cases} \mathbf{AlCl_3} \\ \mathbf{AlCl_3} \end{cases}$.

Aluminic oxide ...... $\mathbf{Al_2O_3}$ or $\begin{cases} \mathbf{AlO} \\ \mathbf{AlO} \end{cases} O$.

Aluminic hydrate.⎫
 (*Gibbsite.*) ......⎬ $\mathbf{Al_2Ho_6}$ or $\begin{cases} \mathbf{AlHo_3} \\ \mathbf{AlHo_3} \end{cases}$.

Aluminic oxydi-hydrate. (*Diaspore.*) .... $Al_2O_2Ho_2$ or $\begin{cases} AlOHo \\ AlOHo \end{cases}$

Aluminic sulphide ... $Al_2S''_3$ or $\begin{cases} AlS'' \\ AlS''S'' \end{cases}$

Dipotassic aluminate $Al_2O_2Ko_2$ or $\begin{cases} AlOKo \\ AlOKo \end{cases}$

Magnesic aluminate. (*Spinelle.*) Al_2O_2Mgo'' or $\begin{cases} Al \\ Al \end{cases}O_2Mgo''$

Aluminic sulphate ... $S_3O_6('Al'''_2O_6)^{vi}, 18OH_2$ or

$$\begin{matrix} SO_2 \\ SO_2 \\ SO_2 \end{matrix}\Bigg]-('Al'''_2O_6)^{vi}, 18OH_2.$$

Aluminic sulphate tetrahydrate. (*Aluminite.*) ... $SO_2('Al'''_2O_2Ho_4)'', 7OH_2.$

Allophane .... $SiHo_2('Al'''_2Ho_4O_2)'', (2 \text{ or } 4)OH_2.$

Prehnite .... $Si_3Ho_2Cao_2''('Al'''_2O_6)^{vi}.$

Zoisite.................. $Si_4Cao''_3\left(\begin{cases} 'Al'''_2O_5 \\ O \\ 'Al'''_2O_5 \end{cases}\right)^x.$

Spodumene........... $Si_{15}O_{15}Lio_6('Al'''_2O_6)^{vi}_4.$

Petalite .... $Si_{30}O_{45}Nao_2Lio_4('Al'''_2O_6)^{vi}_4.$

wi Sou 3

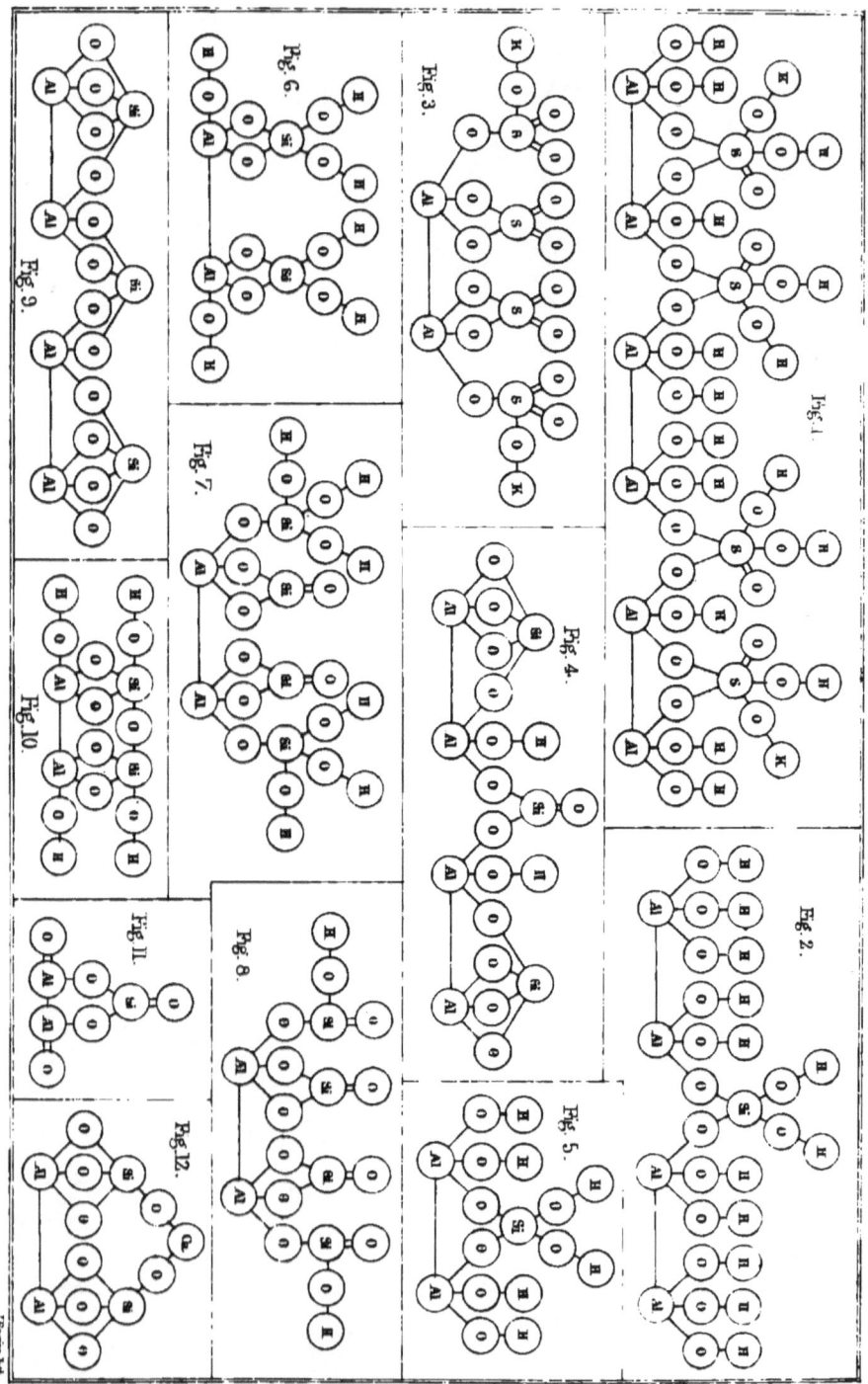

Alunite, alum- stone. Fig. 1.	$\mathbf{S}OHoKo$ $('Al'''Ho_3O_3)'''$. $\mathbf{S}OHo_2$ $('Al'''_2Ho_4O_2)''$. $\mathbf{S}OHo_2$ $('Al'''_2Ho_4O_2)''$. $\mathbf{S}OHoKo$ $('Al'''_2Ho_3O_3)'''$.
Collyrite. Fig. 2....	$\mathbf{S}iHo_2('Al'''_2Ho_5O)_2,4\mathbf{O}H_2$.
Dipotassic alumi- nic tetrasul- phate. (*Common* *alum.*) Fig. 3.	$\mathbf{S}O_2Ko\!\!-\!\!\!\rceil$ $\mathbf{S}O_2\!\!-\!\!\!-$ $\mathbf{S}O_2\!\!-\!\!\!-('Al'''_2O_6)^{vi},24\mathbf{O}H_2$. $\mathbf{S}O_2Ko\!\!-\!\!\!\rfloor$
Wörthite. Fig. 4....	$\mathbf{S}i$ $\mathbf{S}iO\!\begin{Bmatrix}('Al'''_2HoO_5)^v\\('Al'''_2HoO_5)^v\end{Bmatrix}$ $\mathbf{S}i$
Miloschine. Fig. 5.	$\mathbf{S}iHo_2('Al'''_2Ho_4O_2)''$.
Porcelain clay of Passau. Fig. 6.	$\mathbf{S}iHo_2('Al'''_2Ho_2O_4)^{iv}$. $\mathbf{S}iHo_2$
Cimolite, kaolin of Ellenbogen. Fig. 7.............	$\mathbf{S}iHo_3\!\!-\!\!\!\rceil$ $\mathbf{S}iO\!\!-\!\!\!-$ $\mathbf{S}iO\!\!-\!\!\!-('Al'''_2O_6)^{vi}$. $\mathbf{S}iHo_3\!\!-\!\!\!\rfloor$
Agalmatolite. Fig. 8.	$\mathbf{S}iOHo\!\!-\!\!\!\rceil$ $\mathbf{S}iO\!\!-\!\!\!-$ $\mathbf{S}iO\!\!-\!\!\!-('Al'''_2O_6)^{vi}$. $\mathbf{S}iOHo\!\!-\!\!\!\rfloor$
Buchholzite, xeno- lite. Fig. 9. ...	$\mathbf{S}i\!('Al'''_2O_6)^{vi}$ $\mathbf{S}i\!('Al'''_2O_6)^{vi}$ $\mathbf{S}i\!('Al'''_2O_6)^{vi}$.
Porcelain clay. Fig. 10.	$\Big\{\begin{array}{l}\mathbf{S}iHo\!\!-\!\!\!\rceil\\O\quad('Al'''_2Ho_2O_4)^{iv}.\\\mathbf{S}iHo\!\!-\!\!\!\rfloor\end{array}$
Andalusite, chias- tolite, cyanite, fibrolite, sillima- nite. Fig. 11.	$\mathbf{S}iO('Al'''_2O_4)''$.
Wernerite. Fig. 12.	$\mathbf{S}i_2Cao''('Al'''_2O_6)^{vi}$.

Saponite. Fig. 13. $\mathbf{Si_7Mgo''_6Ho_{10}('Al'''_2O_6)^{vi}}$.

Lepidolite. Fig. 14. $\mathbf{Si_9O_8Ko_2Lio_4('Al'''_2O_6)^{vi}_2('Al'''_2F_4O_2)''}$.

Analcime. Fig. 15. $\left\{ \begin{array}{l} \mathbf{Si}Ho_2Nao \\ O \\ \mathbf{Si} \\ \mathbf{Si} \\ O \\ \mathbf{Si}Ho_2Nao \end{array} \right.$ $('Al'''_2O_6)^{vi}$.

Razoumoffskin. Fig. 16. .... } $\begin{array}{l} \mathbf{Si}Ho_2 \\ \mathbf{Si}Ho_2 \\ \mathbf{Si}Ho_2 \end{array}$ $('Al'''_2O_6)^{vi}$.

Malthacite. Fig. 17. $\mathbf{Si_3O_{11}Ho_4('Al'''_2O_6)^{vi}}$.

Albite. Fig. 18. ... $\left\{ \begin{array}{l} \mathbf{Si}ONao \\ \mathbf{Si}O \\ O \\ \mathbf{Si}O \\ \mathbf{Si}O \\ O \\ \mathbf{Si}O \\ \mathbf{Si}ONao \end{array} \right.$ $('Al'''_2O_6)^{vi}$.

CHAPTER XXII.

TETRAD ELEMENTS.

SECTION III.

PLATINUM, Pt.

Atomic weight $=197\cdot4$. *Molecular weight unknown. Sp. gr.* $21\cdot5$. *Atomicity'' and* ^{iv}.

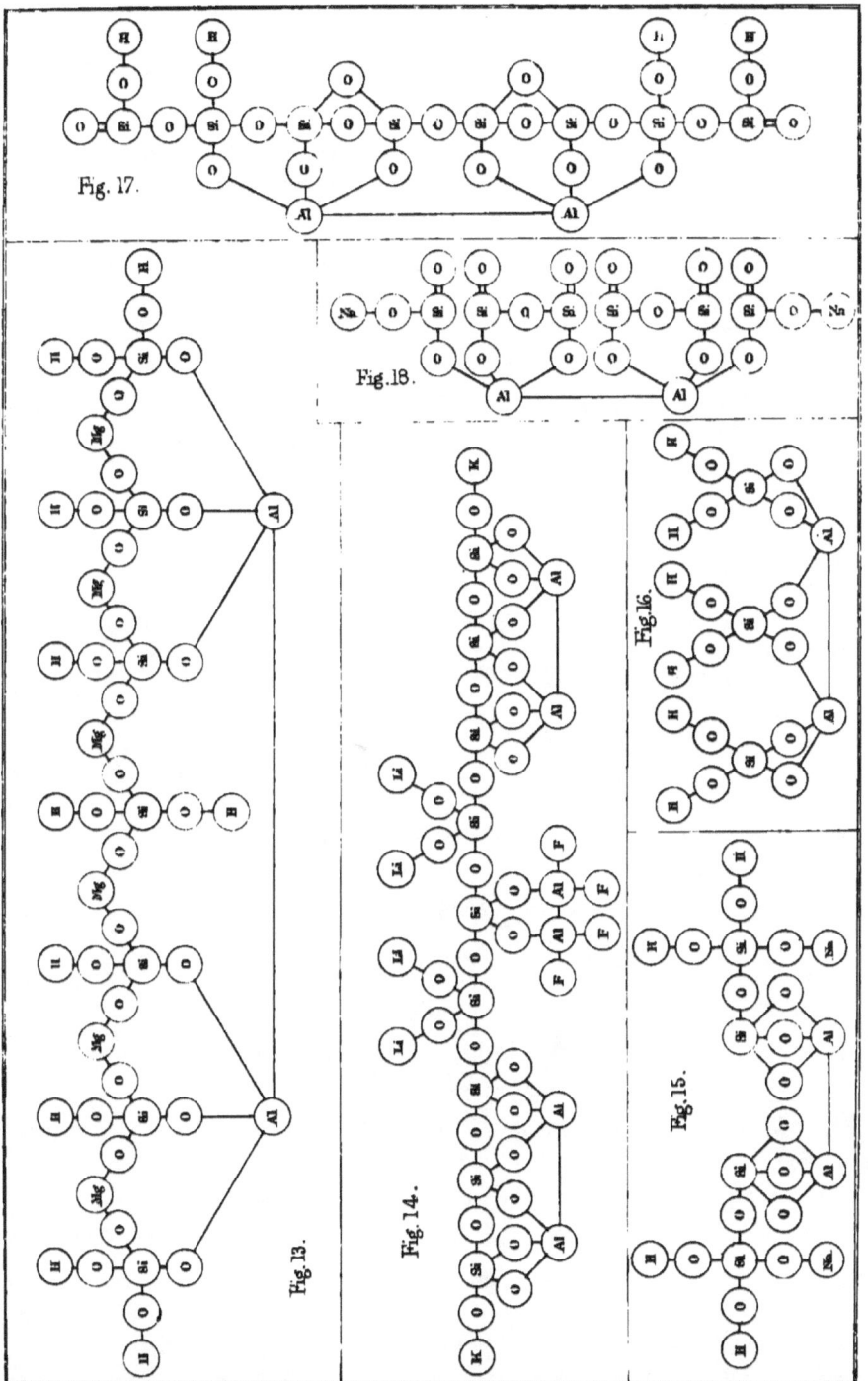

Fig. 17.

Fig. 18.

Fig. 16.

Fig. 15.

Fig. 14.

Fig. 13.

The following compounds will serve to illustrate the atomicity and general character of this metal :—

Platinous chloride ... **PtCl₂**.

Platinic chloride **PtCl₄**.

Platinous oxide ...... **PtO**.

Platinic oxide......... **PtO₂**.

Platinous hydrate ... **PtHo₂**.

Platinic hydrate...... **PtHo₄**.

Platinosodiammonic dichloride. (*Green salt of Magnus.*) ......

$$\left\{ \begin{array}{l} \mathbf{NH_3Cl} \\ \mathbf{Pt''} \\ \mathbf{NH_3Cl} \end{array} \right..$$

White compound of Reiset ....

$$\left\{ \begin{array}{l} \mathbf{NH_2(N^vH_4)Cl} \\ \mathbf{Pt''} \\ \mathbf{NH_2(N^vH_4)Cl} \end{array} \right..$$

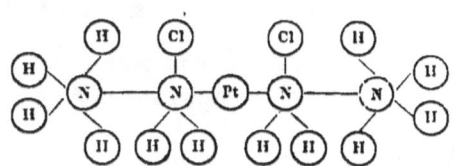

Platoso-diammon di-
ammonium dihy-
drate .... $\left\{\begin{array}{l}\mathbf{N}H_2(N^vH_4)Ho \\ Pt'' \\ \mathbf{N}H_2(N^vH_4)Ho\end{array}\right.$

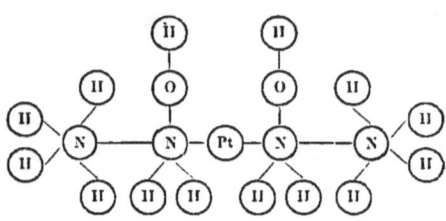

Diplatosammonic ox-
ide .... $\left\{\begin{array}{l}\mathbf{N}H_3\rceil \\ Pt''_3 O. \\ \mathbf{N}H_3\rfloor\end{array}\right.$

Platinous sulphide ... $\mathbf{Pt}S''.$

Platinic sulphide $\mathbf{Pt}S''_2.$

CHAPTER XXIII.

TETRAD ELEMENTS.

SECTION IV.

LEAD, Pb.

Atomic weight $=207$. *Molecular weight unknown. Sp. gr.*
11·445. *Fuses at* 335°. *Boils at a white heat. Atomicity* " *and* ᶦᵛ. *Also sometimes pseudo-triatomic.*

The following list contains the names and formulæ of the most interesting compounds of this metal :—

Plumbic chloride $\mathbf{Pb}Cl_2.$

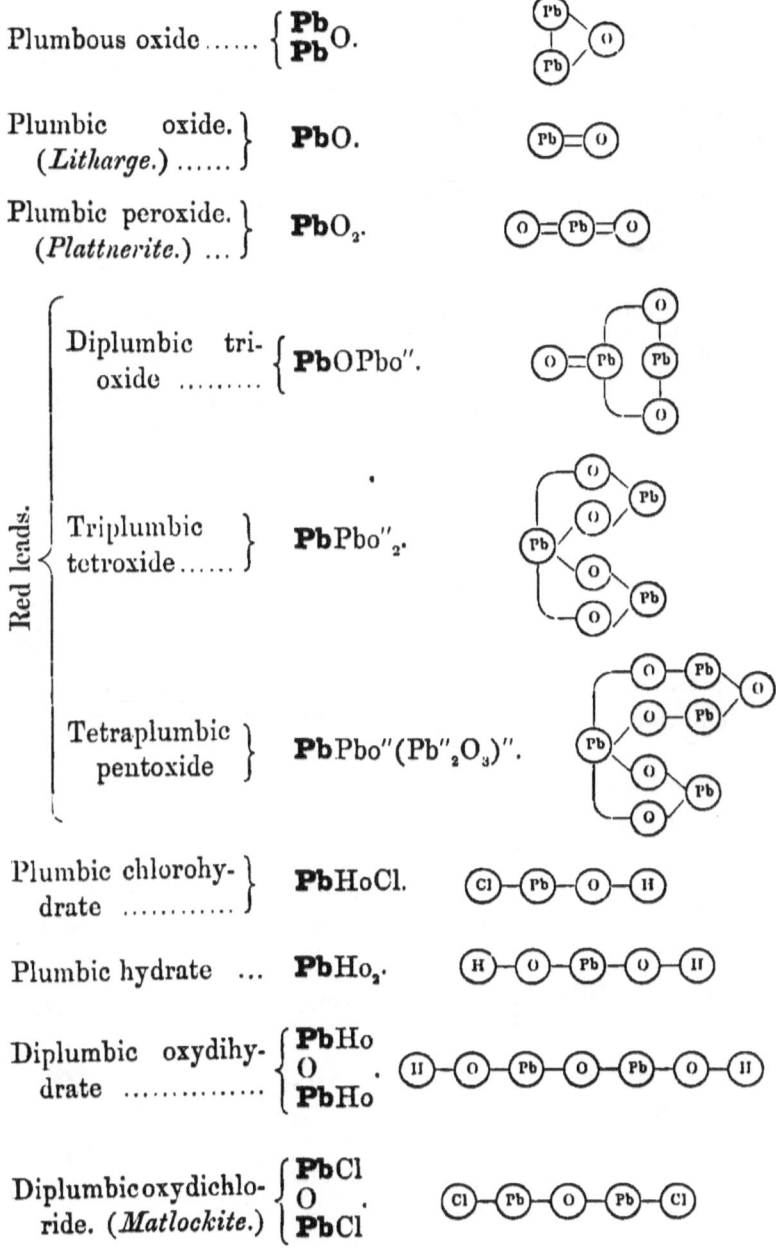

Plumbous oxide ...... { **Pb** **Pb**O.

Plumbic oxide. }
(*Litharge.*) ...... } **Pb**O.

Plumbic peroxide. }
(*Plattnerite.*) ... } **Pb**O₂.

Red leads.

Diplumbic tri- } **Pb**OPbo″.
oxide .... {

Triplumbic } **Pb**Pbo″₂.
tetroxide ...... }

Tetraplumbic } **Pb**Pbo″(Pb″₂O₃)″.
pentoxide }

Plumbic chlorohy- } **Pb**HoCl.
drate .... }

Plumbic hydrate ... **Pb**Ho₂.

Diplumbic oxydihy- { **Pb**Ho
drate .... { O .
{ **Pb**Ho

Diplumbic oxydichlo- { **Pb**Cl
ride. (*Matlockite.*) { O .
{ **Pb**Cl

Triplumbic oxydi-
chloride. (*Mendi-*
pite.) .... $\left\{ \begin{array}{l} \textbf{Pb}\text{Cl} \\ \text{O} \\ \text{Pb}'' \\ \textbf{Pb}\text{Cl} \end{array} \right.$.

(Cl)—(Pb)—(Pb)—(O)—(Pb)—(Cl)

Diplumbic oxychlo-
rohydrate .... $\left\{ \begin{array}{l} \textbf{Pb}\text{Ho} \\ \text{O} \\ \textbf{Pb}\text{Cl} \end{array} \right.$.

(Cl)—(Pb)—(O)—(Pb)—(O)—(H)

Octoplumbic hept-
oxydichloride ... $\Big\}$ $\textbf{Pb}_8\text{O}_7\text{Cl}_2$.

(O)—(Pb)—(O)—(Pb)—(O)—(Pb)—(O)—(Pb)—(O)—(Pb)—(O)—(Pb)—(O)
    | |
(Pb)   (Pb)
(Cl)   (Cl)

Plumbic sulphide.
(*Galena.*) ...... $\Big\}$ $\textbf{Pb}\text{S}''$.

Diplumbic sulphodi-
chloride .... $\left\{ \begin{array}{l} \textbf{Pb}\text{Cl} \\ \text{S}'' \\ \textbf{Pb}\text{Cl} \end{array} \right.$.

(Cl)—(Pb)—(S)—(Pb)—(Cl)

Plumbic sulphate.
(*Lead vitriol.*) $\Big\}$ $\textbf{SO}_2\text{Pbo}''$.

Plumbic dinitrite ... $\left\{ \begin{array}{l} \textbf{NO} \\ \text{Pbo}'' \\ \textbf{NO} \end{array} \right.$.

Plumbic nitrite hy-
drate .... $\Big\}$ $\textbf{N}\text{O}(\text{Pb}''\text{HoO})$.

Diplumbic nitrite-
hydrate .. .. $\Big\}$ $\textbf{N}\text{Pbo}''(\text{Pb}''\text{HoO})$.

(H)—(O)—(Pb)—(O)—(N)⟨ (O) (Pb) (O) ⟩

Plumbic dinitrate ...
$$\left\{\begin{array}{l}\mathbf{N}O_2 \\ \mathbf{P}bo''. \\ \mathbf{N}O_2\end{array}\right.$$

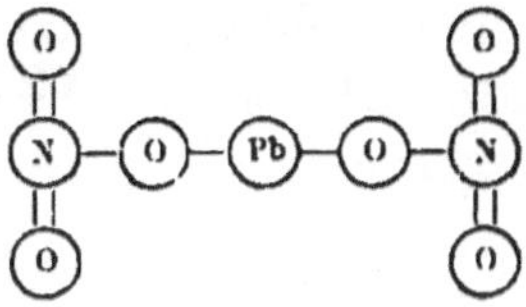

Hydric plumbic nitrate .... } $\mathbf{N}OHoPbo''.$

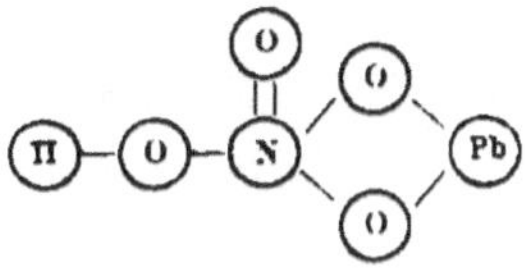

Dihydric diplumbic nitrate hydrate .... } $\mathbf{N}Ho_2Pbo''(PbHoO).$

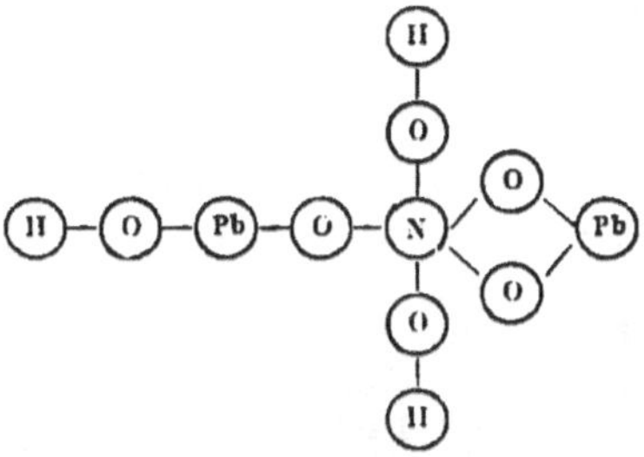

Dihydric diplumbic nitrate nitrite.
(*Basic hyponitrate of lead.*)
$$\left\{\begin{array}{l}\mathbf{N}Ho_2Pbo'' \\ Pbo'' \\ \mathbf{N}O\end{array}\right.$$

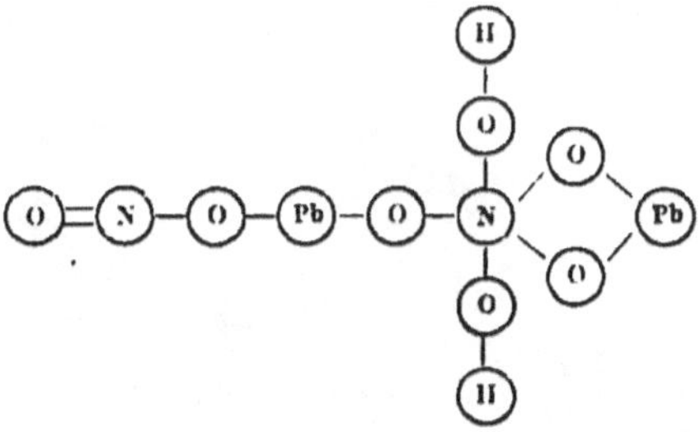

Plumbic carbonate
(*Lead spar, white lead ore.*) ...... } $\mathbf{C}OPbo''.$

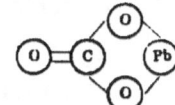

Triplumbic dihy-⎫ $\mathbf{CO}$(OPbHo)⎫
drate dicarbonate⎭ $\mathbf{CO}$(OPbHo)⎭Pbo″;

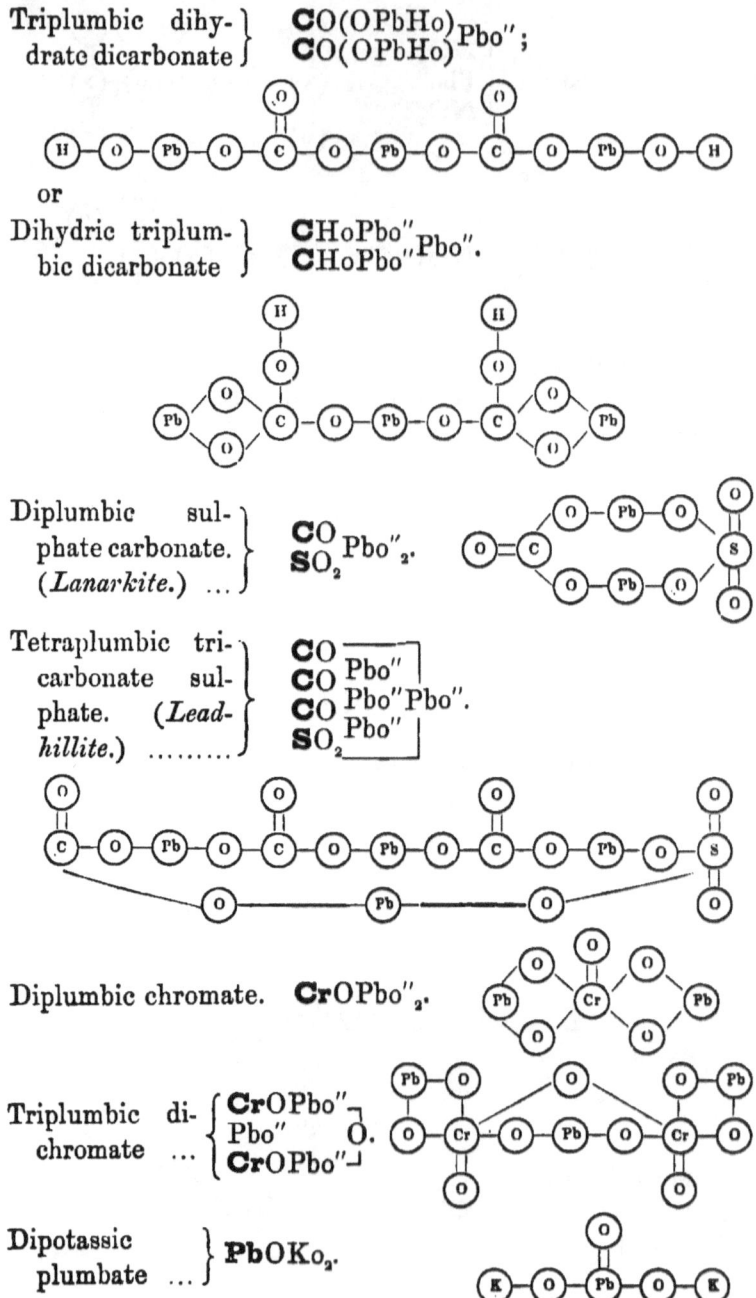

or

Dihydric triplum-⎫ $\mathbf{C}$HoPbo″⎫
bic dicarbonate⎭ $\mathbf{C}$HoPbo″⎭Pbo″.

Diplumbic   sul-⎫
phate carbonate. ⎬ $\mathbf{CO}$⎫Pbo″₂.
(*Lanarkite.*) ...⎭ $\mathbf{SO_2}$⎭

Tetraplumbic tri-⎫ $\mathbf{CO}$⎤
carbonate sul- ⎪ $\mathbf{CO}$⎪Pbo″⎤
phate. (*Lead-* ⎬ $\mathbf{CO}$⎪Pbo″⎬Pbo″.
hillite.) .... ⎭ $\mathbf{SO_2}$⎦Pbo″⎦

Diplumbic chromate. $\mathbf{Cr}$OPbo″₂.

Triplumbic di-⎧ $\mathbf{Cr}$OPbo″⎤
chromate ...⎨   Pbo″   ⎥O.
       ⎩ $\mathbf{Cr}$OPbo″⎦

Dipotassic⎫ $\mathbf{PbOKo_2}$.
plumbate ...⎭

CHAPTER XXIV.

HEXAD ELEMENTS.

SECTION IV.

CHROMIUM, Cr.

Atomic weight $=52\cdot5$. *Molecular weight unknown. Sp. gr.* 6.
Atomicity ", ^{iv}, *and* ^{vi}; *also a pseudo-triad (and a pseud-octad).*
Evidence of atomicity :—See the annexed compounds.

The following are the most important compounds of this
metal :—

Chromous chloride　　$CrCl_2$.

Chromic chloride ... $\begin{cases} CrCl_3. \\ CrCl_3. \end{cases}$

Chromic perfluoride. CrF_6.

Chromous oxide...... CrO.

Chromous hydrate... $CrHo_2$.

Chromic oxide ...... $\begin{cases} CrO \\ CrO \end{cases} O.$

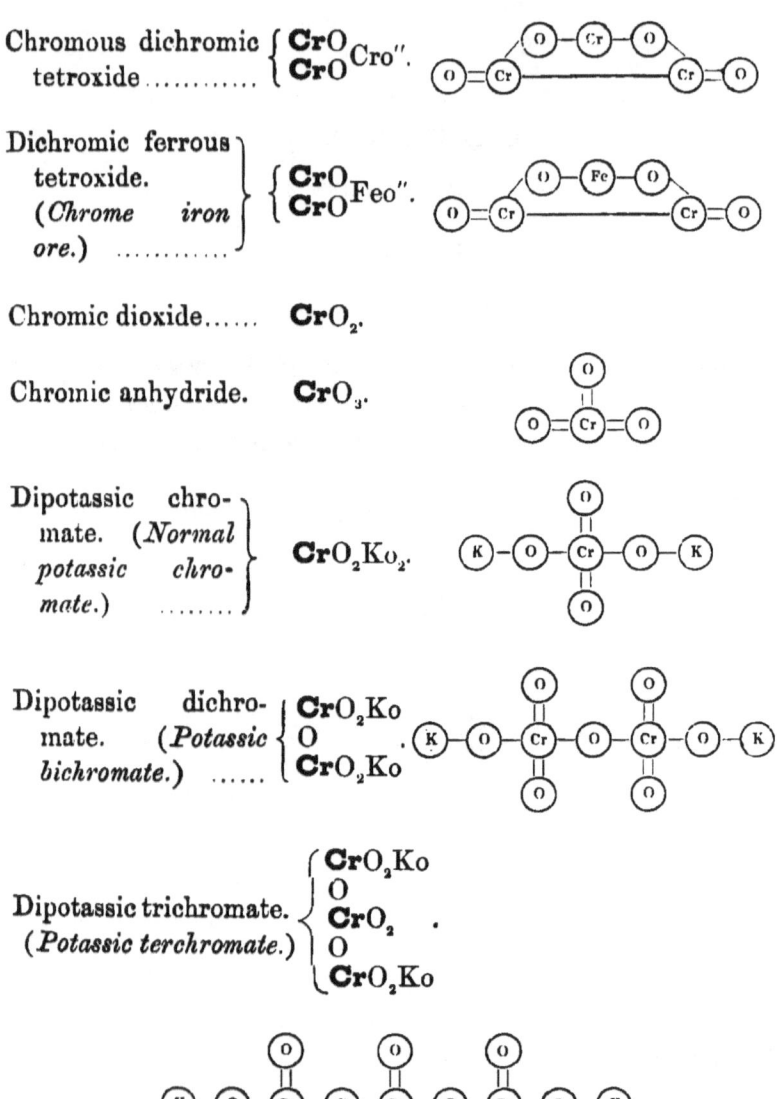

Chromous dichromic tetroxide .... $\left\{\begin{array}{l}\textbf{CrO} \\ \textbf{CrO}\end{array}\right.$ Cro″.

Dichromic ferrous tetroxide. (*Chrome iron ore.*) .... $\left.\right\} \left\{\begin{array}{l}\textbf{CrO} \\ \textbf{CrO}\end{array}\right.$ Feo″.

Chromic dioxide ...... $\textbf{CrO}_2$.

Chromic anhydride. $\textbf{CrO}_3$.

Dipotassic chromate. (*Normal potassic chromate.*) .... $\left.\right\}$ $\textbf{CrO}_2\textbf{Ko}_2$.

Dipotassic dichromate. (*Potassic bichromate.*) ...... $\left\{\begin{array}{l}\textbf{CrO}_2\textbf{Ko} \\ \textbf{O} \\ \textbf{CrO}_2\textbf{Ko}\end{array}\right.$

Dipotassic trichromate. (*Potassic terchromate.*) $\left\{\begin{array}{l}\textbf{CrO}_2\textbf{Ko} \\ \textbf{O} \\ \textbf{CrO}_2 \\ \textbf{O} \\ \textbf{CrO}_2\textbf{Ko}\end{array}\right.$.

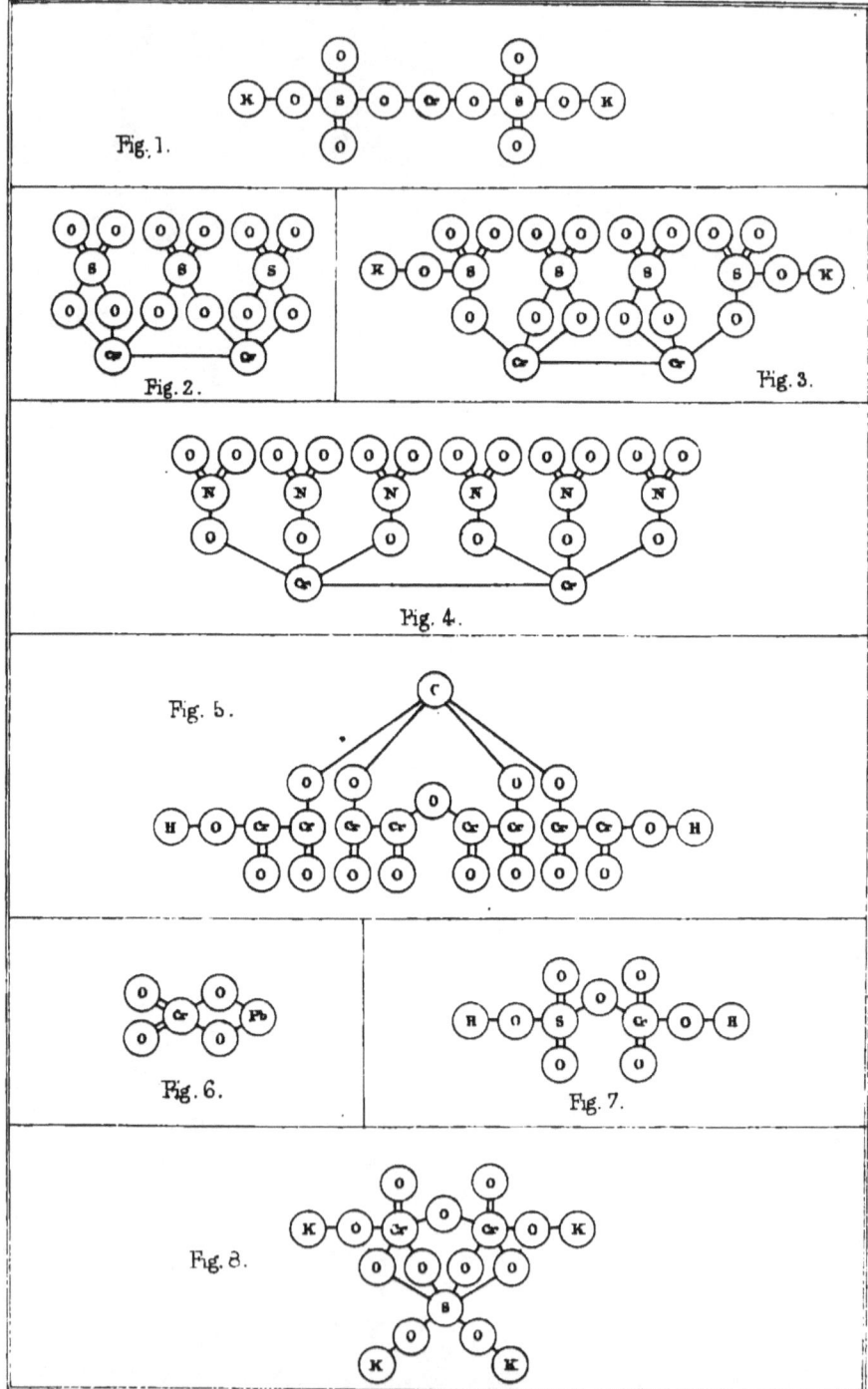

Fig. 1.

Fig. 2.

Fig. 3.

Fig. 4.

Fig. 5.

Fig. 6.

Fig. 7.

Fig. 8.

Dipotassic chromous disulphate. Fig. 1. } $\begin{array}{l}SO_2Ko\\SO_2Ko\end{array}Cro'', 6OH_2.$

Dichromic trisulphate. Fig.2. ...... } $\begin{array}{l}SO_2\rceil\\SO_2-\\SO_2\rfloor\end{array}('Cr'''_2O_6)^{vi}, 15OH_2.$

Dipotassic dichromic tetrasulphate. (*Potassium chrome alum.*) Fig. 3....... } $\begin{array}{l}SO_2Ko\rceil\\SO_2-\\SO_2-\\SO_2Ko\rfloor\end{array}('Cr'''_2O_6)^{vi}, 24OH_2.$

Dichromic hexanitrate. Fig. 4. ...... } $N_6O_{12}('Cr'''_2O_6)^{iv}.$

Octochromic carbonate dihydrate. Fig. 5. .... } $C('Cr'''_8O_{13}Ho_2)^{iv}.$

Plumbic chromate. (*Red lead ore, crocoisite.*) Fig. 6. } $CrO_2Pbo''.$

Sulphochromic acid. (*Dihydric sulphate chromate.*) Fig. 7. ... $\left\{\begin{array}{l}SO_2Ho\\O\\CrO_2Ho\end{array}\right.$

Tetrapotassic dichromosulphate. Fig.8.... $\left\{\begin{array}{l}CrOKo-O_2\rceil\\O\\CrOKo-O_2\rfloor\end{array}S^{vi}Ko_2.\right.$

Chromic hydrate ... $\left\{\begin{array}{l}CrHo_3\\CrHo_3\end{array}, 6OH_2.\right.$

Perchromic acid...... $\left\{\begin{array}{l}CrO_2(OHo)\\CrO_2(OHo)\end{array}\right.$

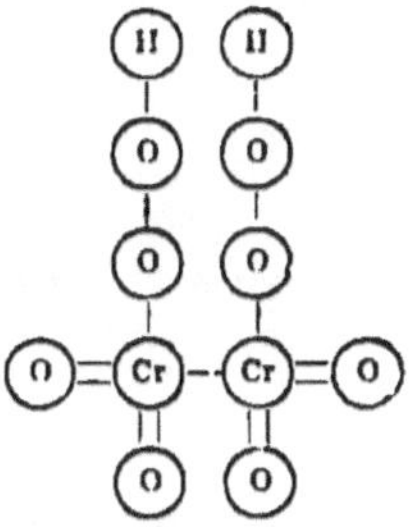

Potassic chlorochromate .... } $CrO_2ClKo.$

Chromic dichlorodioxide. (*Chlorochromic acid.*) ...... } $CrO_2Cl_2.$

Chromous sulphide.... $CrS''.$

Dichromic trisulphide .... $\begin{cases} CrS''S'' \\ CrS''S''. \end{cases}$

Sulphochromic anhydride .... } $CrS''_3.$

MANGANESE, Mn.

Atomic weight =55. *Molecular weight unknown. Sp. gr.* 7 *to* 8. *Atomicity* ", ᶦᵛ, *and* ᵛᶦ; *also a pseudo-triad and a pseud-octad. Evidence of atomicity :—*

Manganic perfluoride .... $Mn^{vi}F_6.$
Analogy with chromium.

The following are some of the more important compounds of this element :—

Manganous chloride. $MnCl_2.$

Manganic chloride... $MnCl_4.$

Dimanganic hexachloride .... $\begin{cases} MnCl_3 \\ MnCl_3. \end{cases}$

Manganous hydrate. $MnHo_2.$

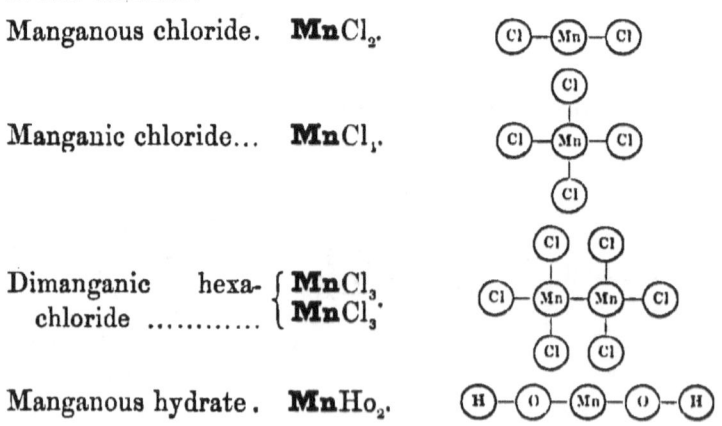

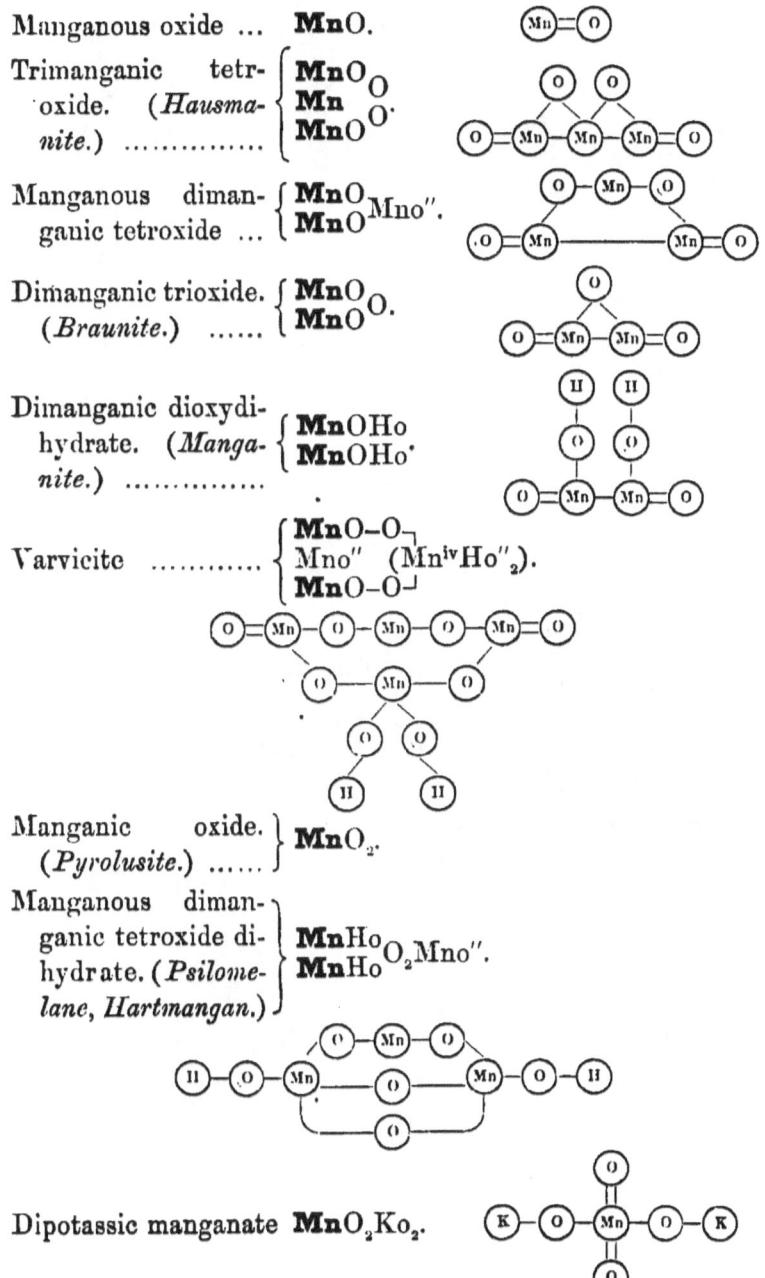

Manganous oxide ... **MnO**.

Trimanganic tetroxide. (*Hausmanite.*) ....
$$\begin{cases} \mathbf{MnO}_O \\ \mathbf{Mn} \\ \mathbf{MnO}^O \end{cases}.$$

Manganous dimanganic tetroxide ...
$$\begin{cases} \mathbf{MnO} \\ \mathbf{MnO} \end{cases} Mno''.$$

Dimanganic trioxide. (*Braunite.*) ......
$$\begin{cases} \mathbf{MnO}_O \\ \mathbf{MnO}^O \end{cases}.$$

Dimanganic dioxydihydrate. (*Manganite.*) ....
$$\begin{cases} \mathbf{MnO}Ho \\ \mathbf{MnO}Ho \end{cases}.$$

Varvicite ....
$$\begin{cases} \mathbf{MnO}\text{-}O \\ Mno'' \ (\mathbf{Mn^{iv}Ho''_2}). \\ \mathbf{MnO}\text{-}O \end{cases}$$

Manganic oxide. (*Pyrolusite.*) ......
$$\left.\right\} \mathbf{MnO}_2.$$

Manganous dimanganic tetroxide dihydrate. (*Psilomelane, Hartmangan.*)
$$\left.\right\} \begin{matrix} \mathbf{MnHo}_O \\ \mathbf{MnHo}^{O_2}Mno''. \end{matrix}$$

Dipotassic manganate **MnO₂Ko₂**.

Dipotassic per-manganate ...... } $\left\{\begin{array}{l} \mathbf{Mn}O_2(OKo) \\ \mathbf{Mn}O_2(OKo) \end{array}\right.$ ·

Manganous sul-phide. (*Man-ganese blende.*). } $\mathbf{Mn}S.$

Disulphopotassic trimanganous disulphide ...... } $\begin{array}{l} \mathbf{Mn}Ks_{S''} \\ \mathbf{Mn}_{S'''} \\ \mathbf{Mn}Ks \end{array}$ ·

Manganous carbo-nate. (*Manga-nese spar.*) ...... } $\mathbf{C}OMno''.$

Dihydric manga-nous sulphate... } $\mathbf{S}OHo_2Mno'',(3, 4\ or\ 6\mathbf{O}H_2).$

Dipotassic manga-nous disulphate } $\begin{array}{l} \mathbf{S}O_2Ko \\ \mathbf{S}O_2Ko \end{array}Mno'',6\mathbf{O}H_2.$

Aluminic manga-nous tetrasul-phate. (*Manga-nese aluminium alum.*) .... } $\left\{\begin{array}{l} \mathbf{S}O_2 \\ \mathbf{S}O_2 \\ Mno''\ (\,'Al'''_2O_6)^{vi},\ 24\mathbf{O}H_2. \\ \mathbf{S}O_2 \\ \mathbf{S}O_2 \end{array}\right.$

Dimanganic trisul-phate .... } $\begin{array}{l} \mathbf{S}O_2 \\ \mathbf{S}O_2-(\,'Mn'''_2O_6)^{vi}. \\ \mathbf{S}O_2 \end{array}$

Dipotassic diman-ganic tetrasul-phate. (*Potas-sium manganese alum.*) .... } $\begin{array}{l} \mathbf{S}O_2Ko \\ \mathbf{S}O_2 \\ \mathbf{S}O_2-(\,'Mn'''_2O_6)^{vi},\ 24\mathbf{O}H_2. \\ \mathbf{S}O_2Ko \end{array}$

Dihydric dimanga-
nous diphosphate } $P_2O_2Ho_2Mno_2''$.

Manganous sili-
cate. (*Silicife-
rous manganese,
red manganese,
rother Mangan-
kiesel, Roth-
braunsteinerz.*) } $SiOMno''$.

Dimanganous sili-
cate. (*Tephro-
ite.*) .... } $SiMno''_2$.

Dihydric dimanga-
nic silicate dihy-
drate. (*Schwar-
zer Mangankie-
sel.*) .... } $SiHo_2(OMnHo)_2$.

Hexmanganic mo-
nosilicate. (*He-
terocline.*) ...... } $Si('Mn'''_6O_{11})^{iv}$.

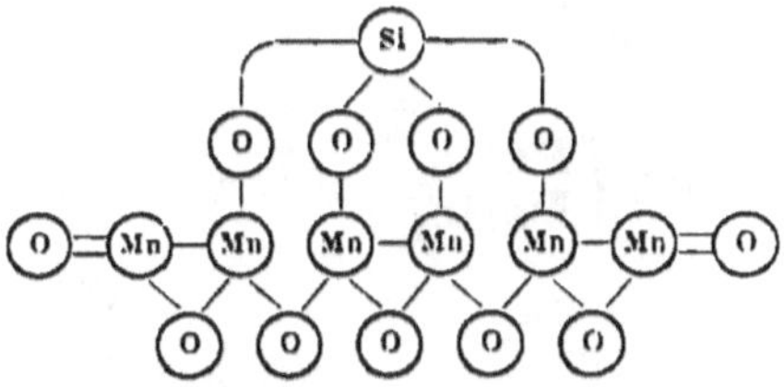

Triglucinic tetraman-
ganous trisilicate
sulphide. (*Hel-
vine.*) ....

$\left\{\begin{array}{l} SiGo'' \underline{\hspace{2cm}} \\ Mno'' \\ SiGo'' \\ Mno'' \\ SiGo'' \underline{\hspace{2cm}} \end{array} \left(\left\{\begin{array}{l} Mn''O \\ S'' \\ Mn''O \end{array}\right\}\right)''\right.$.

Aluminic manganous
disilicate...........

$\left\{\begin{array}{l} Si \underline{\hspace{1cm}} \\ Mno''('Al'''_2O_6)^{vi}, 2OH_2. \\ Si \underline{\hspace{1cm}} \end{array}\right.$

IRON, Fe.

Atomic weight =56.   *Molecular weight unknown.*   *Sp. gr.* 7·8.   *Atomicity* ", ^{iv}, *and* ^{vi}.   *Evidence of atomicity* :—*Analogy with chromium.*

The following is a list of the chief compounds of iron :—

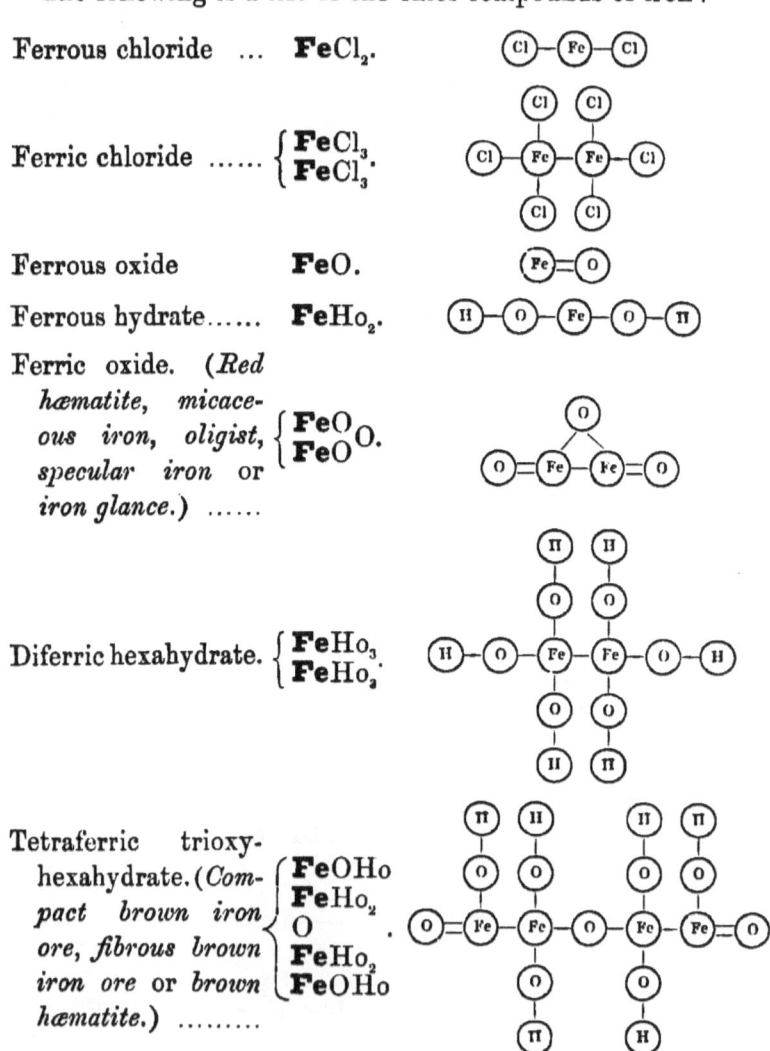

Ferrous chloride ...   **FeCl$_2$**.

Ferric chloride ...... $\begin{cases} \textbf{FeCl}_3. \\ \textbf{FeCl}_3. \end{cases}$

Ferrous oxide   **FeO**.

Ferrous hydrate...... **FeHo$_2$**.

Ferric oxide. (*Red hæmatite, micaceous iron, oligist, specular iron or iron glance.*) ...... $\begin{cases} \textbf{FeO} \\ \textbf{FeO} \end{cases}\!\!\textbf{O}.$

Diferric hexahydrate. $\begin{cases} \textbf{FeHo}_3. \\ \textbf{FeHo}_2. \end{cases}$

Tetraferric trioxy-hexahydrate. (*Compact brown iron ore, fibrous brown iron ore or brown hæmatite.*) .... $\begin{cases} \textbf{FeOHo} \\ \textbf{FeHo}_2 \\ \textbf{O} \\ \textbf{FeHo}_2 \\ \textbf{FeOHo} \end{cases}\!\!.$

Diferric dioxy-dihy-
drate. (*Needle* $\begin{cases} \mathbf{Fe}OHo \\ \mathbf{Fe}OHo^{\cdot} \end{cases}$
iron ore, brown
iron ore.)

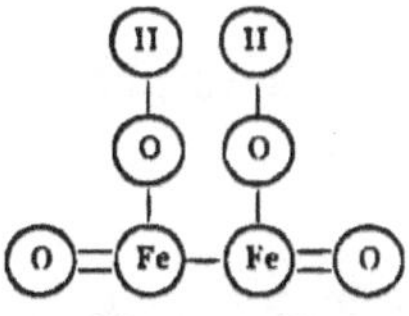

Diferric oxy-tetra-
hydrate. (Another $\begin{cases} \mathbf{Fe}Ho_2O. \\ \mathbf{Fe}Ho_2 \end{cases}$
variety of *brown*
iron ore.)

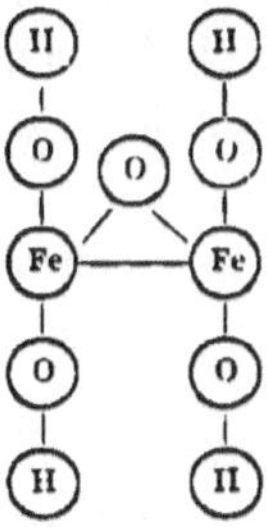

Triferric tetroxide ... $\begin{cases} \mathbf{Fe}O_O \\ \mathbf{Fe}\;O \\ \mathbf{Fe}O^O \end{cases};$

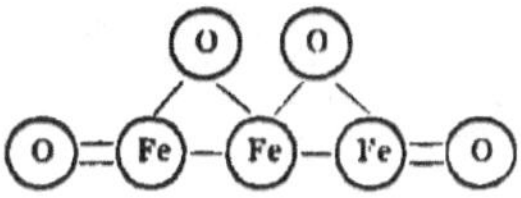

or

Ferrous diferric te-
troxide. (*Magnetic* $\begin{cases} \mathbf{Fe}O \\ \mathbf{Fe}O \end{cases} Feo''.$
iron ore.)

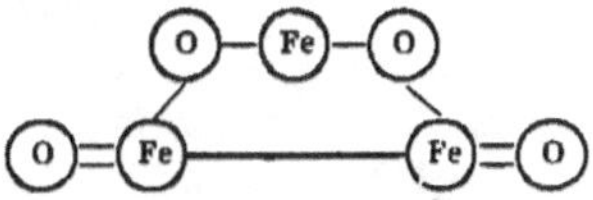

Potassic ferrate ...... $\mathbf{Fe}O_2Ko_2.$

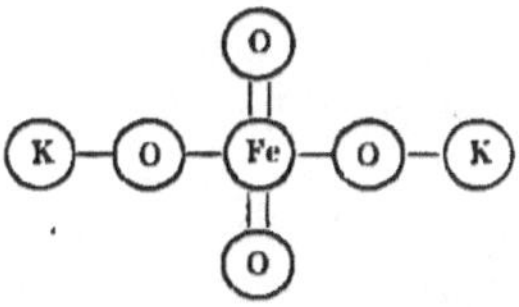

Octoferrous sulphide $^{xiv}(\mathbf{Fe}_8)''S''.$

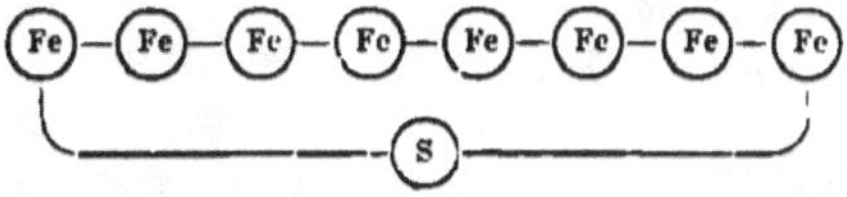

Diferrous sulphide... $\begin{cases} \mathbf{Fe} \\ \mathbf{Fe} \end{cases} S''.$

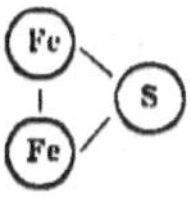

Ferrous sulphide $\mathbf{Fe}S''.$

Ferric disulphide.
(*Iron pyrites,* $\mathbf{Fe}S''_2.$
martial pyrites.)

K

Diferric trisulphide $\begin{cases} \textbf{FeS}''_2\textbf{S}'' \\ \textbf{FeS}''_2\textbf{S}''. \end{cases}$

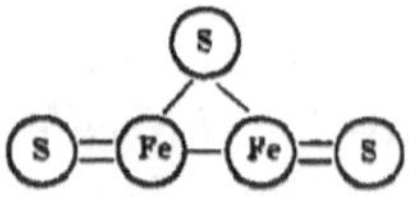

Diferric dicupric tetrasulphide. (*Copper pyrites.*) $\begin{cases} \textbf{FeS} \\ \textbf{FeS}(Cu_2S''_2)''. \end{cases}$

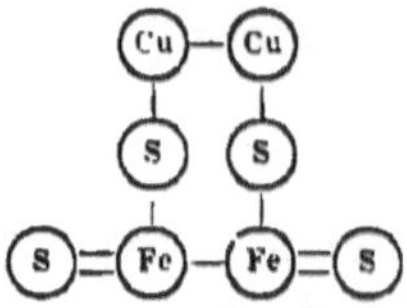

Diferric hexanitrate. Fig. 1... $\}$ $\textbf{N}_6O_{12}('Fe'''_2O_6)^{vi}.$

Hexahydric diferric diphosphate dihydrate. Fig. 2 .... $\}$ $\textbf{P}_2Ho_6('Fe'''_2Ho_2O_4)^{iv}.$

Ferrous sulphate. Fig. 3 .... $\}$ $\textbf{S}OHo_2Feo'',6OH_2.$

Dipotassic ferrous disulphate. Fig. 4. $\begin{cases} \textbf{S}O_2Ko \\ Feo'''_2 &,6OH_2. \\ \textbf{S}O_2Ko. \end{cases}$

Diferric trisulphate. (*Coquimbite.*) Fig. 5. $\}$ $\begin{aligned} &\textbf{S}O_2- \\ &\textbf{S}O_2-('Fe'''_2O_6)^{vi},9OH_2. \\ &\textbf{S}O_2- \end{aligned}$

Dipotassic diferric tetrasulphate. (*Potassium iron alum.*) Fig. 6. $\}$ $\begin{aligned} &\textbf{S}O_2Ko \\ &\textbf{S}O_2- \\ &\textbf{S}O_2- ('Fe'''_2O_6)^{vi},24OH_2. \\ &\textbf{S}O_2Ko \end{aligned}$

Tetraferric sulphate. Fig. 7. $\}$ $\textbf{S}O_2('Fe'''_4O_7)'',6OH_2.$

Tetrahydric tetraferric sulphate octohydrate. (*Vitriol ochre.*) Fig. 8 .... $\}$ $\textbf{S}Ho_4('Fe'''_4OHo_8O_2)''.$

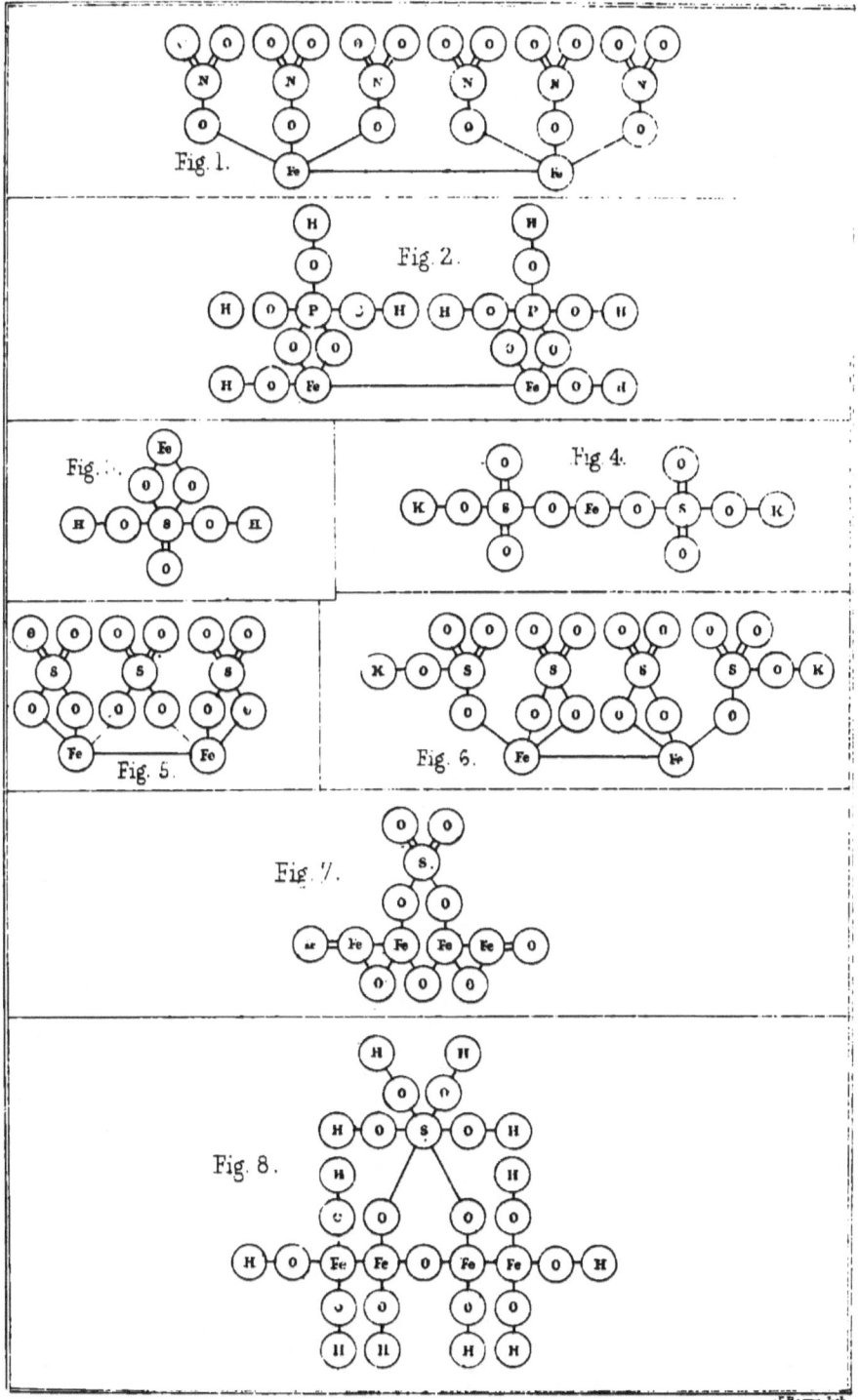

Fig. 1.

Fig. 2.

Fig. 3.

Fig. 4.

Fig. 5.

Fig. 6.

Fig. 7.

Fig. 8.

J. Brown, lith.

Heptaferric octo-
sulphide. (*Mag-
netic pyrites*.) ... } $\mathbf{Fe_7S''_8}$.

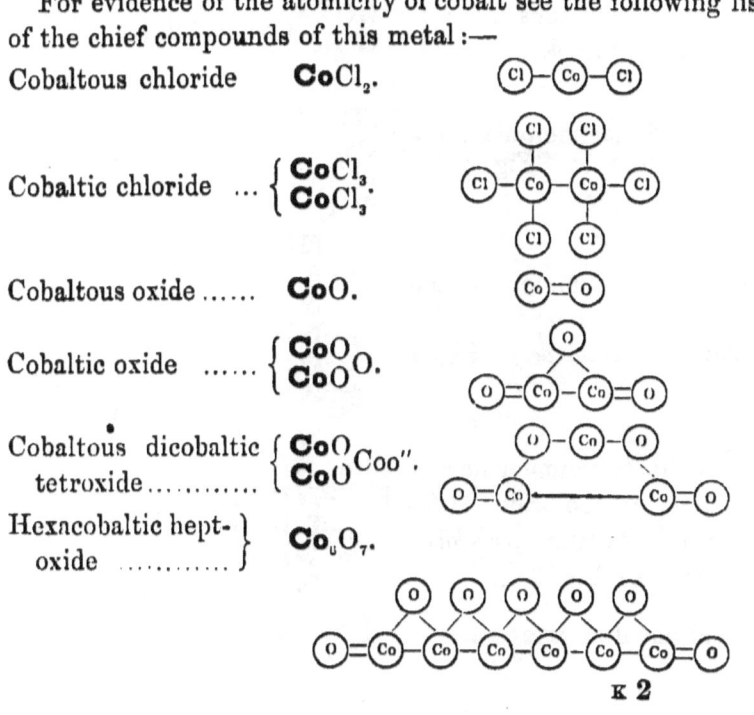

Ferrous carbonate.
(*Spathic iron
ore*.) } $\mathbf{C}$OFeo''.

Ferrous nitrate ...... $\left\{ \begin{array}{l} \mathbf{NO_2} \\ \text{Feo}'' \\ \mathbf{NO_2}. \end{array} \right.$

COBALT, Co.

Atomic weight = 58·8. *Molecular weight unknown.* *Sp.gr.* 8·5.
Atomicity '', ^{iv}, *and* ^{vi} ? *also pseudo-triatomic.*

For evidence of the atomicity of cobalt see the following list
of the chief compounds of this metal :—

Cobaltous chloride $\mathbf{CoCl_2}$.

Cobaltic chloride ... $\left\{ \begin{array}{l} \mathbf{CoCl_3} \\ \mathbf{CoCl_3}. \end{array} \right.$

Cobaltous oxide ...... $\mathbf{CoO}$.

Cobaltic oxide ...... $\left\{ \begin{array}{l} \mathbf{CoO}_O. \\ \mathbf{CoO} \end{array} \right.$

Cobaltous dicobaltic
tetroxide .... $\left\{ \begin{array}{l} \mathbf{CoO}_{Coo''.} \\ \mathbf{CoO} \end{array} \right.$

Hexacobaltic hept-
oxide .... } $\mathbf{Co_6O_7}$.

Cobaltous hydrate .... $CoHo_2$.   (H)—(O)—(Co)—(O)—(H)

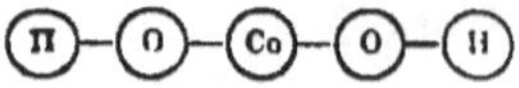

Cobaltic oxy-dihydrate ...... $\begin{cases} CoOHo \\ CoOHo \end{cases}$

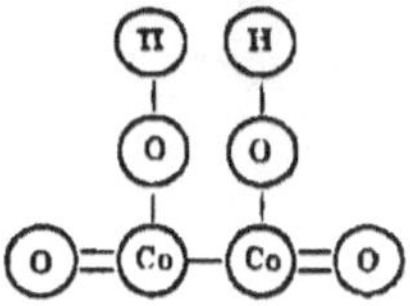

Cobaltous sulphide. ......:.... CoS''.

Cobaltic sulphide. (*Cobalt pyrites.*) .... $\begin{cases} CoS''S'' \\ CoS'''S'' \end{cases}$

Dipotassic cobaltous disulphate. Fig. 1 .... $\begin{matrix} SO_2Ko \\ SO_2Ko \end{matrix} Coo'',6OH_2$.

Dihydric pentacobaltous dicarbonate tetrahydrate. Fig. 2 .... $\begin{matrix} CHo(OCo''Ho)_2 \\ CHo(OCo''Ho)_2 \end{matrix} Coo'',OH_2$.

Dicobaltous carbonate dihydrate. Fig. 3. .... $CO(OCo''Ho)_2$.

Dipotassic cobaltous dicarbonate. Fig. 4. .... $\begin{matrix} COKo \\ COKo \end{matrix} Coo'',10OH_2$.

Cobaltoso-diammon-diammonic dichloride. Fig. 5. ... $\begin{cases} NH_4 \\ NH_2Cl \\ Co'' \\ NH_2Cl \\ NH_4 \end{cases}$

Dicobaltic hexammon-hexammonic hexachloride. (*Luteo-cobalt chloride.*) Fig. 6. $\begin{cases} Co\begin{matrix} -NH_2(N^vH_4)Cl \\ -NH_2(N^vH_4)Cl \\ -NH_2(N^vH_4)Cl \end{matrix} \\ Co\begin{matrix} -NH_2(N^vH_4)Cl \\ -NH_2(N^vH_4)Cl \\ -NH_2(N^vH_4)Cl \end{matrix} \end{cases}$

Dicobaltic tetrammon-hexammonic hexachloride. (*Roseo-cobalt* or *purpureo-cobalt chloride.*) Fig. 7............. $\begin{cases} Co\begin{matrix} -NH_2(N^vH_4)Cl \\ -NH_2(N^vH_4)Cl \\ -NH_3Cl \end{matrix} \\ Co\begin{matrix} -NH_3Cl \\ -NH_2(N^vH_4)Cl \\ -NH_2(N^vH_4)Cl \end{matrix} \end{cases}$

Cobaltous dinitrate .... $\begin{matrix} NO_2 \\ NO_2 \end{matrix} Coo'',6OH_2$.

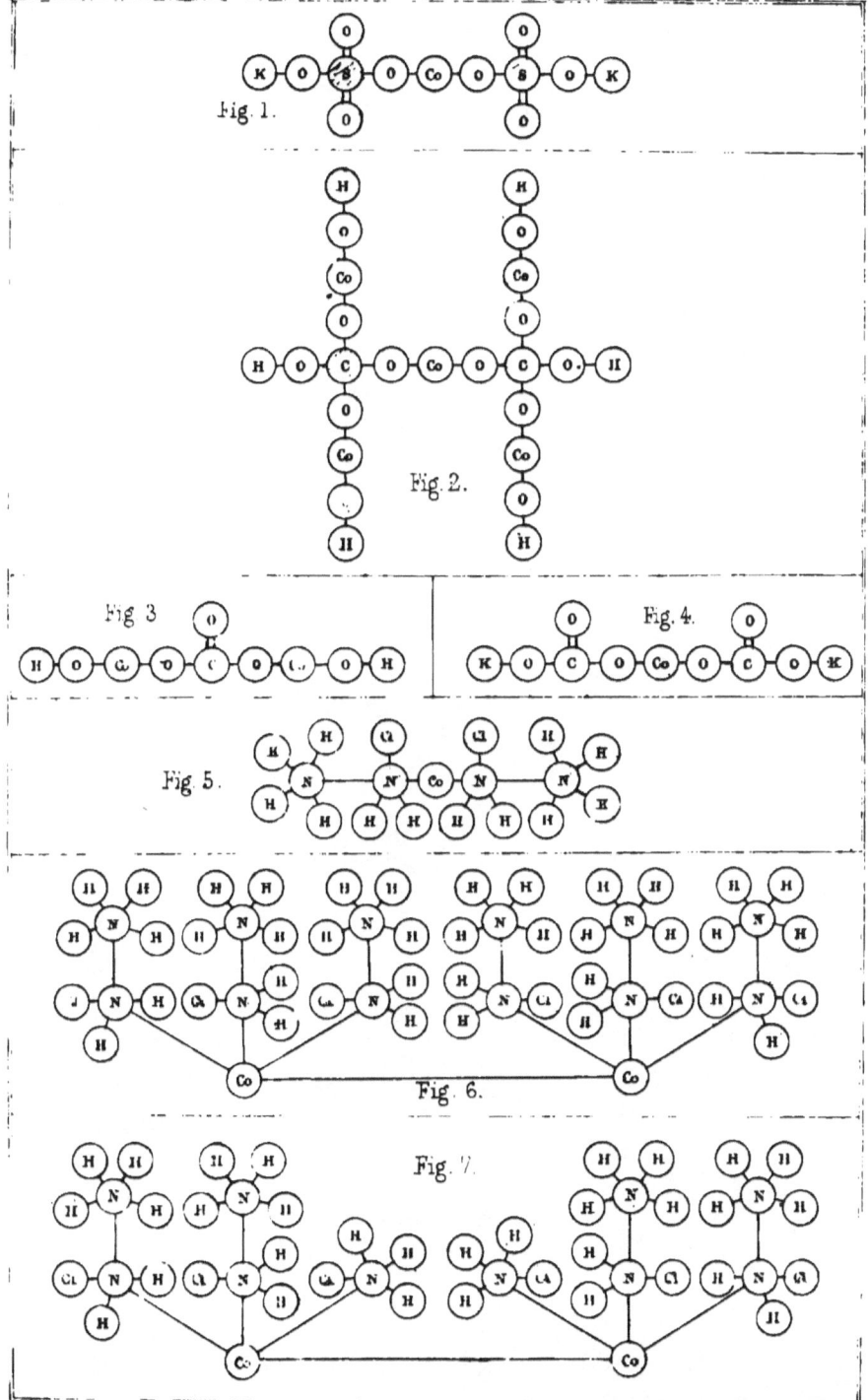

Fig. 1.

Fig. 2.

Fig 3

Fig. 4.

Fig 5.

Fig. 6.

Fig. 7.

Cobaltic disulphide .... **CoS**$_2$.

Dicobaltous oxysulphide...... **Co Co**OS″.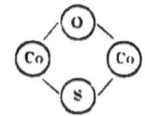

Dihydric cobaltous sulphate.. **S**OHo$_2$**Coo**″,6**O**H$_2$.

NICKEL, Ni.

Atomic weight =58·8. *Molecular weight unknown. Sp. gr.* 8·7. *Atomicity* ″, ^{iv}, *and* ^{vi}? *also pseudo-triatomic.*

Annexed is a list of the chief compounds of this metal :—

Nickelous chloride ...... **NiCl**$_2$. ⓒⓛ—Ⓝⓘ—ⓒⓛ

Nickelous oxide .... **NiO.**

Nickelic oxide .... { **NiO NiO**O.

Nickelic hydrate .... { **NiHo**$_3$ **NiHo**$_3$.

Nickelous hydrate ...... **NiHo**$_2$.

Dinickelous sulphide ... { **Ni Ni**S″.

Nickelous sulphide.
(*Capillary pyrites, hair nickel.*) .... } **NiS**″.

Nickelic disulphide ...... **NiS**″$_2$. ⓢ=Ⓝⓘ=ⓢ

Dihydric nickelous sul-
phate .... } **S**OHo$_2$**Nio**″, 6**O**H$_2$.

Dipotassic nickelous di-
sulphate ....
$\left\{\begin{array}{l}\textbf{SO}_2\textbf{Ko} \\ \text{Nio}'' \\ \textbf{SO}_2\textbf{Ko}\end{array}\right.,6\textbf{OH}_2.$

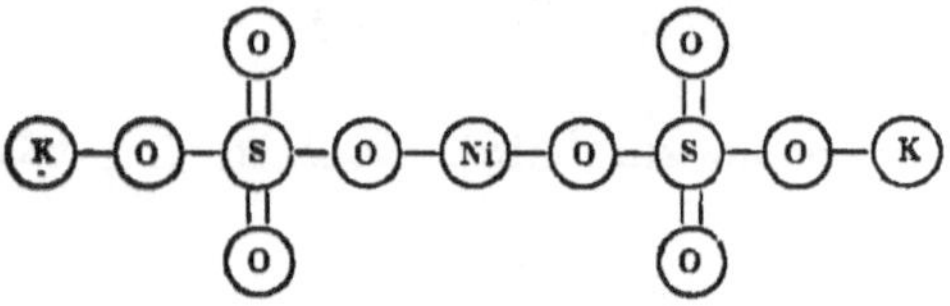

Nickelous dinitrate ......
$\left\{\begin{array}{l}\textbf{NO}_2 \\ \text{Nio}'',6\textbf{OH}_2. \\ \textbf{NO}_2\end{array}\right.$

Dihydric pentanickelous
dicarbonate   tetrahy-
drate ....
$\left\{\begin{array}{l}\textbf{C}\text{Ho(Ni}''\text{OHo)}_2 \\ \text{Nio}'' \\ \textbf{C}\text{Ho(Ni}''\text{OHo)}_2\end{array}\right.,2\textbf{OH}_2.$

Dipotassic nickelous tri-
carbonate ....
$\left\{\begin{array}{l}\textbf{C}\text{OKo} \\ \text{O} \\ \textbf{C}\text{Nio}'',10\textbf{OH}_2. \\ \text{O} \\ \textbf{C}\text{OKo}\end{array}\right.$

Nickelous diarsenide.
(*Kupfernickel.*) ...
$\left.\begin{array}{l}\ \end{array}\right\}$ $''\textbf{As}'_{''}\textbf{As}'\text{Ni}'';$

or

$''\left\{\begin{array}{l}\textbf{As}' \\ \textbf{As}'\end{array}\right.\text{Ni}''.$

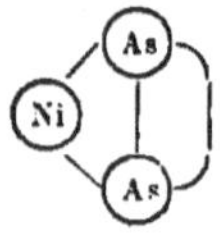

Nickelic tetrarsenide.
$\begin{array}{l}''\textbf{As}'{-} \\ ''\textbf{As}' \\ ''\textbf{As}'\text{Ni}^{\text{iv}}. \\ ''\textbf{As}'{-}\end{array}$

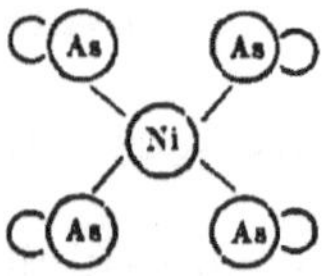

Nickelous sulphide di-
arsenide. (*Grey nickel
ore, nickel glance.*)
$\left\{\begin{array}{l}\textbf{As} \\ \textbf{As}\end{array}\right.\text{S}''\text{Ni}''.$

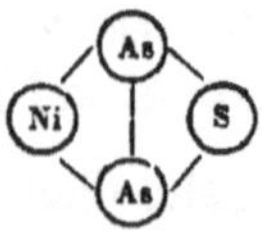

CHAPTER XXV.

ORGANIC CHEMISTRY.

INTRODUCTORY.

THE number of elements entering into the composition of organic compounds is usually very small.

The really essential element is carbon.

A large number of compounds consist of carbon united with only *one* other element, either hydrogen or nitrogen.

The members of another very large class contain only *three* elements,—carbon, hydrogen, and oxygen; carbon, hydrogen, and nitrogen; or carbon, nitrogen, and oxygen.

Others contain *four* elements, carbon, hydrogen, oxygen, and nitrogen.

Some contain sulphur, chlorine, bromine, iodine, silicon, phosphorus, boron, and metals; but these are comparatively rare, and those containing sulphur are alone found in nature.

Although the number of the elements entering into the composition of organic compounds is much smaller than that taking part in the formation of minerals, yet the complexity of the former class of compounds is frequently much greater.

There is no instance of an organic compound containing but two single atoms, and only one containing three (hydrocyanic acid, HCN).

On the other hand, the number of atoms in a single organic compound is sometimes very great.   Thus:—

Spermaceti contains	98	atoms.
Stearin	„	173	„
Margarin	„	217	„
Albumen	„	226	„

In most, possibly in all, organic compounds *carbon is a tetrad.* It therefore forms with 1 atom of hydrogen a triad compound radical, which exists in chloroform $(\mathbf{C^{iv}H})'''\mathrm{Cl}_3$; a dyad compound radical with 2 atoms of hydrogen, as in methylenic dichloride $(\mathbf{C^{iv}H_2})''\mathrm{Cl}_2$; and a monad compound radical with 3 atoms of hydrogen, as in methylic iodide $(\mathbf{C^{iv}H_3})\mathrm{I}$. But in these compounds it will be seen that the carbon is always saturated and always a tetrad.

This is in conformity with the following general law:—" In every molecule of a chemical compound the sum of the bonds is always an even number," because every number, whether odd or even, when multiplied by 2 gives an even number. To this may be added the following law, to which no exception is known:—"The sum of the bonds in any molecule is at least twice as great as the active atomicity of the most polyad element in the compound."

A compound organic radical consists, therefore, of one or more atoms of carbon in which one or more bonds are unsatisfied, and it is either a monad, dyad, &c. radical, according to the number of monad atoms required to complete its active atomicity.

Such a radical, when a monad, triad, or pentad, cannot exist as a separate atom; like hydrogen or nitrogen, when isolated, it combines with itself, forming a diatomic molecule. It is only by the union of two atoms that the vacated bonds can be satisfied.

The following are two of the principal series of compound organic radicals:—

Monads.			Dyads.	
Methyl... $\left\{\begin{array}{l}\mathbf{CH_3}\\\mathbf{CH_3}\end{array}\right.$	or	$\left\{\begin{array}{l}\mathrm{Me}\\\mathrm{Me}\end{array}\right.$	Methylene... $\mathbf{CH_2}$ or Me''*.	
Ethyl ... $\left\{\begin{array}{l}\mathbf{C_2H_5}\\\mathbf{C_2H_5}\end{array}\right.$	or	$\left\{\begin{array}{l}\mathrm{Et}\\\mathrm{Et}\end{array}\right.$	Ethylene ... $\mathbf{C_2H_4}$ or Et''.	
Propyl... $\left\{\begin{array}{l}\mathbf{C_3H_7}\\\mathbf{C_3H_7}\end{array}\right.$	or	$\left\{\begin{array}{l}\mathrm{Pr}\\\mathrm{Pr}\end{array}\right.$	Propylene... $\mathbf{C_3H_6}$ or Pr''.	
Butyl ... $\left\{\begin{array}{l}\mathbf{C_4H_9}\\\mathbf{C_4H_9}\end{array}\right.$	or	$\left\{\begin{array}{l}\mathrm{Bu}\\\mathrm{Bu}\end{array}\right.$	Butylene... $\mathbf{C_4H_8}$ or Bu''.	
Amyl ... $\left\{\begin{array}{l}\mathbf{C_5H_{11}}\\\mathbf{C_5H_{11}}\end{array}\right.$		$\left\{\begin{array}{l}\mathrm{Ay}\\\mathrm{Ay}\end{array}\right.$	Amylene ... $\mathbf{C_5H_{10}}$ or Ay''.	

* Only known in combination

Organic radicals are the analogues of the monad, dyad, and triad elements of mineral chemistry.

Such being the constitution of the organic radicals, it will now be necessary to investigate their functions in organic compounds, and to examine the general plan according to which these compounds are built up.

We shall be assisted in this investigation if we reduce the formulæ of these compounds to a few types or fundamental forms. In doing this it is necessary to avoid, as far as possible, all empirical grouping of atoms. Our formulæ must express, as exactly as possible, how the elements are combined with each other. Thus in a compound containing $C_xH_yO_z$, the formula ought to show, first, whether the hydrogen is combined with the carbon or with the oxygen, or if combined with both, it should indicate how many atoms are united with the carbon, and how many with the oxygen. Secondly, the formula ought to show whether the oxygen be united with the carbon or with the hydrogen, or partly with the one and partly with the other, or, lastly, whether it be performing the function of linking hydrogen to carbon.

This information is most completely given, in notation, by making carbon the dominant or grouping element in non-nitrogenous compounds, and nitrogen in the remaining organic compounds.

Non-nitrogenous organic compounds, exclusive of organo-metallic bodies, can be conveniently considered under the two following types, viz. :—

1. The monadelphic, or marsh-gas type .... }

2. The diadelphic, or methyl type...

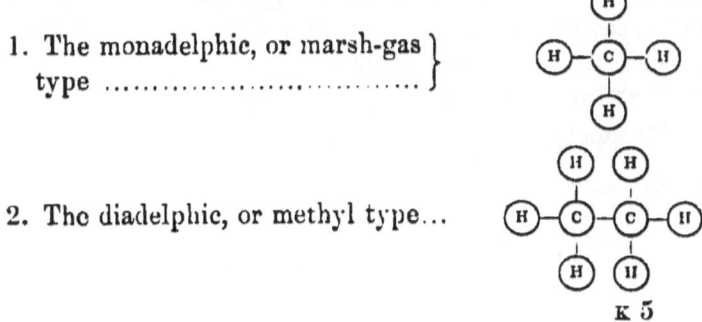

K 5

The nitrogenous organic compounds arrange themselves in the most convenient manner under the two following types:—

3. The ammonia type ....

4. The ammonic chloride type ......

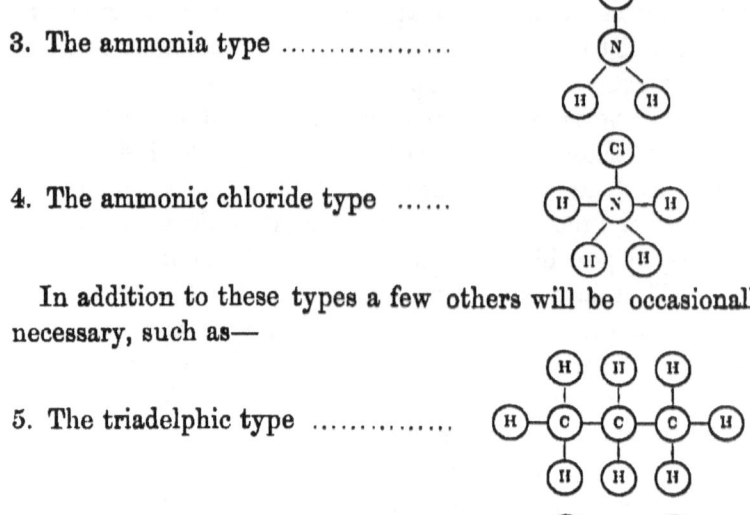

In addition to these types a few others will be occasionally necessary, such as—

5. The triadelphic type ....

6. The double monadelphic type ...

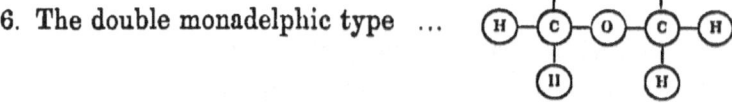

7. The condensed diadelphic or } olefine type.....................

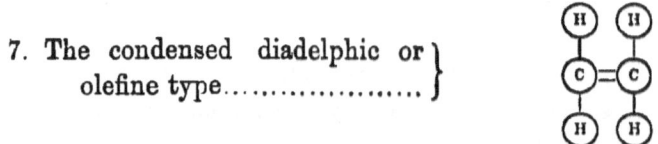

Double, and sometimes even treble ammonia and ammonic-chloride types are required for certain compounds, but they do not need special description here.

The above types are written symbolically as follows:—

1. Monadelphic type .... CH_4.

2. Diadelphic type .... $\begin{cases} CH_3 \\ CH_3 \end{cases}$.

3. Ammonia type........................ $N'''H_3$.

4. Ammonic-chloride type .... N^vH_4Cl.

5. Triadelphic type .... $\left\{\begin{array}{l} \mathbf{CH_3} \\ \mathbf{CH_2}. \\ \mathbf{CH_3} \end{array}\right.$

6. Double monadelphic type .... $\left\{\begin{array}{l} \mathbf{CH_3} \\ \mathbf{O} \\ \mathbf{CH_3} \end{array}\right.$

7. Condensed diadelphic type ......" $\left\{\begin{array}{l} \mathbf{CH_2} \\ \mathbf{CH_2}. \end{array}\right.$

In order to facilitate the use of these symbolic types, it is advisable to become familiarized with the symbols of the following analogues of hydroxyl, in addition to those already given for inorganic compound radicals at p. 28, and for the monad and dyad organic radicals at p. 200 :—

Methoxyl .... $\mathbf{CH_3O}$ or Meo.
Ethoxyl .... $\mathbf{C_2H_5O}$ or Eto.
Propoxyl .... $\mathbf{C_3H_7O}$ or Pro.
Butoxyl .... $\mathbf{C_4H_9O}$ or Buo.
Amoxyl .... $\mathbf{C_5H_{11}O}$ or Ayo.

Formulæ written on the triadelphic type can be reduced to the diadelphic type, and those on the latter to the monadelphic type as follows :—

Propylic hydride. $\left\{\begin{array}{l} \mathbf{CH_3} \\ \mathbf{CH_2} \\ \mathbf{CH_3} \end{array}\right.$ = $\left\{\begin{array}{l} \mathbf{CMeH_2} \\ \mathbf{CH_3} \end{array}\right.$ = $\mathbf{CMe_2H_2}$ or $\mathbf{CEtH_3}$.

Triadelphic.　　　Diadelphic.　　　Monadelphic.

The above alternative monadelphic formulæ, although apparently different, are in reality identical, as can be easily proved by expressing both graphically, thus :—

$\mathbf{CMe_2H_2}$ =

$$\mathbf{CEtH_2} \;=\; \begin{array}{c} \textcircled{\scriptsize H} \\ | \\ \textcircled{\scriptsize H}-\textcircled{\scriptsize C}-\textcircled{\scriptsize H} \\ | \\ \textcircled{\scriptsize H}-\textcircled{\scriptsize C}-\textcircled{\scriptsize H} \\ | \\ \textcircled{\scriptsize H}-\textcircled{\scriptsize C}-\textcircled{\scriptsize H} \\ | \\ \textcircled{\scriptsize H} \end{array}$$

On the other hand, monadelphic formulæ, if they contain a sufficient number of carbon atoms, can be expanded into diadelphic, triadelphic, &c. formulæ in the following manner:—

$$\mathbf{CBuPrH_2} = \left\{ \begin{array}{l} \mathbf{CPrH_2} \\ \mathbf{CPrH_2} \end{array} \right. = \left\{ \begin{array}{l} \mathbf{CEtH_2} \\ \mathbf{CH_2} \\ \mathbf{CH_2} \\ \mathbf{CEtH_2} \end{array} \right. = \left\{ \begin{array}{l} \mathbf{CMeH_2} \\ \mathbf{CH_2} \\ \mathbf{CH_2} \\ \mathbf{CH_2} \\ \mathbf{CH_2} \\ \mathbf{CMeH_2} \end{array} \right. = \left\{ \begin{array}{l} \mathbf{CH_3} \\ \mathbf{CH_2} \\ \mathbf{CH_2} \\ \mathbf{CH_2} \\ \mathbf{CH_2} \\ \mathbf{CH_2} \\ \mathbf{CH_2} \\ \mathbf{CH_3} \end{array} \right.$$

Normal butyl, or propylated methyl. (Molecule.)

This development of normal butyl depends on the following facts:—1st, that normal butyl is propylated methyl; 2nd, that normal propyl is ethylated methyl; and 3rd, that ethyl is methylated methyl.

Secondary and tertiary organic radicals (see p. 207) cannot be completely developed vertically:—

$$\left\{ \begin{array}{l} \mathbf{CEtMeH} \\ \mathbf{CEtMeH} \end{array} \right. = \left\{ \begin{array}{l} \mathbf{CMeH_2} \\ \mathbf{CMeH} \\ \mathbf{CMeH} \\ \mathbf{CMeH_2} \end{array} \right. = \left\{ \begin{array}{l} \mathbf{CH_3} \\ \mathbf{CH_2} \\ \mathbf{CMeH} \\ \mathbf{CMeH} \\ \mathbf{CH_2} \\ \mathbf{CH_3} \end{array} \right. = \left\{ \begin{array}{l} \mathbf{CH_3} \\ \mathbf{CH_2} \\ \mathbf{CH(CH_3)} \\ \mathbf{CH(CH_3)} \\ \mathbf{CH_2} \\ \mathbf{CH_3} \end{array} \right. :$$

Methylo-ethylated methyl. (Molecule.)

$$\left\{ \begin{array}{l} \mathbf{CMe_3} \\ \mathbf{CMe_3} \end{array} \right. = \left\{ \begin{array}{l} \mathbf{CH_3} \\ \mathbf{CMe_2} \\ \mathbf{CMe_2} \\ \mathbf{CH_3} \end{array} \right. = \left\{ \begin{array}{l} \mathbf{CH_3} \\ \mathbf{C(CH_3)(CH_3)} \\ \mathbf{C(CH_3)(CH_3)} \\ \mathbf{CH_3} \end{array} \right. \cdot$$

Trimethylated methyl. (Molecule.)

It will be seen, on comparing the above formulæ with the graphic representation of the respective compounds, that both express the same facts. Thus in the developed symbolic formula of normal butyl it is evident that the two extreme carbon atoms are united with three atoms of hydrogen and one of carbon, and all the intermediate carbon atoms with two of hydrogen and two of carbon, exactly as shown in the following graphic representation :—

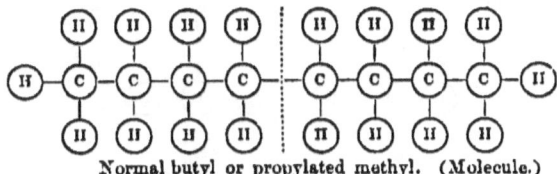

Normal butyl or propylated methyl. (Molecule.)

From the developed symbolic formula of methylo-ethylated methyl, it is evident that the two central carbon atoms are united with three atoms of carbon and one of hydrogen, that the two extreme carbon atoms are united with one atom of carbon and three of hydrogen, that the intermediate carbon atoms are joined to two hydrogen and two carbon atoms, and that the remaining carbon atoms are each combined with one atom of carbon and three of hydrogen, thus ·—

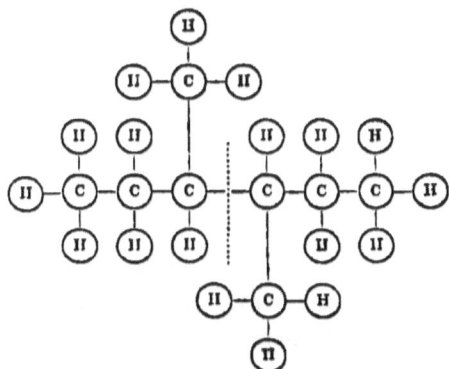

Methylo-ethylated methyl. (Molecule.)

Lastly, in the developed symbolic formula for trimethylated methyl, the two central atoms of carbon are each united with four carbon atoms, and all the remaining atoms of carbon are united with three of hydrogen, and one of carbon, thus :—

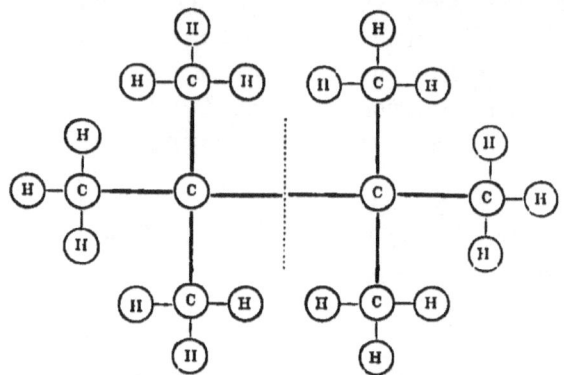

Trimethylated methyl.   (Molecule.)

CLASSIFICATION OF ORGANIC COMPOUNDS.

The most important organic compounds can be conveniently divided into the following twelve families :—

1. Organic radicals.
2. Hydrides of organic radicals.
3. Alcohols.
4. Ethers.
5. Haloid ethers.
6. Aldehydes.
7. Acids.
8. Anhydrides.
9. Ketones.
10. Ethereal salts.
11. Organic compounds containing triad or pentad nitrogen.
12. Organo-metallic bodies.

CHAPTER XXVI.

ORGANIC RADICALS.

This family of organic compounds is divided into two classes :—

Class I. Basylous or positive radicals.
Class II. Chlorous or negative radicals.

CLASS I.

BASYLOUS OR POSITIVE RADICALS.

Monads.

Methyl or $(C_nH_{2n+1})_2$ Series.
Vinyl or $(C_nH_{2n-1})_2$ Series.
Phenyl or $(C_nH_{2n-7})_2$ Series.

Dyads.

Ethylene or C_nH_{2n} Series.
Acetylene or C_nH_{2n-2} Series.
Phenylene or C_nH_{2n-8} Series.

Triads.

Glyceryl or $(C_nH_{2n-1})'''_2$ Series.

CLASS I.

BASYLOUS RADICALS.

MONADS.

METHYL or $(C_nH_{2n+1})_2$ SERIES.

These radicals are divided into three sections, viz. **Normal,** Secondary, and Tertiary :—

General formula.

1. Normal Radicals......... $\begin{cases} C(C_nH_{2n+1})H_2 \\ C(C_nH_{2n+1})H_2 \end{cases}$

General formulæ.

2. Secondary Radicals   ... $\left\{ \begin{array}{l} \mathbf{C}(C_nH_{2n+1})_2H \\ \mathbf{C}(C_nH_{2n+1})_2H' \end{array} \right.$

3. Tertiary Radicals   ...... $\left\{ \begin{array}{l} \mathbf{C}(C_nH_{2n+1})_3 \\ \mathbf{C}(C_nH_{2n+1})_3 \end{array} \right.$

In the first of the above formulæ n may $=0$, but in the others it must be a positive integer.

Examples of the secondary and tertiary series of radicals may be seen in the secondary and tertiary fatty acids. They have not yet been isolated.

It is evident that, besides the three series of radicals shown above, three other series, containing, in the same molecule, normal and secondary, normal and tertiary, and secondary and tertiary radicals, may exist; but up to the present time they have not been obtained.

1. *Normal Radicals.*

This series contains the radicals of the methylic series of alcohols.

These radicals also enter into the composition of the normal series of fatty acids.

The following list contains all the radicals of this section that have been hitherto obtained:—

Boiling-points.

Methyl   ...... $\left\{ \begin{array}{l} Me \\ Me' \end{array} \right.$ or $\left\{ \begin{array}{l} \mathbf{CH_3} \\ \mathbf{CH_3} \end{array} \right.$ .... not known.

Ethyl   .... $\left\{ \begin{array}{l} Et \\ Et' \end{array} \right.$ or $\left\{ \begin{array}{l} \mathbf{CMeH_2} \\ \mathbf{CMeH_2'} \end{array} \right.$, or $\left\{ \begin{array}{l} \mathbf{CH_3} \\ \mathbf{CH_2} \\ \mathbf{CH_2} \\ \mathbf{CH_3} \end{array} \right.$ ...... about $-23°$.

Propyl .... $\left\{ \begin{array}{l} Pr \\ Pr' \end{array} \right.$ or $\left\{ \begin{array}{l} \mathbf{CEtH_2} \\ \mathbf{CEtH_2'} \end{array} \right.$, or $\left\{ \begin{array}{l} \mathbf{CH_3} \\ \mathbf{CH_2} \\ \mathbf{CH_2} \\ \mathbf{CH_2} \\ \mathbf{CH_3} \end{array} \right.$ .... $+68°$.

Boiling-
points.

Butyl .... $\left\{ \begin{array}{l} \text{Bu} \\ \text{Bu}' \end{array} \right.$ or $\left\{ \begin{array}{l} \mathbf{C}\text{PrH}_2 \\ \mathbf{C}\text{PrH}_2 \end{array} \right.$ .... 119°.

Amyl .... $\left\{ \begin{array}{l} \text{Ay} \\ \text{Ay}' \end{array} \right.$ or $\left\{ \begin{array}{l} \mathbf{C}\text{BuH}_2 \\ \mathbf{C}\text{BuH}_2 \end{array} \right.$ .... 159°.

Caproyl ...... $\left\{ \begin{array}{l} \text{Cp} \\ \text{Cp}' \end{array} \right.$ or $\left\{ \begin{array}{l} \mathbf{C}\text{AyH}_2 \\ \mathbf{C}\text{AyH}_2 \end{array} \right.$ .... 202°.

Preparation.—1. By the action of zinc on the iodides of the normal radicals :—

$$2\mathbf{C}(C_nH_{2n+1})I \;+\; Zn \;=\; \mathbf{ZnI}_2 \;+\; \left\{ \begin{array}{l} \mathbf{C}(C_nH_{2n+1})H_2 \\ \mathbf{C}(C_nH_{2n+1})H_2 \end{array} \right.$$

Part of the liberated radical is at the same time decomposed into the hydride of the radical and the corresponding dyad radical :—

$$\left\{ \begin{array}{l} \mathbf{C}(C_nH_{2n+1})H_2 \\ \mathbf{C}(C_nH_{2n+1})H_2 \end{array} \right. \;=\; \left\{ \begin{array}{l} \mathbf{C}(C_nH_{2n+1})H_2 \\ H \end{array} \right. \;+\; [\mathbf{C}(C_nH_{2n+1})H]''.$$

A remarkable special method for preparing ethyl consists in exposing mercury and ethylic iodide to the influence of sunlight :—

$$2\text{EtI} \;+\; Hg \;=\; \mathbf{HgI}_2 \;+\; \text{Et}_2.$$
Ethylic   Mercuric Ethyl.
iodide.   iodide.

2. By the electrolysis of the salts of the normal fatty acids. In this process nascent oxygen acts upon the fatty acid, converting its oxatyl into carbonic anhydride, the positive radical being set free :—

$$2\left\{ \begin{array}{l} \mathbf{C}(C_nH_{2n+1})H_2 \\ \text{COHo} \end{array} \right. \;+\; O \;=\; \left\{ \begin{array}{l} \mathbf{C}(C_nH_{2n+1})H_2 \\ \mathbf{C}(C_nH_{2n+1})H_2 \end{array} \right.$$
  Fatty acid.   Radical.

$$+\; 2\mathbf{CO}_2 \;+\; \mathbf{OH}_2.$$
   Carbonic Water.
   anhydride.

3. By acting with zinc upon the iodides of two radicals simultaneously, the so-called double or mixed radicals are produced:—

$$\text{MeI} \;+\; \text{EtI} \;+\; Zn \;=\; \mathbf{ZnI}_2 \;+\; \left\{ \begin{array}{l} \text{Me} \\ \text{Et} \end{array} \right.$$
 Methylic Ethylic   Zincic   Methyl
 iodide.   iodide.   iodide.   ethyl.

ETHYL.

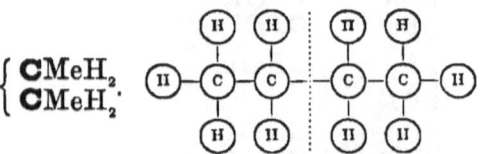

$$\left\{ \begin{array}{l} \mathbf{C}\text{MeH}_2 \\ \mathbf{C}\text{MeH}_2 \end{array} \right.$$

Molecular weight $=58$. *Molecular volume* ⬚. *1 litre of
ethyl gas weighs* 29 *criths. Boils at about* $-23°$ C.

Preparation.—By digesting together at 120° ethylic iodide
and zinc, the reaction being exactly similar to that between
hydriodic acid and zinc:—

$$\text{Zn} \quad + \quad 2\text{HI} \quad = \quad \mathbf{ZnI}_2 \quad + \quad \left\{ \begin{array}{l} \text{H} \\ \text{H} \end{array} \right.$$

Hydriodic   Zincic   Hydrogen.
acid.   iodide.

$$\text{Zn} \quad + \quad 2\text{EtI} \quad = \quad \mathbf{ZnI}_2 \quad + \quad \left\{ \begin{array}{l} \text{Et} \\ \text{Et} \end{array} \right.$$

Ethylic   Zincic   Ethyl.
iodide.   iodide.

VINYL SERIES.

General formula... $\left\{ \begin{array}{l} \mathbf{C}(\text{C}_n\text{H}_{2n-1})\text{H}_2 \\ \mathbf{C}(\text{C}_n\text{H}_{2n-1})\text{H}_2 \end{array} \right.$

The first member of this series—vinyl—has not yet been
isolated.

ALLYL.

$$\left\{ \begin{array}{l} \mathbf{C}_3\text{H}_5 \\ \mathbf{C}_3\text{H}_5 \end{array} \right. \text{or} \left\{ \begin{array}{l} \mathbf{C}(\text{CMe}''\text{H})\text{H}_2 \\ \mathbf{C}(\text{CMe}''\text{H})\text{H}_2 \end{array} \right.$$

Molecular weight $=82$. *Molecular volume* ⬚. *1 litre of
allyl vapour weighs* 41 *criths. Sp. gr.* 0·684. *Boils at* 59°.

Preparation.—By digesting allylic iodide with sodium, and then distilling :—

$$Na_2 + 2C(CMe''H)H_2I = \begin{cases} C(CMe''H)H_2 \\ C(CMe''H)H_2 \end{cases} + 2NaI.$$

Allylic iodide.   Allyl.   Sodic iodide.

Character.—Bromine and iodine unite directly with allyl, producing allylic tetrabromide and tetriodide. In these compounds the molecule of allyl plays the part of a tetrad radical.

In allylic tetrabromide, four latent carbon bonds in the molecule of allyl have become active, and are united with four atoms of bromine :—

$$\begin{cases} C(CMe''H)H_2 \\ C(CMe''H)H_2 \end{cases} + Br_4 = \begin{cases} C[C(CH_2Br)BrH]H_2 \\ C[C(CH_2Br)BrH]H_2 \end{cases}$$

Allyl.   Allylic tetrabromide.

An analogous case is met with in ferric chloride, where two tetrad atoms, united by one bond of each, become together hexatomic :—

$$\begin{cases} Fe \\ Fe \end{cases} Cl_6.$$

Allylic monobromide can only be obtained by the action of phosphorous tribromide on allylic alcohol :—

$$3C(CMe''H)H_2Ho + \dot{P}Br_3 = 3C(CMe''H)H_2Br$$

Allylic alcohol.   Phosphorous tribromide.   Allylic monobromide.

$$+ POHHo_2.$$

Phosphorous acid.

PHENYL SERIES.

General formula... $\begin{cases} C(C_nH_{2n-7})H_2 \\ C(C_nH_{2n-7})H_2 \end{cases}$

These radicals are but very imperfectly known, phenyl alone having been investigated.

PHENYL.

$$\begin{cases} \mathbf{C}(C_5H_3)H_2 \\ \mathbf{C}(C_5H_3)H_2 \end{cases}$$

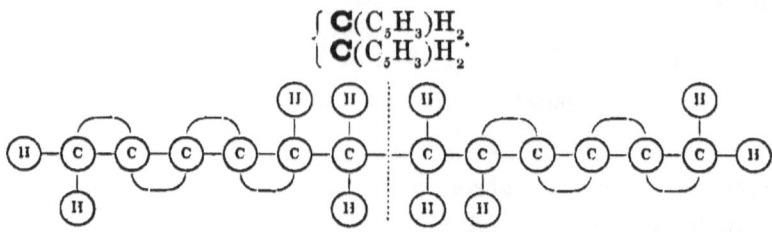

Molecular weight $= 154$. *Molecular volume* ▭. *1 litre of phenyl vapour weighs 77 criths. Fuses at 70°. Boils at 240°.*

Preparation.—By the action of sodium on phenylic bromide :—

$$2\mathbf{C}(C_5H_3)H_2Br \;+\; Na_2 \;=\; \begin{cases} \mathbf{C}(C_5H_3)H_2 \\ \mathbf{C}(C_5H_3)H_2 \end{cases} \;+\; 2NaBr.$$

Phenylic bromide.   Phenyl.   Sodic bromide.

Reaction. — By treatment with bromine, phenyl produces bromphenyl and hydrobromic acid :—

$$\begin{cases} \mathbf{C}(C_5H_3)H_2 \\ \mathbf{C}(C_5H_3)H_2 \end{cases} \;+\; Br_4 \;=\; \begin{cases} \mathbf{C}(C_5H_3)HBr \\ \mathbf{C}(C_5H_3)HBr \end{cases} \;+\; 2HBr.$$

Phenyl.   Bromphenyl.   Hydrobromic
                                                        acid.

CHAPTER XXVII.

BASYLOUS RADICALS.

DYADS.

ETHYLENE or C_nH_{2n} *SERIES.*

Preparation.—These compounds are produced as follows :—

1. In many cases of destructive distillation, where, however, the reaction cannot be traced.

2. By the abstraction of the elements of water from the normal monacid alcohols of the methylic series, as for intance: —

$$\left\{ \begin{array}{l} \mathbf{CH_3} \\ \mathbf{CH_2Ho} \end{array} \right. = \, '' \left\{ \begin{array}{l} \mathbf{CH_2} \\ \mathbf{CH_2} \end{array} \right. + \, \mathbf{OH_2}.$$

Ethylic alcohol.   Ethylene.   Water.

3. By passing the vapours of the haloid compounds of the normal monad radicals of the C_nH_{2n+1} series over heated lime, thus:—

$$\left\{ \begin{array}{l} \mathbf{CH_3} \\ \mathbf{CH_2Cl} \end{array} \right. = \, '' \left\{ \begin{array}{l} \mathbf{CH_2} \\ \mathbf{CH_2} \end{array} \right. + \, \mathbf{HCl}.$$

Ethylic   Ethylene.   Hydrochloric
chloride.   acid.

4. By the transformation of the monad radicals at the moment of liberation from their compounds, when they split into dyad radicals and the hydrides of monad radicals :—

$$\left\{ \begin{array}{l} \mathbf{C(CH_3)H_2} \\ \mathbf{C(CH_3)H_2} \end{array} \right. = \, '' \left\{ \begin{array}{l} \mathbf{CH_2} \\ \mathbf{CH_2} \end{array} \right. + \, \left\{ \begin{array}{l} \mathbf{CH_3} \\ \mathbf{CH_3} \end{array} \right.$$

Ethyl.   Ethylene.   Ethylic
hydride.

5. By the action of the iodide of a monad radical on the sodium compound of a monad radical :—

$$\left\{ \begin{array}{l} \mathbf{CH_3} \\ \mathbf{CH_2I} \end{array} \right. + \left\{ \begin{array}{l} \mathbf{CH_3} \\ \mathbf{CH_2Na} \end{array} \right. = \, \mathbf{NaI} \, + '' \left\{ \begin{array}{l} \mathbf{CH_2} \\ \mathbf{CH_2} \end{array} \right. + \left\{ \begin{array}{l} \mathbf{CH_3} \\ \mathbf{CH_3} \end{array} \right.$$

Ethylic   Sodic   Sodic   Ethylene.   Ethylic
iodide.   ethide.   iodide.   hydride.

Character.—The lower members of this series of dyad radicals are gaseous, the higher solid, and the intermediate ones liquid.

The following list includes the known dyad radicals of this series, together with their fusing- and boiling-points :—

		Fusing-point.	Boiling-point.
Ethylene	C_2H_4	——	——
Propylene	C_3H_6	——	$-17 \cdot 8$
Butylene	C_4H_8	——	$+35 \cdot 0$
Amylene	C_5H_{10}	——	$55 \cdot 0$
Hexylene	C_6H_{12}	——	$39 \cdot 0$
Heptylene or Œnanthylene	C_7H_{14}	——	$55 \cdot 0$
Octylene or Caprylene	C_8H_{16}	——	$95 \cdot 0$

		Fusing-point.	Boiling-point.
Nonylene	C_9H_{18}	——	125°
Paramylene	$C_{10}H_{20}$	——	——
Cetene	$C_{16}H_{32}$	——	——
Cerotene	$C_{27}H_{54}$	57°	275
Melene	$C_{30}H_{60}$	62	375.

Reactions.—1. The dyad radicals of this series all unite directly with chlorine, bromine, and iodine, producing compounds which, in the case of ethylene, are represented by

$$\left\{\begin{array}{l} CH_2Cl \\ CH_2Cl \end{array}\right. \qquad \left\{\begin{array}{l} CH_2Br \\ CH_2Br \end{array}\right. \qquad \left\{\begin{array}{l} CH_2I \\ CH_2I \end{array}\right.$$

Ethylenic   Ethylenic   Ethylenic
dichloride.   dibromide.   diiodide.

These compounds, when treated with alcoholic solution of potassic hydrate, furnish one molecule of a hydracid, thus:—

$$\left\{\begin{array}{l} CH_2Cl \\ CH_2Cl \end{array}\right. + KHo = KCl + {''}\left\{\begin{array}{l} CHCl \\ CH_2 \end{array}\right. + OH_2.$$

Ethylenic   Potassic   Potassic   Vinylic chloride,   Water.
dichloride.   hydrate.   chloride.   or chlorinated
                                                ethylene.

The monochlorinated radical thus obtained again unites with two atoms of chlorine, producing chlorinated ethylenic dichloride,

$$\left\{\begin{array}{l} CHClCl \\ CH_2Cl \end{array}\right. ,$$

which, by further treatment with alcoholic potash, yields dichlorinated ethylene; and so, by alternate treatments with chlorine and potassic hydrate, ethylene becomes transformed into tetrachlorinated ethylene. The following formulæ show the first, intermediate, and final compounds:—

$$_{''}\left\{\begin{array}{l} CH_2 \\ CH_2 \end{array}\right. \left\{\begin{array}{l} CH_2Cl \\ CH_2Cl \end{array}\right. {}_{''}\left\{\begin{array}{l} CHCl \\ CH_2 \end{array}\right. \left\{\begin{array}{l} CHClCl \\ CH_2Cl \end{array}\right. {}_{''}\left\{\begin{array}{l} CHCl \\ CHCl \end{array}\right.$$

$$\left\{\begin{array}{l} CHClCl \\ CHClCl \end{array}\right. {}_{''}\left\{\begin{array}{l} CCl_2 \\ CHCl \end{array}\right. \left\{\begin{array}{l} CCl_2Cl \\ CHClCl \end{array}\right. {}_{''}\left\{\begin{array}{l} CCl_2 \\ CCl_2 \end{array}\right.$$

Tetrachlorinated ethylene absorbs two additional atoms of chlorine, producing the solid dicarbonic hexachloride :—

$$\left\{ \begin{array}{l} CCl_3. \\ CCl_3 \end{array} \right.$$

Dicarbonic hexachloride.

2. The dyad radicals of the ethylene series can be transformed into the monad radicals from which they are derived. If ethylene be digested with hydriodic acid for 50 hours at 100° C., it is transformed into ethylic iodide :—

$$'' \left\{ \begin{array}{l} CH_2 \\ CH_2 \end{array} \right. + \; HI \; = \; \left\{ \begin{array}{l} CH_3 \\ CH_2I. \end{array} \right.$$

Ethylene.   Hydriodic   Ethylic
    acid.   iodide.

From this, ethyl may be prepared, as shown at p. 210.

Isomerism of ethylene and ethylidene compounds.—The chlorides of the dyad radicals are isomeric :—

1. With the chlorides of the monochlorinated normal monad radicals.

2. With the chlorides derived from the aldehydes, which, however, are identical with the chlorides of the monochlorinated normal monad radicals :—

$$\left\{ \begin{array}{l} CH_2Cl \\ CH_2Cl \end{array} \right. \qquad \left\{ \begin{array}{l} CH_3 \\ CHCl_2 \end{array} \right. \qquad \left\{ \begin{array}{l} CH_3 \\ CHCl_2. \end{array} \right.$$

Ethylenic   Ethylidenic dichlo-   Monochlorinated
dichloride.   ride (obtained from   ethylic chloride.
    aldehyde.)

These substances, when treated with alcoholic potash, all yield the same vinylic chloride :—

$$\left\{ \begin{array}{l} CH_2Cl \\ CH_2Cl \end{array} \right. + \; KHo \; = \; '' \left\{ \begin{array}{l} CHCl \\ CH_2 \end{array} \right. + \; OH_2 \; + \; KCl.$$

Ethylenic   Potassic   Vinylic   Water.   Potassic
dichloride.   hydrate.   chloride.   chloride.

$$\left\{ \begin{array}{l} CH_3 \\ CHCl_2 \end{array} \right. + \; KHo \; = \; '' \left\{ \begin{array}{l} CHCl \\ CH_2 \end{array} \right. + \; OH_2 \; + \; KCl.$$

Monochlorinated   Potassic   Vinylic   Water.   Potassic
ethylic chloride, or   hydrate.   chloride.   chloride.
Ethylidenic dichlo-
ride.

But certain compounds of ethylene yield paralactic acid,

whilst the corresponding compounds of ethylidene give lactic acid. The boiling-points of their chlorides also differ, ethylenic dichloride boiling at 85°, whilst ethylidenic dichloride boils at 64°; on the other hand, ethylenic oxide boils at 13°·5, whilst ethylidenic oxide (aldehyde) boils at 20°.

The oxides of the dyad radicals are isomeric—

1. With the corresponding aldehydes.

2. With the alcohols of the vinylic or $C_nH_{2n-1}Ho$ series.

The nature of this isomerism is seen from the following formulæ:—

$\begin{cases} \mathbf{CH_2} \\ \mathbf{CH_2} \end{cases}O$	$\begin{cases} \mathbf{CH_3} \\ \mathbf{COH} \end{cases}$	$,, \begin{cases} \mathbf{CH_2} \\ \mathbf{CHHo^{\cdot}} \end{cases}$
Ethylenic oxide.	Acetic aldehyde.	Vinylic alcohol.

ETHYLENE.

$,, \begin{cases} \mathbf{CH_2} \\ \mathbf{CH_2} \end{cases}.$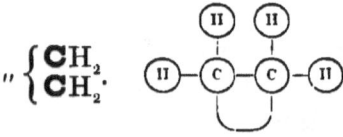

Molecular weight =28. *Molecular volume* ☐☐. 1 *litre weighs* 14 *criths*.

Preparation.—See general methods (p. 213).

Reactions.—1. Decomposed into carbon and marsh-gas by passing through a red-hot tube:—

$$,, \begin{cases} \mathbf{CH_2} \\ \mathbf{CH_2} \end{cases} = \mathbf{CH_4} + \mathbf{C}.$$

Ethylene.   Marsh-gas.

2. Burns in chlorine with deposition of carbon:—

$$,, \begin{cases} \mathbf{CH_2} \\ \mathbf{CH_2} \end{cases} + 2Cl_2 = C_2 + 4HCl.$$

Ethylene.   Hydrochloric acid.

Ethylidene, the isomer of ethylene, has not yet been isolated, unless the hydrocarbon C_2H_4 derived from the transformation of ethyl be this body. The formula of ethylidene is doubtless

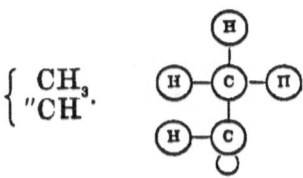

$$\left\{ \begin{array}{l} CH_3 \\ ''CH \end{array} \right.$$

ACETYLENE or C_nH_{2n-2} SERIES.

Acetylene is the only radical belonging to this series which is well known. The series comprises the following members :—

Acetylene........................ C_2H_2.
Allylene .... C_3H_4.
Crotonylene................... C_4H_6.

These radicals stand in the same relation to the alcohols of the vinylic series as ethylene bears to ethylic alcohol. They are also probably capable of assuming tetrad functions.

ACETYLENE.

$$'''C'_2H_2 \text{ or } \left\{ \begin{array}{l} ''CH \\ ''CH \end{array} \right.$$

Molecular weight $=26$. *Molecular volume* ▢. 1 *litre weighs* 13 *criths*.

Preparation.—1. By synthesis from its elements. When an electric arc from a powerful voltaic battery passes between carbon poles in an atmosphere of hydrogen, acetylene is produced.

2. By the action of water on potassic carbide :—

$$\underset{\substack{\text{Potassic}\\\text{carbide.}}}{C_2K_2} + \underset{\substack{\text{Water.}}}{2OH_2} = \underset{\substack{\text{Acetylene.}}}{C_2H_2} + \underset{\substack{\text{Potassic}\\\text{hydrate.}}}{2KHo.}$$

L

3. By the action of heat upon olefiant gas or the vapours of alcohol, ether, or wood-spirit, or by passing induction sparks through marsh-gas:—

$$2CH_4 = C_2H_2 + 3H_2.$$

Marsh-gas.　　　Acetylene.

4. By heating the vapour of methylic chloride to low redness:—

$$2CH_3Cl = C_2H_2 + 2HCl + H_2.$$

Methylic　　　Acetylene.　　Hydrochloric
chloride.　　　　　　　　　　acid.

5. By passing the vapour of chloroform over ignited copper:—

$$2CHCl_3 + Cu_6 = C_2H_2 + 3'Cu'_2Cl_2.$$

Chloroform.　　　　　　Acetylene.　　Cuprous
　　　　　　　　　　　　　　　　　chloride.

6. By the action of calcic carbide upon water:—

$$C_2Ca'' + 2OH_2 = C_2H_2 + CaHo_2.$$

Calcic　　　Water.　　Acetylene.　　Calcic
carbide.　　　　　　　　　　　　hydrate.

7. From vinylic bromide, one of the derivatives of ethylene, acetylene may be obtained by the action of alcoholic potash:—

$$C_2H_3Br + KHo = C_2H_2 + KBr + OH_2.$$

Vinylic　　　Potassic　　Acetylene.　　Potassic　　Water.
bromide.　　hydrate.　　　　　　　　bromide.

8. By the incomplete combustion of bodies containing carbon and hydrogen:—

$$4CH_4 + 3O_2 = 2C_2H_2 + 6OH_2:$$

Marsh-gas.　　　　　　Acetylene.　　Water.

$$2C_2H_4 + O_2 = 2C_2H_2 + 2OH_2.$$

Olefiant gas.　　　　Acetylene.　　Water.

The crude acetylene, obtained by any of these processes, is best purified by passing it through an ammoniacal solution of cuprous chloride, with which it forms a red precipitate containing

taining $\left\{ \begin{array}{l} C_2'Cu'_2H \\ O \\ C_2'Cu'_2H \end{array} \right.$

$$2'Cu'_2Cl_2 \; + \; 2C_2H_2 \; + \; OH_2 \; = \; \begin{cases} C_2'Cu'_2H \\ O \\ C_2'Cu'_2H \end{cases} + \; 4HCl.$$

Cuprous chloride.   Acetylene.   Cuprosovinylic ether. (Acetylide of copper.)   Hydrochloric acid

If ethylene have been present in the crude acetylene, the liquid containing the red precipitate is next heated to boiling, in order to decompose a compound which ethylene forms with copper. The cuprosovinylic ether is then collected upon a filter and washed. On heating cuprosovinylic ether with hydrochloric acid, pure acetylene is evolved :—

$$\begin{cases} C_2'Cu'_2H \\ O \\ C_2'Cu'_2H \end{cases} + \; 4HCl \; = \; 2C_2H_2 \; + \; 2'Cu'_2Cl_2 \; + \; OH_2.$$

Cuprosovinylic ether. (Acetylide of copper.)   Hydrochloric acid.   Acetylene.   Cuprous chloride.   Water.

Reactions.—1. When cuprosovinylic ether is heated with zinc and dilute ammonia, the nascent hydrogen evolved by the action of the zinc upon the ammonia unites with acetylene, producing ethylene :—

$$\begin{cases} C_2'Cu'_2H \\ O \\ C_2'Cu'_2H \end{cases} + \; 2H_2 \; = \; 2C_2H_2 \; + \; 4Cu \; + \; OH_2 ;$$

Cuprosovinylic ether.   Acetylene.   Water.

$$C_2H_2 \; + \; H_2 \; = \; C_2H_4.$$

Acetylene.   Ethylene.

2. Acetylene is absorbed by sulphuric acid, producing vinyl-sulphuric acid :—

$$C_2H_2 \; + \; SO_2Ho_2 \; = \; SO_2(C_2H_3O)Ho.$$

Acetylene.   Sulphuric acid.   Vinyl-sulphuric acid.

3. Acetylene unites with bromine, forming acetylenic dibromide :—

$$'''C'_2H_2 \; + \; Br_2 \; = \; ''C''_2H_2Br_2.$$

Acetylene.   Acetylenic dibromide.

BROMACETYLENE.

$C_2HBr.$

By boiling together dibromethylenic dibromide with alcoholic potash, a spontaneously inflammable gas is evolved, which is bromacetylene.

$$\left\{ \begin{array}{l} \mathbf{CHBrBr} \\ \mathbf{CHBrBr} \end{array} \right. = \quad HBr \quad + \quad Br_2 \quad + \quad \mathbf{C_2HBr.}$$

Dibromethylenic   Hydrobromic   Bromacetylene.
dibromide.   acid.

PHENYLENE or C_nH_{2n-8} SERIES.

The dyad radicals of this series are very little known. The following has alone been isolated :—

Phenylene, $C_6H_4.$

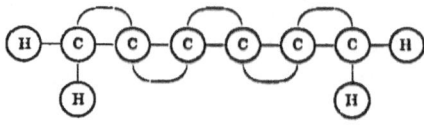

BASYLOUS RADICALS.

TRIADS.

These radicals are unknown in the separate state, unless they are identical with the dyad radicals of the acetylene series :—

$$\mathbf{C_2H_2} \quad = \quad \left\{ \begin{array}{l} \mathbf{(CH)}''' \\ \mathbf{(CH)}'''' \end{array} \right.$$

Acetylene.   Formyl.

They are, however, well known in a numerous class of compounds belonging to families which will be studied hereafter.

CHAPTER XXVIII.

ORGANIC RADICALS.

CLASS II.

CHLOROUS OR NEGATIVE RADICALS.

Every positive or basylous radical may be looked upon as the source of a chlorous or negative radical, which is generated by displacing a portion of the hydrogen of the former by oxygen. Thus :—

Ethyl	$\begin{cases} C_2H_5 \\ C_2H_5 \end{cases}$	yields acetyl	$\begin{cases} C_2H_3O \\ C_2H_3O \end{cases}$	
Allyl	$\begin{cases} C_3H_5 \\ C_3H_5 \end{cases}$	,, acryl	$\begin{cases} C_3H_3O \\ C_3H_3O \end{cases}$	
Ethylene	$(C_2H_4)''$	,, glycolyl	$(C_2H_2O)''$.	
Propylene	$(C_3H_6)''$	,, lactyl	$(C_3H_4O)''$.	
,,		,, malonyl	$(C_3H_2O_2)''$.	

The constitution of the so-called compounds of these negative radicals may, however, be more simply explained from another point of view; and, in fact, it will only be necessary for us to establish the existence of two negative radicals, in order to investigate the whole range of chlorous organic compounds. These are—

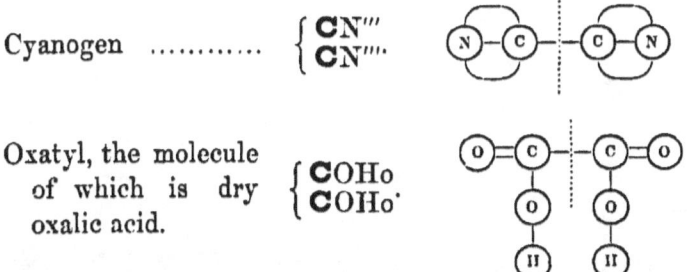

Cyanogen .... $\begin{cases} CN''' \\ CN'''' \end{cases}$

Oxatyl, the molecule of which is dry oxalic acid. $\begin{cases} COHo \\ COHo \end{cases}$

These two radicals are the acidifying principles of nearly all organic acids; they are therefore highly important compounds. The atom of each consists of an atom of carbon, one bond of which is free to combine with other elements or groups of elements, the other three bonds being saturated, in cyanogen by triad nitrogen, and in oxatyl by one atom of oxygen and one of hydroxyl. In the molecules of both, the two free bonds of the carbon saturate each other.

These radicals are also closely related to each other. Thus, if cyanogen be dissolved in water, it is soon transformed into ammonic oxalate :—

$$\begin{cases} \mathbf{CN}''' \\ \mathbf{CN}''' \end{cases} + 4\mathbf{OH}_2 = \begin{cases} \mathbf{C}O(N^vH_4O) \\ \mathbf{C}O(N^vH_4O) \end{cases}$$

Cyanogen.   Water.   Ammonic oxalate.

In the presence of potassic hydrate, cyanogen evolves ammonia and produces potassic oxalate :—

$$\begin{cases} \mathbf{CN}''' \\ \mathbf{CN}''' \end{cases} + 2\mathbf{KHo} + 2\mathbf{OH}_2 = \begin{cases} \mathbf{C}OKo \\ \mathbf{C}OKo \end{cases} + 2\mathbf{NH}_3.$$

Cyanogen.   Potassic hydrate.   Water.   Potassic oxalate.   Ammonia.

From these salts oxalic acid, or the molecule of oxatyl, may of course be readily obtained by the action of sulphuric acid.

In the converse manner, oxatyl may be converted into cyanogen, by transforming it into ammonic oxalate and submitting the salt to the action of heat :—

$$\begin{cases} \mathbf{C}O(N^vH_4O) \\ \mathbf{C}O(N^vH_4O) \end{cases} = 4\mathbf{OH}_2 + \begin{cases} \mathbf{CN}''' \\ \mathbf{CN}''' \end{cases}$$

Ammonic oxalate.   Water.   Cyanogen.

CYANOGEN.

$$\begin{cases} \mathbf{CN}''' \\ \mathbf{CN}''' \end{cases} \text{ or } Cy_2.$$

Molecular weight=52. *Molecular volume* ▭. 1 *litre weighs* 26 *criths. Fuses at* —34°. *Boils at* —20°·7.

Occurrence.—Amongst the gases of blast furnaces,—a proof of its withstanding an extremely high temperature.

Preparation.—By the action of heat on mercuric cyanide :—

$$HgCy_2 = Hg + Cy_2.$$

Mercuric
cyanide.   Cyanogen.

This equation only partially expresses the reaction, as a brown non-volatile compound (paracyanogen), Cy_n, is simultaneously produced.

Reaction.—Cyanogen unites directly with potassium :—

$$Cy_2 + K_2 = 2KCy.$$

Cyanogen.   Potassic
cyanide.

HYDROCYANIC ACID.

$$\left\{ \begin{array}{l} CN''' \\ H \end{array} \right. \text{ or } HCy.$$

Molecular weight $=27$. *Molecular volume* $\boxed{}$. 1 *litre of hydrocyanic acid vapour weighs* 13·5 *criths. Sp. gr. of liquid* 0·7058. *Fuses at* $-15°$. *Boils at* $26°·5$.

Preparation.—1. In the anhydrous condition, by passing hydrosulphuric acid over mercuric cyanide :—

$$HgCy_2 + SH_2 = HgS'' + 2HCy.$$

Mercuric   Sulphuretted   Mercuric   Hydrocyanic
cyanide.   hydrogen.   sulphide.   acid.

2. By distilling potassic cyanide, or ferrocyanide, with dilute sulphuric acid :—

$$2KCy + SO_2Ho_2 = 2HCy + SO_2Ko_2.$$

Potassic   Sulphuric   Hydrocyanic   Potassic
cyanide.   acid.   acid.   sulphate.

3. By passing nitrogen over an ignited mixture of potassic carbonate and carbon :—

$$COKo_2 + C_4 + N_2 = 2CN'''K + 3CO.$$

Potassic   Potassic   Carbonic
carbonate.   cyanide.   oxide.

The potassic cyanide thus formed is then treated according to process No. 2.

Reactions.—1. Hydrocyanic acid in contact with water slowly passes, partly into ammonic oxalate as mentioned at p. 222, and partly into ammonic formate:—

$$\mathbf{CN'''H} \ + \ \mathbf{2OH_2} \ = \ \left\{ \begin{array}{l} \mathbf{H} \\ \mathbf{CO(N'H_4O)'} \end{array} \right.$$

Hydrocyanic   Water.   Ammonic
   acid.   formate.

2. If hydrocyanic acid be mixed with concentrated hydrochloric acid, formic acid and ammonic chloride are produced:—

$$\mathbf{CN'''H} \ + \ \mathbf{2OH_2} \ + \ \mathbf{HCl} \ = \ \left\{ \begin{array}{l} \mathbf{H} \\ \mathbf{COHo} \end{array} \right. \ + \ \mathbf{N'H_4Cl.}$$

Hydrocyanic   Water.   Hydrochloric   Formic   Ammonic
  acid.   acid.   acid.   chloride.

3. The displacement of the hydrogen in hydrocyanic acid by metals gives rise to a very extensive series of single and double cyanides. The following is a list of the most important of these compounds:—

Single Cyanides.

Potassic cyanide	$KCy.$
Zincic cyanide	$\mathbf{Zn}Cy_2.$
Cadmic cyanide	$\mathbf{Cd}Cy_2.$
Nickelous cyanide	$\mathbf{Ni}Cy_2.$
Argentic cyanide	$AgCy.$
Mercuric cyanide	$\mathbf{Hg}Cy_2.$
Aurous cyanide	$AuCy.$
Cuprous cyanide	$\mathbf{'Cu'_2}Cy_2.$
Ferrous cyanide	$\mathbf{Fe}Cy_2.$
Cobaltous cyanide	$\mathbf{Co}Cy_2.$

Double Cyanides.

Dipotassic zincic tetracyanide	$K_2Zn'', Cy_4.$
Dipotassic cadmic tetracyanide	$K_2Cd'', Cy_4.$
Dipotassic nickelous tetracyanide	$K_2Ni'', Cy_4.$
Potassic argentic dicyanide	$KAg, Cy_2.$
Potassic aurous dicyanide	$KAu, Cy_2.$
Potassic auric tetracyanide	$KAu'''Cy_4.$
Dipotassic cuprous tetracyanide	$K_2'Cu'_2, Cy_4.$
Dipotassic platinous tetracyanide	$K_2Pt'', Cy_4.$

Tetrapotassic diplatinic decacyanide ...... $K_4{}'Pt'''_2Cy_{10}.$
Tetrapotassic ferrous hexacyanide. (Potassic ferrocyanide.) .... $K_4, Fe''Cy_6.$
Hexapotassic diferric dodecacyanide. (Potassic ferricyanide.) .... $K_6, 'Fe'''_2Cy_{12}.$
Hexapotassic dicobaltic dodecacyanide. (Potassic cobalticyanide.)................ $K_6, 'Co'''_2Cy_{12}.$
Hexapotassic dichromic dodecacyanide ... $K_6, 'Cr'''_2Cy_{12}.$
Hexapotassic dimanganic dodecacyanide... $K_6'Mn'''_2, Cy_{12}.$

The cyanides of the alkali metals when fused in contact with air, absorb oxygen, producing cyanates:—

$$KCy + O = CyKo.$$
Potassic cyanide.   Potassic cyanate.

Some of the single cyanides, as potassic cyanide, are readily decomposed by acids; others, as ferrous and aurous cyanides, may be boiled with moderately strong acids without decomposition.

Most of the insoluble single cyanides dissolve in solutions of the alkaline cyanides, forming double cyanides. Some of these double compounds, when acted upon by hydrochloric acid, evolve hydrocyanic acid, producing chlorides of both metals, as in the case of dipotassic zincic tetracyanide. These are called *easily decomposable cyanides*, and are indicated in the above Table by the comma being placed between the cyanogen and the metals.

Other double cyanides do not evolve hydrocyanic acid under the influence of hydrochloric acid, but produce a chloride of one of the metals, the remaining elements of the compound uniting with hydrogen to form a complex acid. In the above Table the double cyanides of this class are indicated by the comma being placed between the metals.

The most important of these double cyanides are the potassic ferrocyanide $K_4, Fe''Cy_6$, and the potassic ferricyanide $K_6, 'Fe'''_2Cy_{12}.$

POTASSIC FERROCYANIDE.

$$K_4, Fe''Cy_6 \text{ or } K_4Cfy.$$

Preparation.—1. By placing a mixture of iron filings with solution of potassic cyanide in contact with the air, oxygen is absorbed and potassic ferrocyanide produced :—

$$Fe + 6KCy + OH_2 + O = K_4Fe''Cy_6 + 2KHo.$$

       Potassic   Water.   Potassic   Potassic
       cyanide.   ferrocyanide.   hydrate.

2. By digesting potassic cyanide with ferrous sulphide :—

$$FeS'' + 6KCy = K_4Fe''Cy_6 + SK_2.$$

      Ferrous   Potassic   Potassic   Potassic
      sulphide.   cyanide.   ferrocyanide.   sulphide.

3. On a manufacturing scale it is prepared by fusing nitrogenous animal matter with potassic carbonate and iron filings in iron vessels, lixiviating the resulting mass with water, and crystallizing.

Reactions.—1. Potassic ferrocyanide, when fused with potassic carbonate, forms potassic cyanide and cyanate :—

$$Fe''Cy_6K_4 + COKo_2 = 5KCy + CyKo + Fe + CO_2.$$

   Potassic   Potassic   Potassic   Potassic   Carbonic
   ferrocyanide.   carbonate.   cyanide.   cyanate.   anhydride.

2. By mixing solution of potassic ferrocyanide with ether and hydrochloric acid, hydroferrocyanic acid is precipitated :—

$$Fe''Cy_6K_4 + 4HCl = 4KCl + Fe''Cy_6H_4.$$

   Potassic   Hydrochloric   Potassic   Hydroferrocyanic
   ferrocyanide.   acid.   chloride.   acid.

3. Potassic ferrocyanide produces, with solutions of ferrous salts, a light-blue precipitate, which rapidly becomes dark blue in contact with the air :—

$$Fe''Cy_6K_4 + SO_2Feo'' = Fe''Cy_6Fe''K_2 + SO_2Ko_2.$$

   Potassic   Ferrous   Light-blue   Potassic
   ferrocyanide.   sulphate.   precipitate.   sulphate.

4. With ferric salts it gives prussian blue :—

$$3Fe''Cy_6K_4 \; + \; 2\mathbf{Fe_2Cl_6} \; = \; 3Fe''Cy_2,'Fe'''_4Cy_{12} \; + \; 12KCl.$$

Potassic ferrocyanide.   Ferric chloride.   Prussian blue.   Potassic chloride.

5. With cupric salts it gives a red precipitate of cupric ferrocyanide :—

$$K_4Fe''Cy_6 \; + \; 2SO_2Cuo'' \; = \; Cu''_2Fe''Cy_6 \; + \; 2SO_2Ko_2.$$

Potassic ferrocyanide.   Cupric sulphate.   Cupric ferrocyanide.   Potassic sulphate.

POTASSIC FERRICYANIDE.

$K_6'Fe'''_2Cy_{12}$ or $K_6Cfdy.$

Preparation.—By the action of oxidizing substances, such as chlorine and nitric acid, on potassic ferrocyanide :—

$$2K_4Fe''Cy_6 \; + \; Cl_2 \; = \; K_6'Fe'''_2Cy_{12} \; + \; 2KCl.$$

Potassic ferrocyanide.   Potassic ferricyanide.   Potassic chloride.

Reaction.—Potassic ferricyanide produces no precipitate with solutions of ferric salts, but causes a deep-blue precipitate with ferrous compounds :—

$$K_6'Fe'''_2Cy_{12} \; + \; 3SO_2Feo'' \; = \; Fe''_3'Fe'''_2Cy_{12} \; + \; 3SO_2Ko_2.$$

Potassic ferricyanide.   Ferrous sulphate.   Turnbull's blue.   Potassic sulphate.

OTHER COMPOUNDS OF CYAYOGEN.

There are three isomeric chlorides of cyanogen :—

$CyCl.$ $Cy_2Cl_2.$ $Cy_3Cl_2.$

Gaseous.   Liquid.   Solid.

The molecular volume of all three cyanogen chlorides is ⬜⬜.

1 *litre of gaseous cyanogen chloride weighs* ....   61·5 *criths.*

1 *litre of vapour of liquid cyanogen chloride weighs* 123   „

1 *litre of vapour of solid cyanogen chloride weighs* 184·5   „

Cyanogen produces, with hydroxyl, three isomeric acids and an isomeric neutral body :—

Cyanic acid .... $CyOH$ or $CyHo$.
Cyanuric acid .... $Cy_3O_3H_3$ or Cy_3Ho_3.
Fulminuric acid .... $Cy_3O_3H_3$ or Cy_3Ho_3.
Cyamelide .... $Cy_nO_nH_n$ or Cy_nHo_n.

When potassic cyanide is boiled with sulphur, the latter is dissolved and the solution contains potassic sulphocyanate:—

$$CyK \quad + \quad S \quad = \quad CyKs.$$

Potassic   Potassic
cyanide.   sulphocyanate.

This compound produces with ferric salts a blood-red colour.

OXATYL.

$$\begin{cases} \mathbf{COHo} \\ \mathbf{COHo} \end{cases}$$

This radical, in the isolated condition, constitutes dry oxalic acid; and in combination with hydrogen and other radicals it enters into the composition of nearly all organic acids. Acids containing one atom of oxatyl are monobasic, those containing two atoms are dibasic, and those containing three are tribasic.

The relations between methyl, oxatyl, and cyanogen are very simple:—

In methyl the two carbon atoms are united together by one bond of each, the remaining three bonds of each atom being saturated by three atoms of hydrogen. In cyanogen the carbon atoms are united in the same manner, but the three remaining bonds of each carbon atom are saturated by triad nitrogen; whilst in oxatyl the three remaining bonds are saturated with the dyad oxygen and the monad radical hydroxyl.

Oxatyl has not been united with chlorine to produce oxatylic chloride ($COHoCl$); nor has its hydroxyl been replaced by chlorine to form $\begin{cases} COCl \\ COCl \end{cases}$. When treated with phosphoric chloride, it yields carbonic oxide and carbonic anhydride :—

$$\begin{cases} COHo \\ COHo \end{cases} + PCl_5 = CO + CO_2 + 2HCl + POCl_3.$$

| Oxatyl. | Phosphoric chloride. | Carbonic oxide. | Carbonic anhydride. | Hydrochloric acid. | Phosphoric oxytrichloride. |

OXALIC ACID.

$$\begin{cases} COHo \\ COHo \end{cases}, 2OH_2 \text{ (crystallized)}.$$

Occurrence.—In the form of the hydric potassic salt in *Oxalis acetosella*, and in the form of different salts in many other plants, and also in the animal organism.

Preparation.—1. From its elements through the medium of cyanogen. (See p. 222.)

2. By the oxidation of a large number of organic compounds. Most organic substances are converted by oxidizing agents into oxalic acid before their final transformation into carbonic anhydride and water: thus sugar is transformed into oxalic acid by the action of nitric acid.

3. By heating sawdust with a mixture of potash and soda, oxalates of these bases are formed.

Transformations.—1. By the action of heat, oxalic acid is transformed into carbonic anhydride and oxatylic hydride, or formic acid :—

$$\begin{cases} COHo \\ COHo \end{cases} = CO_2 + \begin{cases} H \\ COHo \end{cases}$$

Oxalic acid.　　Carbonic　　Formic acid.
　　　　　　　anhydride.

A portion of the formic acid is at the same time decomposed into water and carbonic oxide :—

$$\begin{cases} H \\ COHo \end{cases} = OH_2 + CO.$$

Formic acid.　　Water.　　Carbonic oxide.

2. Substances having a strong attraction for water, as sulphuric acid, transform oxalic acid into water, carbonic oxide, and carbonic anhydride :—

$$\begin{cases} COHo \\ COHo \end{cases} = CO + CO_2 + OH_2.$$

Oxalic acid.　　Carbonic oxide.　　Carbonic anhydride.　　Water.

3. Heated with an excess of alkali, oxalic acid (or an oxalate) yields hydrogen and a carbonate :—

$$\begin{cases} COKo \\ COKo \end{cases} + 2KHo = 2COKo_2 + H_2.$$

Potassic oxalate.　　Potassic hydrate.　　Potassic carbonate.

4. Argentic oxalate explodes when heated, producing silver and carbonic anhydride :—

$$\begin{cases} COAgo \\ COAgo \end{cases} = 2CO_2 + Ag_2.$$

Argentic oxalate.　　Carbonic anhydride.

Salts of Oxalic acid.—Oxalic acid forms three series of salts:—

Normal.	Acid.	Superacid.	
$\begin{cases} COKo \\ COKo \end{cases}$	$\begin{cases} COHo \\ COKo \end{cases}$	$\begin{cases} COHo \\ COKo' \end{cases}$	$\begin{cases} COHo \\ COHo \end{cases}$

$$\begin{cases} CO \\ CO \end{cases} Bao''. \qquad \begin{cases} COHo \\ CO \\ Bao'' \\ CO \\ COHo \end{cases}.$$

OXAMIC ACID.

$$\begin{cases} \mathbf{CO(N'''H_2)} \\ \mathbf{COHo} \end{cases} \text{ or } \begin{cases} \mathbf{COAd} \\ \mathbf{COHo} \end{cases}$$

Preparation.—By heating hydric ammonic oxalate to 230°:—

$$\begin{cases} \mathbf{CO(N^vH_4O)} \\ \mathbf{COHo} \end{cases} = \begin{cases} \mathbf{CO(N'''H_2)} \\ \mathbf{COHo} \end{cases} + \mathbf{OH_2}.$$

Hydric ammonic   Oxamic acid.   Water.
oxalate.

Reaction.—By boiling oxamic acid with water it is retransformed into hydric ammonic oxalate.

OXAMIDE.

$$\begin{cases} \mathbf{CO(N'''H_2)} \\ \mathbf{CO(N'''H_2)} \end{cases} \text{ or } \begin{cases} \mathbf{COAd} \\ \mathbf{COAd} \end{cases}$$

Preparation.—1. By distilling normal ammonic oxalate :—

$$\begin{cases} \mathbf{CO(N^vH_4O)} \\ \mathbf{CO(N^vH_4O)} \end{cases} = 2\mathbf{OH_2} + \begin{cases} \mathbf{CO(N'''H_2)} \\ \mathbf{CO(N'''H_2)} \end{cases}.$$

Normal ammonic   Water.   Oxamide.
oxalate.

2. By acting upon ethylic oxalate by ammonia :—

$$\begin{cases} \mathbf{COEto} \\ \mathbf{COEto} \end{cases} + 2\mathbf{NH_3} = \begin{cases} \mathbf{CO(N'''H_2)} \\ \mathbf{CO(N'''H_2)} \end{cases} + 2\mathbf{EtHo}.$$

Ethylic   Ammonia.   Oxamide.   Alcohol.
oxalate.

Reactions.—1. Oxamide, when heated with phosphoric anhydride, evolves cyanogen :—

$$\begin{cases} \mathbf{CO(N'''H_2)} \\ \mathbf{CO(N'''H_2)} \end{cases} = 2\mathbf{OH_2} + \begin{cases} \mathbf{CN'''} \\ \mathbf{CN'''} \end{cases}.$$

Oxamide.   Water.   Cyanogen.

2. Dilute acids convert it into oxalic acid and ammonic salts :—

$$\begin{cases} \mathbf{CO(N'''H_2)} \\ \mathbf{CO(N'''H_2)} \end{cases} + \mathbf{SO_2Ho_2} + 2\mathbf{OH_2} = \begin{cases} \mathbf{COHo} \\ \mathbf{COHo} \end{cases}$$

Oxamide.   Sulphuric acid.   Water.   Oxalic acid.

$$+ \mathbf{SO_2(N^vH_4O)_2}.$$

Ammonic sulphate.

By distilling the oxalates of the compound ammonias instead of ammonic oxalate, compound oxamides are obtained:—

$$\left\{\begin{array}{l} CO(N^vMeH_3O) \\ CO(N^vMeH_3O) \end{array}\right. \; = \; 2OH_2 \; + \; \left\{\begin{array}{l} CO(N'''MeH) \\ CO(N'''MeH) \end{array}\right.$$

    Methylammonic oxalate.   Water.   Dimethyloxamide.

$$\left\{\begin{array}{l} CO(N^vPhH_3O) \\ CO(N^vPhH_3O) \end{array}\right. \; = \; 2OH_2 \; + \; \left\{\begin{array}{l} CO(N'''PhH) \\ CO(N'''PhH) \end{array}\right.$$

    Phenylammonic oxalate.   Water.   Diphenyloxamide.

CHAPTER XXIX.

HYDRIDES OF THE COMPOUND RADICALS.

This family is divided into two classes:—

Class   I. Hydrides of the Basylous or Positive Radicals.
Class II. Hydrides of the Chlorous or Negative Radicals.

CLASS I.

HYDRIDES OF THE BASYLOUS OR POSITIVE RADICALS.

Two series of hydrides belonging to this class are well known; they are:—

1. Hydrides of the Radicals of the Methyl series.
2. Hydrides of the Radicals of the Phenyl series.

1. *HYDRIDES OF THE RADICALS OF THE METHYL SERIES, Marsh-gas, or C_nH_{2n+2} Series.*

There is some difference of opinion as to whether these compounds are identical or isomeric with the radicals of the methyl series. Thus methyl and ethylic hydride both contain C_2H_6, and ethyl and butylic hydride both contain C_4H_{10}. The graphic formulæ exhibit no difference between these pairs of bodies respectively. Thus :—

Methyl or ethylic hydride.   Ethyl or butylic hydride.

These formulæ do not show us whether the molecule of methyl or ethylic hydride will separate at a and so be represented by the formula $\left\{ \begin{matrix} \mathbf{C}(CH_3)H_2 \\ H \end{matrix} \right.$; or at b, and so be written thus, $\left\{ \begin{matrix} CH_3 \\ CH_3 \end{matrix} \right.$; or whether the molecule of ethyl or butylic hydride will separate at c and so be formulated $\left\{ \begin{matrix} \mathbf{C}(C_3H_7)H_2 \\ H \end{matrix} \right.$; or at d, when it should be represented by $\left\{ \begin{matrix} \mathbf{C}(CH_3)H_2 \\ \mathbf{C}(CH_3)H_2 \end{matrix} \right.$. Some experiments in connexion with this subject appear to show that these compounds are isomeric.

A difference between methyl and ethylic hydride can only be conceived on the supposition that the four bonds of carbon have not equal values in combination, an hypothesis which is not altogether unsupported by facts.

Preparation.—1. There is only one process of general application for preparing these hydrides; it consists in bringing water into contact with the zinc compounds of the respective radicals: —

$$\mathbf{Zn}(C_nH_{2n+1})_2 \quad + \quad 2OH_2 \quad = \quad \mathbf{Zn}Ho_2 \quad + \quad 2 \left\{ \begin{matrix} C_nH_{2n+1} \cdot \\ H \end{matrix} \right.$$

Zinc compound of radical.   Water.   Zincic hydrate.   Hydride of radical.

The corresponding compounds containing more positive metals might doubtless be substituted for those of zinc.

2. There are several special processes which may be used for preparing these hydrides. Thus all the hydrides above that of

methyl may be obtained, together with the corresponding dyad radical, by acting upon the iodide of the monad radical by zinc :—

$$ 2 \left\{ \underset{\text{I}}{\text{C}_n\text{H}_{2n+1}} + \text{Zn} = \textbf{ZnI}_2 + \text{C}_n\text{H}_{2n} + \left\{ \underset{\text{H}}{\text{C}_n\text{H}_{2n+1}} \right. \right. $$

Iodide of the monad radical. Zincic iodide. Dyad radical. Hydride of monad radical.

Methylic hydride, or marsh-gas, is produced during putrefaction, and by the distillation of potassic acetate with excess of potassic hydrate.

The destructive distillation of coal and of allied substances also furnishes a large number of the members of this series.

Character.—They are all distinguished by their great chemical indifference, and by their forming substitution compounds containing chlorine, bromine, &c.

The following list contains the hydrides of the monad radicals at present studied :—

		Boiling-points.
Methylic hydride, or Marsh-gas.........	MeH or $C\ H_4$	
Ethylic hydride	EtH or $C_2\ H_6$	
Propylic or tritylic hydride	PrH or $C_3\ H_8$	
Butylic or tetrylic hydride	BuH or $C_4\ H_{10}$	slightly above 0°
Amylic or pentylic hydride	AyH or $C_5\ H_{12}$	30°
Hexylic or caproylic hydride	CpH or $C_6\ H_{14}$	68°
Heptylic hydride	$C_7\ H_{16}$	92– 94°
Octylic hydride	$C_8\ H_{18}$	116–118°
Nonylic hydride.......................	$C_9\ H_{20}$	136–138°
Decatylic hydride	$C_{10}H_{22}$	160–162°
Endecatylic hydride	$C_{11}H_{24}$	180–184°
Dodecatylic hydride	$C_{12}H_{26}$	196–200°
Tridecatylic hydride	$C_{13}H_{28}$	216–218°
Tetradecatylic hydride	$C_{14}H_{30}$	236–240°
Pentadecatylic hydride	$C_{15}H_{32}$	255–260°

METHYLIC HYDRIDE, *Marsh-gas, Light Carburetted Hydrogen, Fire-damp.*

CH_4 or MeH.

Molecular weight $= 16$.   *Molecular volume* $\square$.   1 *litre weighs* 8 *criths*.

Occurrence.—1. As a product of the decomposition of organic substances out of contact with air.

2. Evolved in coal-mines.

3. The gas of the mud-volcano at Bulganak in the Crimea is nearly pure marsh-gas.

Preparation.—1. By the action of water on zincic methide. (See general reaction, p. 233.)

2. By distilling two parts of potassic acetate, two of potassic hydrate, and three of lime :—

$$\left\{ \begin{matrix} CH_3 \\ COKo \end{matrix} \right. + \quad KHo \quad = \quad COKo_2 \quad + \quad CH_4.$$

| Potassic acetate. | Potassic hydrate. | Potassic carbonate. | Methylic hydride. |

3. By the reduction of carbonic chloride or of chloroform with sodium amalgam and water :—

$$CCl_4 + H_8 = 4HCl + CH_4:$$

Carbonic chloride.   Hydrochloric acid.   Methylic hydride.

$$CHCl_3 + H_6 = 3HCl + CH_4.$$

Chloroform.   Hydrochloric acid.   Methylic hydride.

4. By passing carbonic disulphide and hydrosulphuric acid, or carbonic disulphide and steam, over ignited copper :—

$$CS''_2 + 2SH_2 + 4Cu = 4CuS'' + CH_4.$$

Carbonic disulphide.   Sulphuretted hydrogen.   Cupric sulphide.   Methylic hydride.

5. By the destructive distillation of organic substances, such as wood, coal, &c.

Reaction.—When equal volumes of methylic hydride and chlorine are exposed to diffused daylight, methylic chloride is formed :—

$$CH_4 + Cl_2 = HCl + CH_3Cl.$$

Methylic hydride.   Hydrochloric acid.   Methylic chloride.

ETHYLIC HYDRIDE.

C_2H_6 or $CMeH_3$.

Molecular weight $=30$. *Molecular volume* ⊏⊐. 1 *litre weighs*
15 *criths*.

Preparation.—1. By the action of water on zincic ethide
(see p. 233).

2. By the action of ethylic iodide on sodic ethide, ethylene
being simultaneously produced :—

$$CMeH_2Na \quad + \quad CMeH_2I \quad = \quad NaI \quad + \quad C_2H_4 \quad + \quad CMeH_3.$$
Sodic ethide.　　　　Ethylic　　　　　Sodic　　　Ethylene.　　　Ethylic
　　　　　　　　　　iodide.　　　　　iodide.　　　　　　　　　hydride.

Reactions.—1. When equal volumes of ethylic hydride and
chlorine are exposed to diffused daylight, the following action
takes place :—

$$CMeH_3 \quad + \quad Cl_2 \quad = \quad CMeH_2Cl \quad + \quad HCl.$$
Ethylic　　　　　　　　　　　　β Ethylic　　　Hydrochloric
hydride.　　　　　　　　　　　chloride.　　　　　acid.

A small portion of the body $CMeH_2Cl$ is ordinary ethylic
chloride, which is a liquid, boiling at $12°·5$; but the rest is a gas
which does not condense at $-18°$.

2. When a mixture of two volumes of chlorine and one of
ethylic hydride is exposed to the action of diffused daylight, an
oily liquid having the composition of ethylenic dichloride is
formed :—

$$CMeH_3 \quad + \quad 2Cl_2 \quad = \quad C_2H_4Cl_2 \quad + \quad 2HCl.$$
Ethylic　　　　　　　　　　　　　　　　　　Hydrochloric
hydride.　　　　　　　　　　　　　　　　　　acid.

AMYLIC HYDRIDE.

C_5H_{12} or $CBuH_3$.

Molecular weight $=72$. *Molecular volume* ⊏⊐. 1 *litre of*
amylic hydride-vapour weighs 36 *criths*. *Boils at* $30°$.

Occurrence.—In petroleum and coal-oil.

Preparation.—By digesting zinc and amylic iodide with water or alcohol at 100° :—

$$2\mathbf{C}BuH_2I \;+\; 2Zn \;+\; 2OH_2 \;=\; 2\mathbf{C}BuH,$$

Amylic iodide.   Water.   Amylic hydride.

$$+\; \mathbf{ZnHo_2} \;+\; \mathbf{ZnI_2}.$$

Zincic hydrate.   Zincic iodide.

PARAFFIN.

This body is produced, together with numerous other compounds of a like nature, by the destructive distillation of bog-head coal and similar substances. It is also found in petroleum and asphalt. If chlorine be passed into melted paraffin, the latter is slowly attacked, hydrochloric acid being evolved. In this reaction paraffin resembles the hydrides of the monad radicals, and differs from the dyad radicals, to which class it was formerly considered to belong. In the formula C_nH_{2n+2} for paraffin, the value of n has not yet been satisfactorily determined; in fact it is probable that several distinct hydrides of the class now under consideration are confounded under this name.

2. *HYDRIDES OF THE RADICALS OF THE PHENYL SERIES.*

The following five members of this series are known, viz. :—

	Formula.	Boiling-point.	Sp. gr.
Benzol	$\mathbf{C_6H_6}$	80·5	0·85
Toluol	$\mathbf{C_7H_8}$	110·0	0·87
Xylol	$\mathbf{C_8H_{10}}$	128·5	——
Cumol	$\mathbf{C_9H_{12}}$	148·5	0·87
Cymol	$\mathbf{C_{10}H_{14}}$	171·4	0·86

Preparation.—1. These hydrides are produced by the distillation of the alkaline salts of the acids containing the same positive radicals, with excess of potassic hydrate:—

$$\left\{\begin{array}{l} C_nH_{2n-7} \\ COKo \end{array}\right. \quad + \quad KHo \quad = \quad COKo_2 \quad + \quad \left\{\begin{array}{l} C_nH_{2n-7} \\ H \end{array}\right..$$

Potassic salt.   Potassic   Potassic   Hydride of
                      hydrate.   carbonate.   radical.

2. By the destructive distillation of various organic substances, such as coal.

Properties.—These hydrides are distinguished from those of the radicals of the C_nH_{2n+1} series by being less indifferent to chemical agents. By treatment with strong nitric acid they yield nitro-compounds :—

Thus Benzol, C_6H_6, gives nitrobenzol, $C_6H_5 (N^vO_2)$.
„   Toluol, C_7H_8,   „   nitrotoluol, $C_7H_7 (N^vO_2)$.
„   Xylol, C_8H_{10},   „   nitroxylol, $C_8H_9 (N^vO_2)$.
„   Cumol, C_9H_{12},   „   nitrocumol, $C_9H_{11}(N^vO_2)$.
„   Cymol, $C_{10}H_{14}$,   „   nitrocymol, $C_{10}H_{13}(N^vO_2)$.

Under the influence of reducing agents, these nitro-compounds yield aniline and its homologues.

BENZOL, *Benzene, Benzine, Phenylic Hydride,*

Bicarburet of Hydrogen.

C_6H_6 or $C(C_5H_3)H_3$, or PhH.

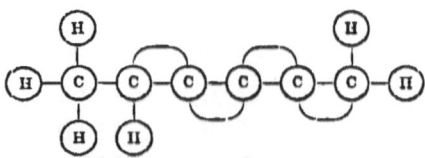

Molecular weight $= 78$. *Molecular volume* ▭. *1 litre of benzol-vapour weighs 39 criths. Fuses at 5°·5. Boils at 80°·5.*

Occurrence.—In Rangoon petroleum and in coal-tar.

Preparation.—1. By heating benzoic acid with excess of lime or baryta :—

$$\left\{ \begin{array}{l} \mathbf{C}(C_5H_3)H_2 \\ \mathbf{COH}o \end{array} \right. + \mathbf{CaO} = \left\{ \begin{array}{l} \mathbf{C}(C_5H_3)H_2 \\ \mathbf{H} \end{array} \right. + \mathbf{COC}ao''.$$

Benzoic acid.   Lime.   Benzol.   Calcic carbonate.

2. By heating the vapour of benzoic acid to redness, when it splits into benzol and carbonic anhydride :—

$$\left\{ \begin{array}{l} \mathbf{C}(C_5H_3)H_2 \\ \mathbf{COH}o \end{array} \right. = \mathbf{CO}_2 + \mathbf{C}(C_5H_3)H_3.$$

Benzoic acid.   Carbonic   Benzol.
anhydride.

3. By heating phthalic acid with lime :—

$$C_8H_6O_4 + 2\mathbf{CaO} = \mathbf{C}_6H_6 + 2\mathbf{COC}ao''.$$

Phthalic   Lime.   Benzol.   Calcic
acid.   carbonate.

4. By passing fats through red-hot tubes.

5. By the destructive distillation of coal.

6. In small quantity, when the vapour of acetic acid or of alcohol is passed through a red-hot tube.

SUBSTITUTION DERIVATIVES OF BENZOL.

I. *Bromo Compounds.*

MONOBROMBENZOL.

$$\mathbf{C}_6H_5Br.$$

Boils at 150°.

Preparation.—By acting with two atoms of bromine on boiling benzol :—

$$\mathbf{C}_6H_6 + Br_2 = \mathbf{C}_6H_5Br + HBr.$$

Benzol.   Monobrombenzol   Hydrobromic
    or phenylic   acid.
    bromide.

DIBROMBENZOL.

$$\mathbf{C}_6H_4Br_2.$$

Fuses at 89°.   *Boils at* 219°.

Preparation.—By treating monobrombenzol with excess of bromine.

TRIBROMBENZOL HYDROBROMATE.

$$C_6H_6Br_6.$$

Preparation.—By exposing a mixture of benzol and bromine to the action of sunlight.

TRIBROMBENZOL.

$$C_6H_3Br_3.$$

Preparation.—By boiling the previous compound with alcoholic potash.

The following graphic formulæ show the probable atomic relations subsisting between benzol, tribrombenzol hydrobromate, and tribrombenzol :—

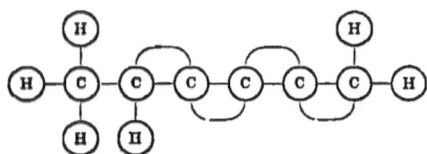

Benzol.

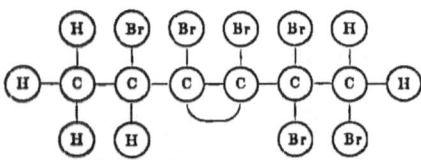

Tribrombenzol hydrobromate.

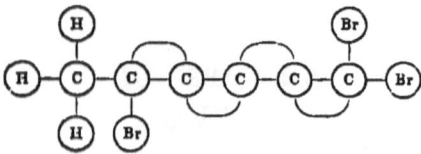

Tribrombenzol.

II. *Chloro-compounds.*

Benzol forms three chloro- substitution compounds, similar to the bromo-compounds just described.

		State of aggregation.	Fusing-point.	Boiling-point.	Sp. gr.
Monochlorbenzol...	C_6H_5Cl,	Liquid...	——	136°.	——
Dichlorbenzol	$C_6H_4Cl_2$,	Solid ...	89°.	——	——
Trichlorbenzol......	$C_6H_3Cl_3$,	Oily ...	——	210°	1·457.

III. *Nitro-compounds.*

Two only have hitherto been produced.

Nitrobenzol ...... $C_6H_5(N^vO_2)$ or $N(C_6H_5)O_2$.

Dinitrobenzol...... $C_6H_4(N^vO_2)_2$ or $N_2(C_6H_4)''O_4$.

NITROBENZOL.

$$N(C_6H_5)O_2 \text{ or } NPhO_2.$$

Molecular weight $=123$. *Molecular volume* □. 1 *litre of nitrobenzol vapour weighs* 61·5 *criths.* *Fuses at* 3°. *Boils at* 220°.

Preparation.—By the action of nitric acid on benzol:—

$$C(C_6H_3)H_3 \ + \ NO_2Ho \ = \ N(C_6H_5)O_2 \ + \ OH_2.$$
Benzol. Nitric acid. Nitrobenzol. Water.

Reactions.—1. By the action of reducing or hydrogenating agents, as zinc and hydrochloric acid, sulphuretted hydrogen, acetic acid and iron, or potassic arsenite, nitrobenzol is converted into aniline:—

$$N(C_6H_5)O_2 \ + \ 3SH_2 \ = \ N(C_6H_5)H_2 \ + \ 2OH_2 \ + \ S_3$$
$$\text{or } C_6H_5(N^vO_2) \ + \ 3SH_2 \ = \ C_6H_5(N'''H_2) \ + \ 2OH_2 \ + \ S_3.$$
Nitrobenzol. Sulphuretted hydrogen. Aniline. Water.

M

The relation between nitrobenzol and aniline will be seen in the following graphic formulæ :—

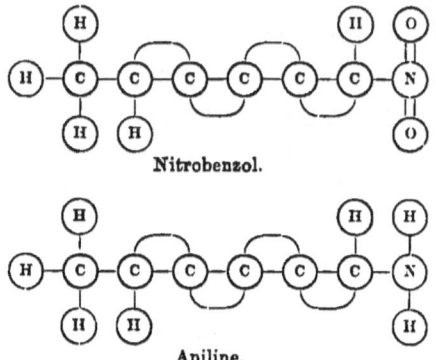

Nitrobenzol.

Aniline.

2. By the action of sodium amalgam and water, nitrobenzol is converted into azobenzol, and finally into benzidine :—

$$2\mathbf{N}^{v}(C_6H_5)O_2 \ + \ H_8 \ = \ {}^{\prime\prime} \left\{ \begin{matrix} \mathbf{N}(C_6H_5) \\ \mathbf{N}(C_6H_5) \end{matrix} \right. \ + \ 4\mathbf{O}H_2.$$

Nitrobenzol.   Azobenzol.   Water.

$$ {}^{\prime\prime} \left\{ \begin{matrix} \mathbf{N}(C_6H_5) \\ \mathbf{N}(C_6H_5) \end{matrix} \right. \ + \ H_2 \ = \ \left\{ \begin{matrix} \mathbf{N}(C_6H_5)H \\ \mathbf{N}(C_6H_5)H \end{matrix} \right. $$

Azobenzol.   Benzidine.

DINITROBENZOL.

$$\mathbf{N}_2(C_6H_4)^{\prime\prime}O_4.$$

Fuses below 100°.

Preparation.—By treating nitrobenzol with a mixture of concentrated nitric and sulphuric acids.

Reaction.—By the action of sulphuretted hydrogen, dinitro-benzol is converted into nitraniline :—

$$\mathbf{N}_2(C_6H_4)^{\prime\prime}O_4 \ + \ 3\mathbf{S}H_2 \ = \ \left\{ \begin{matrix} \mathbf{N}O_2 \\ (C_6H_4)^{\prime\prime} \\ \mathbf{N}H_2 \end{matrix} \right. \ + \ 2\mathbf{O}H_2 \ + \ S_3.$$

Dinitrobenzol.   Sulphuretted   Nitraniline.   Water.
                    hydrogen.

CLASS II.

HYDRIDES OF CHLOROUS OR NEGATIVE RADICALS.

Only two of these are known :—

Cyanic hydride or Hydrocyanic acid.
Oxatylic hydride or Formic acid.

The first has already been considered (p. 223); and the second will be more conveniently studied in connexion with the fatty acids (p. 304).

CHAPTER XXX.

THE ALCOHOLS.

THE alcohols form one of the most important of the families of organic compounds. The simplest member of this family is methylic alcohol, which is derived from marsh-gas by the substitution of one atom of hydroxyl for one of hydrogen.

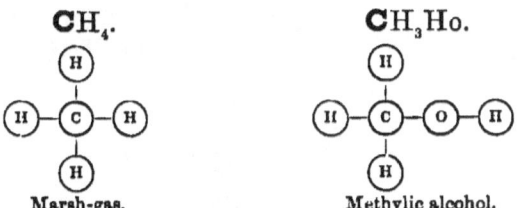

CH_4. CH_3Ho.

Marsh-gas.   Methylic alcohol.

The alcohols have been termed the hydrated oxides of the basylous radicals; but this is erroneous, as they do not contain water. They may more correctly be defined as the compounds of hydroxyl with the basylous organic radicals, whence it follows that each series of basylous radicals forms a corresponding series of alcohols. The alcohols act upon and saturate acids, forming a family of compounds termed *ethereal salts.*

M 2

The acidity or acid-saturating power of the alcohols depends upon the number of atoms of hydroxyl which they contain : the monad radicals give monacid alcohols, the dyad radicals diacid alcohols, &c. We have thus the annexed three principal subdivisions of the alcohol family.

Monacid.	Diacid.	Triacid
Methyl or $C_nH_{2n+1}Ho$ series.	Glycol or $C_nH_{2n}Ho_2$ series.	Glycerin or $C_nH_{2n-1}Ho_3$ series.
Vinyl or $C_nH_{2n-1}Ho$ series.		
Phenyl or $C_nH_{2n-7}Ho$ series.		

The following symbolic and graphic formulæ will exemplify the arrangement of the bonds in these three subdivisions :—

Monacid Alcohols.

Propylic   alcohol.
(*Ethyl series.*) } = C_3H_7Ho or $\begin{cases} C(CH_3)H_2 \\ CH_2Ho \end{cases}$.

Allylic   alcohol.
(*Vinyl series.*) } = C_3H_5Ho or $\begin{cases} C(CH_2)''H \\ CH_2Ho \end{cases}$.

Benzoic   alcohol.
(*Phenyl series.*) } = C_7H_7Ho or $\begin{cases} C(C_5H_3)H_2 \\ CH_2Ho \end{cases}$.

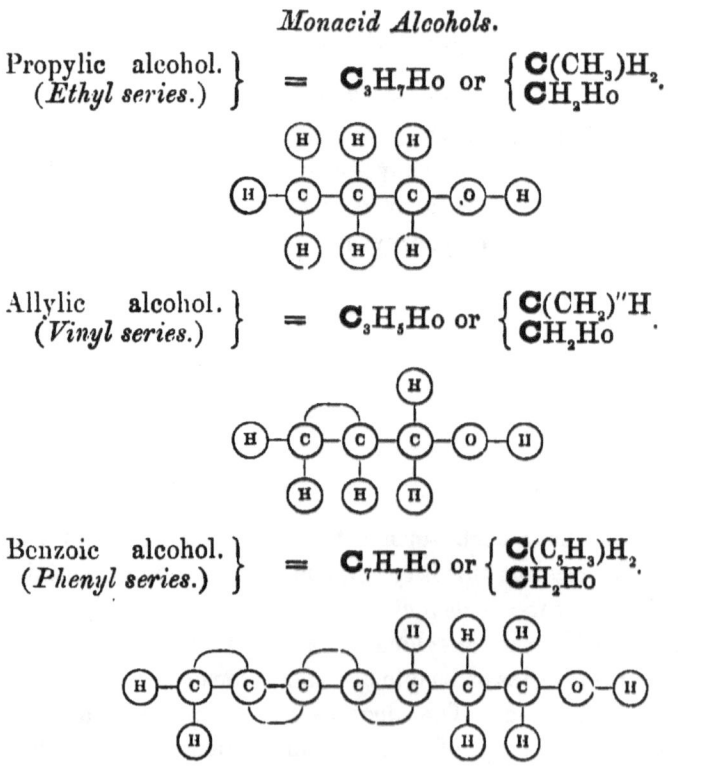

Diacid Alcohols.

Propylenic alcohol, or propylic glycol .... $= C_3H_6Ho_2$ or $\begin{cases} C(CH_3)HHo \\ CH_2Ho \end{cases}$.

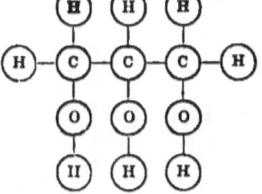

Triacid Alcohols.

Glycerin $= C_3H_5Ho_3$ or $\begin{cases} CH_2Ho \\ CHHo \\ CH_2Ho \end{cases}$.

MONACID ALCOHOLS:

Methyl or $C_nH_{2n+1}Ho$ series.

These alcohols may be divided into three classes viz. :—

1. Monacid normal alcohols $\begin{cases} C(C_nH_{2n+1})H_2 \\ CH_2Ho \end{cases}$.

2.   „   secondary „ $\begin{cases} C(C_nH_{2n+1})H_2 \\ C(C_mH_{2m+1})HHo \end{cases}$.

3.   „   tertiary „ $\begin{cases} C(C_nH_{2n+1})H_2 \\ C(C_mH_{2m+1})_2Ho \end{cases}$.

In the general formula of the normal alcohols n may $=0$, and even the whole radical $C(C_nH_{2n+1})H_2$ may be replaced by hydrogen, as is the case in methylic alcohol. In the formulæ of the secondary and tertiary alcohols n may also $=0$, but m must always be a positive integer.

NORMAL MONACID ALCOHOLS.

General formula $\begin{cases} \mathbf{C}(C_nH_{2n+1})H_2 \\ \mathbf{CH_2Ho} \end{cases}$

The following is a list of the members of this class :—

		Fusing-points.	Boiling-points.
Methylic alcohol	$\begin{cases} H \\ CH_2Ho \end{cases}$	—	66°·5.
Ethylic alcohol	$\begin{cases} CH_3 \\ CH_2Ho \end{cases}$	—	78°·4.
Propylic or tritylic alcohol	$\begin{cases} CMeH_2 \\ CH_2Ho \end{cases}$ or $\begin{cases} C(CH_3)H_2 \\ CH_2Ho \end{cases}$	—	96°.
Butylic or tetrylic alcohol	$\begin{cases} CEtH_2 \\ CH_2Ho \end{cases}$ or $\begin{cases} C(C_2H_5)H_2 \\ CH_2Ho \end{cases}$	—	109°.
Amylic or pentylic alcohol	$\begin{cases} CPrH_2 \\ CH_2Ho \end{cases}$ or $\begin{cases} C(C_3H_7)H_2 \\ CH_2Ho \end{cases}$	−20°.	132°.
Caproylic or hexylic alcohol	$\begin{cases} CBuH_2 \\ CH_2Ho \end{cases}$ or $\begin{cases} C(C_4H_9)H_2 \\ CH_2Ho \end{cases}$	—	—
Œnanthylic or heptylic alcohol	$\begin{cases} CAyH_2 \\ CH_2Ho \end{cases}$ or $\begin{cases} C(C_5H_{11})H_2 \\ CH_2Ho \end{cases}$	—	—
Caprylic or octylic alcohol	$\begin{cases} CCpH_2 \\ CH_2Ho \end{cases}$ or $\begin{cases} C(C_6H_{13})H_2 \\ CH_2Ho \end{cases}$	—	178°.
Cetylic alcohol	$\begin{cases} C(C_{14}H_{29})H_2 \\ CH_2Ho \end{cases}$	49°–50°.	—
Cerotic alcohol	$\begin{cases} C(C_{25}H_{51})H_2 \\ CH_2Ho \end{cases}$	79°.	—
Melissic alcohol	$\begin{cases} C(C_{2n}H_{57})H_2 \\ CH_2Ho \end{cases}$	85°.	—

The lower members of the class are liquid, and the higher solid. There is a rise of about 19° in the boiling-point for every addition of CH_2. They are produced in a variety of operations, such as destructive distillation, fermentation, and animal secretion, but by reactions which cannot usually be traced.

Relations of the normal $C_nH_{2n+1}Ho$ alcohols to the monad C_nH_{2n+1} radicals.

1. The radicals C_nH_{2n+1} which are combined with hydroxyl in the normal alcohols may be separated, by first converting the alcohol into an iodide (see p. 283), and subsequently acting on the iodide by zinc (see p. 209).

2. The radical next lower in the series, than that contained in the alcohol, may be obtained by converting the alcohol into the corresponding fatty acid, and then submitting a salt of this acid to electrolysis (see p. 209).

3. Inversely, the normal alcohols may be obtained by acting upon the radicals with chlorine under the influence of light, when one atom of hydrogen in the radical is displaced by chlorine.

Thus in the case of methyl we have

$$\left\{ \begin{matrix} CH_3 \\ CH_3 \end{matrix} \right. + Cl_2 = \left\{ \begin{matrix} CH_3 \\ CH_2Cl \end{matrix} \right. + HCl :$$

Methyl.   Chlorinated   Hydrochloric
                 methyl.   acid.

by the action of potassic hydrate upon this chlorinated methyl, ethylic alcohol is formed, thus :—

$$\left\{ \begin{matrix} CH_3 \\ CH_2Cl \end{matrix} \right. + KHo = \left\{ \begin{matrix} CH_3 \\ CH_2Ho \end{matrix} \right. + KCl.$$

Chlorinated   Potassic   Ethylic   Potassic
methyl.   hydrate.   alcohol.   chloride.

This reaction requires further investigation (see p. 248).

Relations of the normal C_nH_{2n+1}Ho alcohols to the dyad C_nH_{2n} radicals.

1. The C_nH_{2n} radicals are obtained from the normal C_nH_{2n+1}Ho alcohols by the abstraction of the elements of water :—

$$\left\{ \begin{matrix} CH_3 \\ CH_2Ho \end{matrix} \right. - OH_2 = {}''\left\{ \begin{matrix} CH_2 \\ CH_2 \end{matrix} \right..$$

Ethylic   Water.   Ethylene.
alcohol.

2. Inversely, the normal alcohols are obtained from these radicals by first uniting the latter with hydrochloric, hydrobromic, or hydriodic acid, and then treating the product with potassic hydrate :—

$${}''\left\{ \begin{matrix} CH_2 \\ CH_2 \end{matrix} \right. + HCl = \left\{ \begin{matrix} CH_3 \\ CH_2Cl \end{matrix} \right..$$

Ethylene.   Hydrochloric   Ethylic
             acid.   chloride.

$$\left\{ \begin{matrix} CH_3 \\ CH_2Cl \end{matrix} \right. + KHo = \left\{ \begin{matrix} CH_3 \\ CH_2Ho \end{matrix} \right. + KCl.$$

Ethylic   Potassic   Ethylic   Potassic
chloride.   hydrate.   alcohol.   chloride.

Or by uniting the dyad radicals with sulphuric acid, and distilling the product with water:—

$$SO_2Ho_2 \quad + \quad C_2H_4 \quad = \quad SO_2Ho(C_2H_5O):$$

Sulphuric acid.   Ethylene.   Sulphovinic acid.

$$SO_2EtoHo \quad + \quad OH_2 \quad = \quad SO_2Ho_2 \quad + \quad EtHo.$$

Sulphovinic acid.   Water.   Sulphuric acid.   Ethylic alcohol.

Relations of the normal $C_nH_{2n+1}Ho$ alcohols to the hydrides of the C_nH_{2n+1} radicals.

1. When the alcohols are converted into iodides (see p. 283) and the latter digested with zinc and water at 100°, the corresponding hydrides are produced (see p. 237).

2. When the hydrides of the C_nH_{2n+1} radicals are acted upon by chlorine under the influence of light, they produce the chlorides of the radicals, from which the alcohols may be obtained by the action of potassic hydrate:—

$$EtH \quad + \quad Cl_2 \quad = \quad EtCl \quad + \quad HCl.$$

Ethylic hydride.   Ethylic chloride.   Hydrochloric acid.

$$EtCl \quad + \quad KHo \quad = \quad EtHo \quad + \quad KCl.$$

Ethylic chloride.   Potassic hydrate.   Ethylic alcohol.   Potassic chloride.

The greater quantity of the chlorine compound so formed is isomeric with the chloride of the radical, and possibly gives a corresponding isomeric alcohol.

Relations of the normal $C_nH_{2n+1}Ho$ alcohols to the radical cyanogen. Ascent of the alcohol series. Mendius's reaction.

By the dry distillation of potassic sulphovinate and its homologues with potassic cyanide, the cyanides of the radicals are produced:—

$$SO_2EtoKo \quad + \quad KCy \quad = \quad SO_2Ko_2 \quad + \quad EtCy.$$

Potassic sulphovinate.   Potassic cyanide.   Potassic sulphate.   Ethylic cyanide.

By treatment with nascent hydrogen, this cyanide is converted into propylamine :—

$$\mathbf{NC(CMeH_2)} + \mathbf{H_4} = \mathbf{N[C(CMeH_2)H_2]H_2} \text{ or } \mathbf{NPrH_2}.$$
Ethylic cyanide.   Propylamine.

By the action of nitrous anhydride, propylamine is transformed into propylic alcohol :—

$$2\mathbf{NPrH_2} + \mathbf{N_2O_3} = 2\mathbf{PrHo} + \mathbf{OH_2} + 2\mathbf{N_2}.$$
Propylamine.   Nitrous   Propylic   Water.
                anhydride.   alcohol.

It is obvious that by repeating these reactions on propylic alcohol, butylic alcohol would be obtained, the homologous series of alcohols being ascended one step at each repetition of the process.

METHYLIC ALCOHOL, *Wood Spirit, Pyroxylic Spirit.*
$\mathbf{CH_3Ho}$ or MeHo.

Molecular weight = 32. *Molecular volume* ⊏⊐. 1 *litre of methylic alcohol vapour weighs* 16 *criths.* Sp. gr. 0·798. *Boils at* 66°·5.

Preparation.—1. From marsh-gas, by the action of chlorine and subsequent treatment with potassic hydrate :—

$$\mathbf{CH_4} + \mathbf{Cl_2} = \mathbf{CH_3Cl} + \mathbf{HCl};$$
Marsh-gas.   Methylic   Hydrochlo-
                     chloride.   ric acid.

$$\mathbf{CH_3Cl} + \mathbf{KHo} = \mathbf{CH_3Ho} + \mathbf{KCl}.$$
Methylic   Potassic   Methylic   Potassic
chloride.   hydrate.   alcohol.   chloride.

2. From the essential oil of *Gaultheria procumbens*, by the action of potassic hydrate—

$$\mathbf{C_7H_4OMeoHo} + \mathbf{KHo} = \mathbf{C_7H_4OHoKo} + \mathbf{MeHo}.$$
Oil of *Gaultheria pro-*   Potassic   Potassic salicylate.   Methylic
cumbens. (Metho-   hydrate.   alcohol.
salicylic acid.)

3. By the destructive distillation of wood.

Reactions.—1. Methylic alcohol unites with some salts in the capacity of water of crystallization, as, for instance,—

$$\mathbf{CaCl_2}, 2\mathbf{MeHo}.$$

2. By the action of potassium and sodium, methylates are formed with elimination of hydrogen :—

$$CH_3Ko.$$

Potassic
methylate.

$$CH_3Nao.$$

Sodic
methylate.

3. By oxidation it is transformed into formic acid :—

$$\left\{ \begin{matrix} H \\ CH_2Ho \end{matrix} \right. + O_2 = \left\{ \begin{matrix} H \\ COHo \end{matrix} \right. + OH_2.$$

Methylic
alcohol.

Formic acid.

Water.

4. When distilled with calcic chloro-hypochlorite (*chloride of lime*) and water, chloroform is produced.

$$2CH_3Ho + 4Ca(OCl)Cl = 2CHCl_3 + \left\{ \begin{matrix} CaCl \\ O \\ Ca'' \\ O \\ Ca'' \\ O \\ CaCl \end{matrix} \right. + 3OH_2.$$

Methylic
alcohol.

Calcic chloro-
hypochlorite.

Chloroform.

Calcic oxy-
chloride.

Water.

ETHYLIC ALCOHOL, *Alcohol, Spirit of Wine.*

$$\left\{ \begin{matrix} CH_3 \\ CH_2Ho \end{matrix} \right. \text{ or EtHo.}$$

Molecular weight =46. *Molecular volume* ☐☐. 1 *litre of ethylic alcohol vapour weighs* 23 *criths.* *Sp. gr.* 0·792 *at* 20°. *Boils at* 78°·4.

Preparation.—1. From ethylene (p. 247).

2. By the fermentation of grape-sugar with yeast at about 22° :—

$$C_6H_{12}O_6 = 2C_2H_5Ho + 2CO_2.$$

Grape-sugar.

Ethylic
alcohol.

Carbonic
anhydride.

At the same time, however, other products are formed, but in very small quantities.

Reactions.—1. Treated with potassium or sodium, alcohol forms ethylates :—

$$\left\{ \begin{matrix} CH_3 \\ CH_2Ko \end{matrix} \right.$$

Potassic
ethylate.

$$\left\{ \begin{matrix} CH_3 \\ CH_2Nao \end{matrix} \right.$$

Sodic ethylate.

2. When passed through a red-hot tube, alcohol is decomposed into marsh-gas, hydrogen, and carbonic oxide:—

$$\underset{\substack{\text{Ethylic}\\\text{alcohol.}}}{\mathbf{C_2H_5Ho}} = \underset{\substack{\text{Marsh-}\\\text{gas.}}}{\mathbf{CH_4}} + \mathbf{H_2} + \underset{\substack{\text{Carbonic}\\\text{oxide.}}}{\mathbf{CO}}.$$

Small quantities of ethylene, benzol, and naphthalin are simultaneously produced, whilst carbon is deposited.

3. By oxidation, ethylic alcohol is converted first into aldehyde, and then into acetic acid:—

$$\underset{\substack{\text{Ethylic}\\\text{alcohol.}}}{\left\{\begin{matrix}\mathbf{CH_3}\\\mathbf{CH_2Ho}\end{matrix}\right.} + \mathbf{O} = \underset{\text{Aldehyde.}}{\left\{\begin{matrix}\mathbf{CH_3}\\\mathbf{COH}\end{matrix}\right.} + \underset{\text{Water.}}{\mathbf{OH_2}}.$$

$$\underset{\text{Aldehyde.}}{\left\{\begin{matrix}\mathbf{CH_3}\\\mathbf{COH}\end{matrix}\right.} + \mathbf{O} = \underset{\text{Acetic acid.}}{\left\{\begin{matrix}\mathbf{CH_3}\\\mathbf{COHo}\end{matrix}\right.}$$

4. Distilled with chloride of lime, ethylic alcohol produces chloroform.

Alcoholates are salts containing alcohol in the place of water of crystallization; they are mostly decomposed immediately by water.

The following are known:—

$$\mathbf{ZnCl_2}, 2\mathbf{C_2H_5Ho}.$$
$$\mathbf{CaCl_2}, 4\mathbf{C_2H_5Ho}.$$
$$\mathbf{N_2O_4Mgo''}, 6\mathbf{C_2H_5Ho}.$$

MERCAPTAN, *Sulphur Alcohol, Ethylic sulphhydrate, Hydrosulphate of Ethyl.*

$$\left\{\begin{matrix}\mathbf{CH_3}\\\mathbf{CH_2Hs}\end{matrix}\right. \text{ or EtHs.}$$

Molecular weight =62. *Molecular volume* ☐☐. 1 *litre of mercaptan vapour weighs* 31 *criths.* *Sp. gr.* 0·835. *Boils at* 63°

Preparation.—By distilling potassic sulphovinate with potassic sulphhydrate :—

$$\mathbf{SO_2EtoKo} \; + \; \mathbf{KHs} \; = \; \mathbf{EtHs} \; + \; \mathbf{SO_2Ko_2}.$$

| Potassic sulphovinate. | Potassic sulphhydrate. | Mercaptan. | Potassic sulphate. |

Reactions.—1. By the action of potassium and sodium on mercaptan, an atom of hydrogen is displaced by the metal, producing mercaptides :—

$$\left\{ \begin{array}{l} \mathbf{CH_3} \\ \mathbf{CH_2Ks} \end{array} \right. \qquad\qquad \left\{ \begin{array}{l} \mathbf{CH_3} \\ \mathbf{CH_2Nas} \end{array} \right.$$

Potassic mercaptide.　　　　Sodic mercaptide.

2. Mercaptan acts upon mercuric oxide with great energy, a white crystalline mercuric mercaptide being formed :—

$$2 \left\{ \begin{array}{l} \mathbf{CH_3} \\ \mathbf{CH_2Hs} \end{array} \right. + \; \mathbf{HgO} \; = \; \left\{ \begin{array}{l} \mathbf{CH_3} \\ \mathbf{CH_2} \\ \mathbf{CH_2} \\ \mathbf{CH_3} \end{array} \right\} \mathbf{Hgs''} \; + \; \mathbf{OH_2}.$$

Mercaptan.　　Mercuric oxide.　　Mercuric mercaptide.　　Water.

Propylic alcohol, $\left\{ \begin{array}{l} \mathbf{CMeH_2} \\ \mathbf{CH_2Ho} \end{array} \right.$, is obtained from the fusel oil of the marc brandy of the south of France.

Butylic alcohol, $\left\{ \begin{array}{l} \mathbf{CEtH_2} \\ \mathbf{CH_2Ho} \end{array} \right.$, is contained in the fusel oil produced in the preparation of spirit from the molasses of beet-root sugar.

Amylic alcohol, $\left\{ \begin{array}{l} \mathbf{CPrH_2} \\ \mathbf{CH_2Ho} \end{array} \right.$, is the chief constituent of the fusel oil obtained in the manufacture of alcohol from potatoes or grain.

As far as these alcohols have been studied, they resemble, in their chemical relations, the two previously described.

SECONDARY MONACID ALCOHOLS.

General formula... $\left\{ \begin{array}{l} \mathbf{C}(C_nH_{2n+1})H_2 \\ \mathbf{C}(C_mH_{2m+1})HHo \end{array} \right.$

The secondary alcohols differ from the normal in yielding by oxidation ketones instead of acids.

Three secondary alcohols are at present known :—

Isopropylic alcohol .... $\begin{cases} CH_3 \\ CMeHHo' \end{cases}$

Pseudamylic alcohol .... $\begin{cases} CH_3 \\ CPrHHo' \end{cases}$

Pseudohexylic alcohol .... $\begin{cases} CPrH_2 \\ CMeHHo' \end{cases}$

The first is obtained by the action of nascent hydrogen on acetone :—

$$\begin{cases} CH_3 \\ COMe \end{cases} + H_2 = \begin{cases} CH_3 \\ CMeHHo' \end{cases}$$

Acetone.   Isopropylic alcohol.

The relation existing between ethylic alcohol, propylic alcohol, and isopropylic alcohol, will at once be evident from the following formulæ :—

$$\begin{cases} CH_3 \\ CH_2Ho' \end{cases} \qquad \begin{cases} C(CH_3)H_2 \\ CH_2Ho \end{cases} \qquad \begin{cases} CH_3 \\ C(CH_3)HHo' \end{cases}$$

Ethylic alcohol.   Propylic alcohol.   Isopropylic alcohol.

From these formulæ it is seen that propylic alcohol is ethylic alcohol in which one atom of hydrogen in the methyl (or *non-oxygenated* part of the compound) is displaced by methyl; whereas isopropylic alcohol is ethylic alcohol in which one atom of hydrogen, in the *oxygenated* part of the compound, is displaced by methyl.

Ethylic alcohol boils at ... 78°·4
Propylic alcohol „ ... 96
Isopropylic alcohol „ ... 87

Thus, by substituting an atom of methyl for one of hydrogen in the non-oxygenated part of the alcohol, the addition of CH_2 raises the boiling-point 17°·6; whilst, if an atom of hydrogen in the oxygenated part be similarly displaced, the same addition only raises the boiling-point 8°·6.

Isopropylic alcohol yields by oxidation a ketone, and not an acid. The radical oxatyl being a necessary constituent in organic acids, it will be seen from the following equations that, although propylic alcohol can be converted into an acid without the disruption of its carbon atoms, isopropylic alcohol cannot be so transformed :—

$$\begin{cases} \mathbf{CH_3} \\ \mathbf{CH_2Ho} \end{cases} + \mathbf{O_2} = \begin{cases} \mathbf{CH_3} \\ \mathbf{COHo} \end{cases} + \mathbf{OH_2}.$$

Ethylic alcohol.   Acetic acid.   Water.

$$\begin{cases} \mathbf{C(CH_3)H_2} \\ \mathbf{CH_2Ho} \end{cases} + \mathbf{O_2} = \begin{cases} \mathbf{C(CH_3)H_2} \\ \mathbf{COHo} \end{cases} + \mathbf{OH_2}.$$

Propylic alcohol.   Propionic acid.   Water.

$$\begin{cases} \mathbf{CH_3} \\ \mathbf{C(CH_3)HHo} \end{cases} + \mathbf{O} = \begin{cases} \mathbf{CH_3} \\ \mathbf{CO(CH_3)} \end{cases} + \mathbf{OH_2}.$$

Isopropylic alcohol.   Acetone.   Water.

TERTIARY MONACID ALCOHOLS.

General formula...... $\begin{cases} \mathbf{C(C_nH_{2n+1})H_2} \\ \mathbf{C(C_mH_{2m+1})_2Ho} \end{cases}$

One of these alcohols has recently been obtained, but little is yet known of its reactions.

Pseudobutylic alcohol, $\left\{ \begin{array}{l} \mathbf{CH_3} \\ \mathbf{CMe_2Ho} \end{array} \right.$, has been produced by acting with zincic methide on acetylic chloride, and submitting the product thus obtained to the action of water :—

$$\left\{ \begin{array}{l} \mathbf{CH_3} \\ \mathbf{COCl} \end{array} \right. + 2\mathbf{ZnMe_2} = \left\{ \begin{array}{l} \mathbf{CH_3} \\ \mathbf{CMe_2(Zn''MeO)} \end{array} \right. + \mathbf{ZnMeCl} ;$$

Acetylic   Zincic   Zincic chlor-
chloride.   methide.   methide.

$$\left\{ \begin{array}{l} \mathbf{CH_3} \\ \mathbf{CMe_2(Zn''MeO)} \end{array} \right. + 2\mathbf{OH_2} = \left\{ \begin{array}{l} \mathbf{CH_3} \\ \mathbf{CMe_2Ho} \end{array} \right.$$

                                 Water.   Pseudobutylic
                                                 alcohol.

$$+ \mathbf{CH_4} + \mathbf{ZnHo_2}.$$

Methylic   Zincic
hydride.   hydrate.

or $\left\{ \begin{array}{l} \mathbf{CH_3} \\ \mathbf{C(CH_3)_2Ho} \end{array} \right.$

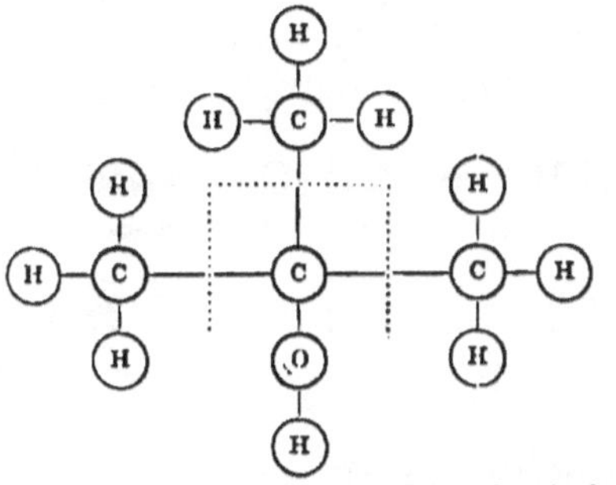

Pseudobutylic alcohol.

CHAPTER XXXI..

MONACID ALCOHOLS:

Vinyl or $C_nH_{2n-1}Ho$ series.

The normal alcohols only of this series are known; and of

these but two have been obtained :—

Vinylic alcohol ..." $\left\{ \begin{array}{l} \mathbf{CH_2} \\ \mathbf{CHHo} \end{array} \right.$ or $\mathbf{CMe''HHo}$.

Allylic alcohol ... $\left\{ \begin{array}{l} \mathbf{CMe''H} \\ \mathbf{CH_2Ho} \end{array} \right.$

Vinylic alcohol.   Allylic alcohol.

VINYLIC ALCOHOL.

Preparation.—By combining acetylene with sulphuric acid and distilling the product with water, in the same manner as in the preparation of ethylic alcohol from ethylene (p. 248) :—

$$\mathbf{SO_2Ho_2} + \mathbf{C_2H_2} = \mathbf{SO_2(C_2H_3O)Ho}.$$

Sulphuric   Acetylene.   Sulphovinylic acid.
acid.

$$\mathbf{SO_2(C_2H_3O)Ho} + \mathbf{OH_2} = \mathbf{SO_2Ho_2} + \mathbf{CMe''HHo}.$$

Sulphovinylic acid.   Water.   Sulphuric   Vinylic
acid.   alcohol.

This alcohol is isomeric with aldehyde and with ethylenic oxide :—

$_{''}\left\{ \begin{array}{l} \mathbf{CH_2} \\ \mathbf{CHHo} \end{array} \right.$ $\left\{ \begin{array}{l} \mathbf{CH_3} \\ \mathbf{COH} \end{array} \right.$ $\left\{ \begin{array}{l} \mathbf{CH_2} \\ \mathbf{CH_2}O. \end{array} \right.$

Vinylic   Aldehyde.   Ethylenic
alcohol.   oxide.

If the above, and not $\left\{ \begin{array}{l} \mathbf{''CH} \\ \mathbf{CH_2Ho} \end{array} \right.$, be the true formula for vinylic alcohol, it is obvious that this body could not yield an acid by oxidation; but if the latter formula represent vinylic alcohol, this alcohol ought to yield on oxidation an acid, $\left\{ \begin{array}{l} \mathbf{''CH} \\ \mathbf{COHo} \end{array} \right.$ homologous with acrylic acid,

ALLYLIC ALCOHOL.

$$\left\{ \begin{array}{l} \mathbf{C}Me''H \\ \mathbf{CH_2}Ho \end{array} \right. \text{ or All Ho.}$$

Boils at 103°.

Preparation.—Glycerin, when submitted to the action of diphosphorous tetriodide, yields allylic iodide :—

$$''\mathbf{P}^{iv}{}_2\mathbf{I}_4 + 2 \left\{ \begin{array}{l} \mathbf{CH_2}Ho \\ \mathbf{CHH}o \\ \mathbf{CH_2}Ho \end{array} \right. = 2 \;''\!\left\{ \begin{array}{l} \mathbf{CH_2} \\ \mathbf{CH} \\ \mathbf{CH_2}I \end{array} \right. + 2\mathbf{P}OHHo_2 + I_2.$$

Diphosphorous   Glycerin.   Allylic   Phosphorous
tetriodide.   iodide.   acid.

The allylic iodide is then decomposed by argentic oxalate when allylic oxalate is formed :—

$$2AllI + \left\{ \begin{array}{l} \mathbf{CO}Ago \\ \mathbf{CO}Ago \end{array} \right. = \left\{ \begin{array}{l} \mathbf{CO}Allo \\ \mathbf{CO}Allo \end{array} \right. + 2AgI.$$

Allylic   Argentic   Allylic   Argentic
iodide.   oxalate.   oxalate.   iodide.

The allylic oxalate is next decomposed by ammonia, when oxamide and allylic alcohol are produced :—

$$\left\{ \begin{array}{l} \mathbf{CO}Allo \\ \mathbf{CO}Allo \end{array} \right. + 2\mathbf{NH_3} = \left\{ \begin{array}{l} \mathbf{CO}(N'''H_2) \\ \mathbf{CO}(N'''H_2) \end{array} \right. + 2AllHo.$$

Allylic   Ammonia.   Oxamide.   Allylic
oxalate.   alcohol.

Reactions.—1. In all ordinary reactions, allylic alcohol behaves like ethylic alcohol. By oxidation it gives acrylic acid :—

$$\left\{ \begin{array}{l} \mathbf{C}Me''H \\ \mathbf{CH_2}Ho \end{array} \right. + O_2 = \left\{ \begin{array}{l} \mathbf{C}Me''H \\ \mathbf{CO}Ho \end{array} \right. + \mathbf{OH_2}.$$

Allylic   Acrylic   Water.
alcohol.   acid.

2. With phosphoric anhydride it yields allylene—

$$= \;''\!\left\{ \begin{array}{l} \mathbf{C}(CH_2)'' \\ \mathbf{CH_2} \end{array} \right.$$

Among the ethereal salts of allylic alcohol, the sulphide and sulphocyanate occur in nature as garlic and mustard oils:—

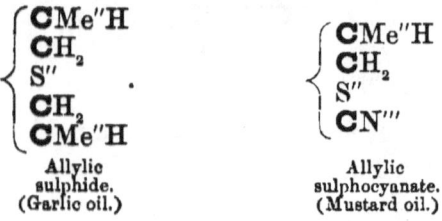

Allylic
sulphide.
(Garlic oil.)

Allylic
sulphocyanate.
(Mustard oil.)

CHAPTER XXXII.

MONACID ALCOHOLS:

Phenyl or C_nH_{2n-7} series.

These alcohols may be divided into a normal and a secondary class. The members of the first class possess the general character of the normal alcohols of the ethyl series, while those of the second class exhibit slightly acid characters.

CLASS I. Normal Alcohols.

General formula............ $\begin{cases} C(C_nH_{2n-7})H_2 \\ CH_2Ho \end{cases}$.

Normal phenylic alcohol... $\begin{cases} C(C_4H)H_2 \\ CH_2Ho \end{cases}$.

Benzoic alcohol*............ $\begin{cases} C(C_5H_3)H_2 \\ CH_2Ho \end{cases}$.

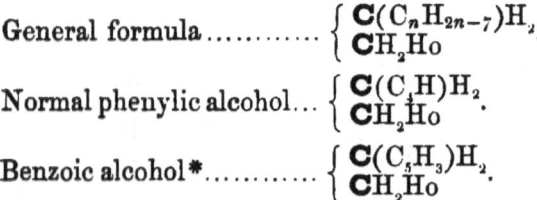

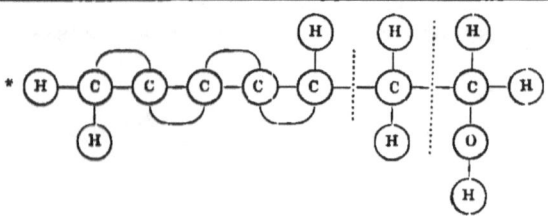

Cumylic alcohol .... $\left\{\begin{array}{l}\mathbf{C}(\mathbf{C_8H_9})\mathbf{H_2} \\ \mathbf{CH_2Ho}\end{array}\right.$

Sycocerylic alcohol......... $\left\{\begin{array}{l}\mathbf{C}(\mathbf{C_{16}H_{25}})\mathbf{H_2} \\ \mathbf{CH_2Ho}\end{array}\right.$

CLASS II. Secondary Alcohols.

General formula............ $\left\{\begin{array}{l}\mathbf{C}(\mathbf{C_mH_{2m+1}})\mathbf{H_2} \\ \mathbf{C}(\mathbf{C_nH_{2n-7}})\mathbf{HHo}\end{array}\right.$

In this formula m may $= 0$, but n must be a positive integer.

Phenylic alcohol.   Carbo-
lic acid* .... $\left\{\begin{array}{l}\mathbf{CH_3} \\ \mathbf{C}(\mathbf{C_4H})\mathbf{HHo}\end{array}\right.$

Cresylic alcohol † .... $\left\{\begin{array}{l}\mathbf{CMeH_2} \\ \mathbf{C}(\mathbf{C_4H})\mathbf{HHo}\end{array}\right.$

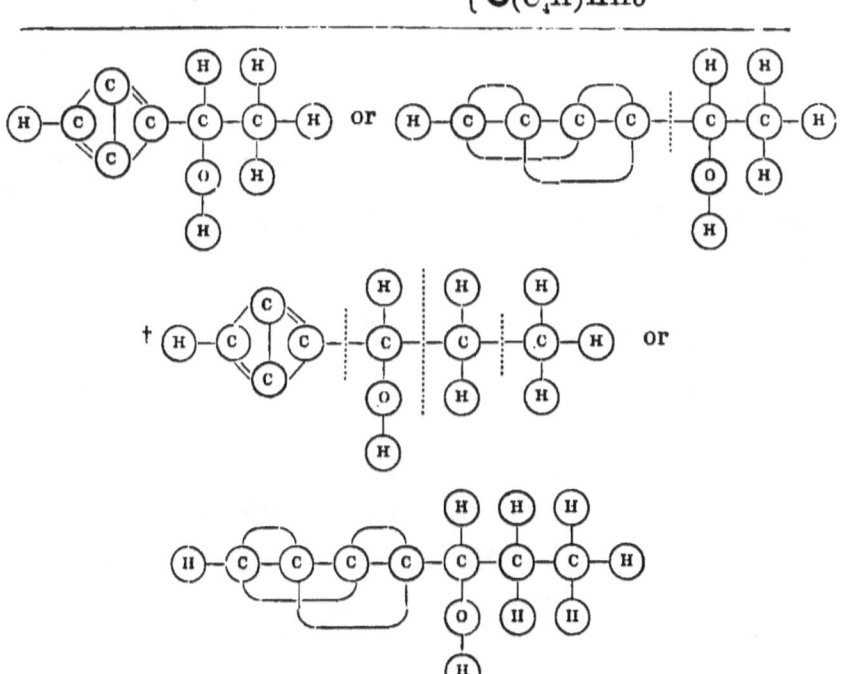

Class I. *NORMAL ALCOHOLS.*

BENZOIC ALCOHOL.

$$\left\{ \begin{array}{l} \mathbf{C}(C_5H_3)H_2 \\ \mathbf{C}H_2Ho \end{array} \right.$$

Boils at 204°.

Preparation.—1. By treating oil of bitter almonds with alcoholic potash :—

$$2\left\{ \begin{array}{l} \mathbf{C}(C_5H_3)H_2 \\ \mathbf{C}OH \end{array} \right. + KHo = \left\{ \begin{array}{l} \mathbf{C}(C_5H_3)H_2 \\ \mathbf{C}H_2Ho \end{array} \right. + \left\{ \begin{array}{l} \mathbf{C}(C_5H_3)H_2 \\ \mathbf{C}OKo \end{array} \right.$$

Benzoic aldehyde.   Potassic   Benzoic alcohol.   Potassic
(Oil of bitter   hydrate.   benzoate.
almonds.)

2. Benzoic alcohol may also be obtained from toluol by first converting the latter into toluylic chloride by the action of chlorine—

$$\left\{ \begin{array}{l} \mathbf{C}(C_5H_3)H_2 \\ \mathbf{C}H_2H \end{array} \right. + Cl_2 = \left\{ \begin{array}{l} \mathbf{C}(C_5H_3)H_2 \\ \mathbf{C}H_2Cl \end{array} \right. + HCl ;$$

Toluol.   Toluylic chloride.   Hydrochloric
(Toluylic hydride.)   acid.

and then submitting the toluylic chloride to the action of potassic hydrate :—

$$\left\{ \begin{array}{l} \mathbf{C}(C_5H_3)H_2 \\ \mathbf{C}H_2Cl \end{array} \right. + KHo = \left\{ \begin{array}{l} \mathbf{C}(C_5H_3)H_2 \\ \mathbf{C}H_2Ho \end{array} \right. + KCl.$$

Toluylic   Potassic   Benzoic   Potassic
chloride.   hydrate.   alcohol.   chloride.

Class II. *SECONDARY ALCOHOLS.*

PHENYLIC ALCOHOL, *Carbolic Acid, Phenylic Acid.*

$$\left\{ \begin{array}{l} \mathbf{C}H_3 \\ \mathbf{C}(C_4H)HHo \end{array} \right. \text{ or PhHo.}$$

Molecular weight =94. *Molecular volume* ☐☐. *1 litre of phenylic alcohol vapour weighs* 47 *criths. Sp. gr.* 1·065 *at* 18°. *Fuses at* 34°. *Boils at* 188°.

Occurrence.—In coal-tar, and in small quantity in the urine of man, of the cow, and of the horse.

Preparation.—1. By the distillation of salicylic acid with baryta or lime:—

$$\left\{ \begin{matrix} \mathbf{C}[C(C_4H)H_2]HHo \\ \mathbf{C}OHo \end{matrix} \right. = \left\{ \begin{matrix} \mathbf{CH_3} \\ \mathbf{C}(\dot{C}_4H)HHo \end{matrix} \right. + \mathbf{CO_2}.$$

Salicylic acid.   Phenylic alcohol.   Carbonic anhydride.

2. It is also produced in the destructive distillation of numerous organic substances.

3. Phenylic alcohol is formed when the vapour of ethylic alcohol or acetic acid is passed through a red-hot tube. In this manner phenylic compounds may be obtained from their elements; for both acetic acid and alcohol may be built up from purely mineral sources.

4. Phenylic alcohol is formed when aniline hydrochlorate is treated with potassic nitrite:—

$$\mathbf{N}PhH_3Cl + \mathbf{N}OKo = PhHo + KCl + \mathbf{OH_2} + \mathbf{N_2}.$$

Aniline hydrochlorate.   Potassic nitrite.   Phenylic alcohol.   Potassic chloride.   Water.

Reactions.—Treated with chlorine, bromine, or nitric acid, phenylic alcohol produces a series of substitution products, of which the following are examples:—

Dichlorphenylic acid............ $\left\{ \begin{matrix} \mathbf{C}HCl_2 \\ \mathbf{C}(C_4H)HHo \end{matrix} \right.$

Trichlorphenylic acid .... $\left\{ \begin{matrix} \mathbf{C}Cl_3 \\ \mathbf{C}(C_4H)HHo \end{matrix} \right.$

Perchlorphenylic acid .... $\left\{ \begin{matrix} \mathbf{C}Cl_3 \\ \mathbf{C}(C_4Cl)ClHo \end{matrix} \right.$

Bromphenylic acid .... $\left\{ \begin{matrix} \mathbf{C}H_2Br \\ \mathbf{C}(C_4H)HHo \end{matrix} \right.$

Nitrophenylic acid .... $\left\{ \begin{matrix} \mathbf{C}H_2(N^vO_2) \\ \mathbf{C}(C_4H)HHo \end{matrix} \right.$

Dinitrophenylic acid............ $\left\{ \begin{matrix} \mathbf{C}H(N^vO_2)_2 \\ \mathbf{C}(C_4H)HHo \end{matrix} \right.$

Trinitrophenylic acid. (*Picric* $\begin{cases} \mathbf{C}(\mathrm{N^vO_2})_3 \\ \mathbf{C}(\mathrm{C_4H})\mathrm{HHo} \end{cases}$
acid.) ....

Amidodinitrophenylic acid. $\begin{cases} \mathbf{C}(\mathrm{N^vO_2})_2(\mathrm{N'''H_2}) \\ \mathbf{C}(\mathrm{C_4H})\mathrm{HHo} \end{cases}$.
(*Picramic acid.*)...............

CRESYLIC ALCOHOL.

$$\begin{cases} \mathbf{C}\mathrm{MeH_2} \\ \mathbf{C}(\mathrm{C_4H})\mathrm{HHo} \end{cases}$$

Boils at 204°.

This alcohol is very little known; it is isomeric with benzoic alcohol.

CHAPTER XXXIII.

DIACID ALCOHOLS or GLYCOLS.

General formula... $\begin{cases} \mathrm{C}_n\mathrm{H}_{2n}\mathrm{Ho} \\ \mathrm{C}_n\mathrm{H}_{2n}\mathrm{Ho} \end{cases}$

The following is a list of the glycols at present known :—

			Boiling-points.
Glycol or Ethylic glycol............	$\mathrm{C_2H_6O_2}$ or	$\begin{cases} \mathbf{C}\mathrm{H_2Ho} \\ \mathbf{C}\mathrm{H_2Ho} \end{cases}$	197°·5.
Propylic glycol...	$\mathrm{C_3H_8O_2}$ or	$\begin{cases} \mathbf{C}\mathrm{MeHHo} \\ \mathbf{C}\mathrm{H_2Ho} \end{cases}$	188°–189°.
Butylic glycol ...	$\mathrm{C_4H_{10}O_2}$ or	$\begin{cases} \mathbf{C}\mathrm{EtHHo} \\ \mathbf{C}\mathrm{H_2Ho} \end{cases}$	183°–184°.
Amylic glycol ...	$\mathrm{C_5H_{12}O_2}$ or	$\begin{cases} \mathbf{C}\mathrm{PrHHo} \\ \mathbf{C}\mathrm{H_2Ho} \end{cases}$	177°.

The existence of normal, secondary, &c. alcohols of this subdivision has not yet been clearly established; but ethylic glycol is probably a normal glycol, whilst propylic, butylic, and amylic glycols are probably secondary glycols, as shown in the above formulæ.

Methylic glycol has not been obtained ; and although a substance of the same composition might exist, yet it would not be strictly homologous with the ethylic and propylic glycols, as will be seen from the following graphic representations :—

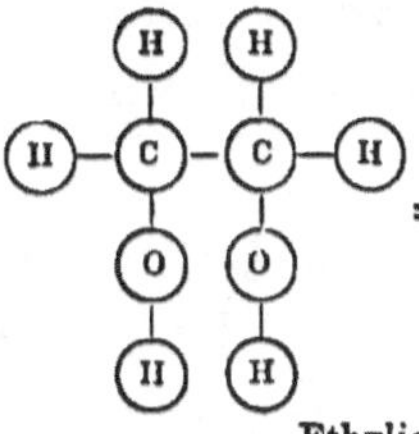

$$\left\{ \begin{array}{l} \mathbf{CH_2Ho} \\ \mathbf{CH_2Ho} \end{array} \right.$$

Ethylic glycol.

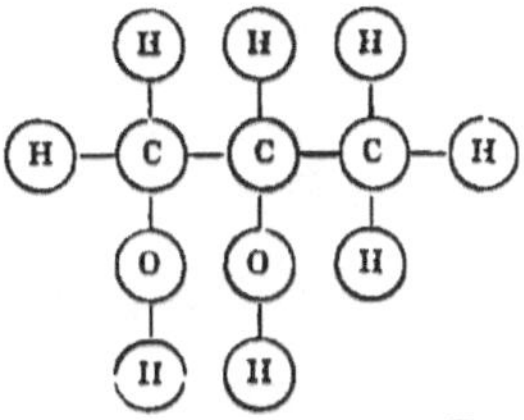

$$\left\{ \begin{array}{l} \mathbf{C(CH_3)HHo} \\ \mathbf{CH_2Ho} \end{array} \right. .$$

Propylic glycol.

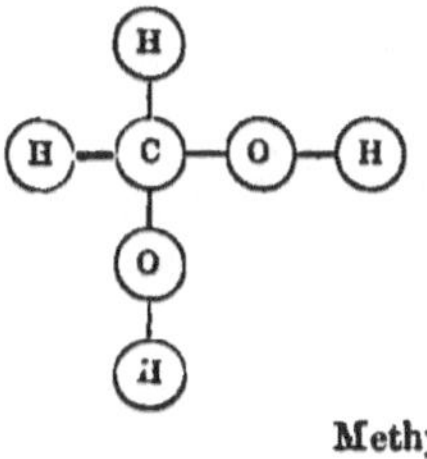

$$= \mathbf{CH_2Ho_2}.$$

Methylic glycol ?

In the ethylic and propylic glycols, the two atoms of hydroxyl are united with different carbon atoms ; in the methylic glycol they would be united with the same carbon atom.

It will be observed that the boiling-points of the glycols differ from each other in a direction inversely to that previously noticed in the case of the normal monacid alcohols : the more complex substances boil at a lower temperature than the simpler ones.

GLYCOL. ETHYLIC GLYCOL, *Ethylenic Alcohol.*

$$\left\{ \begin{array}{l} CH_2Ho \\ CH_2Ho \end{array} \right.$$

Molecular weight = 62. *Molecular volume* ☐☐. 1 *litre of ethylic glycol vapour weighs* 31 *criths.* *Sp. gr.* 1·125. *Boils at* 197°·5.

Preparation.—Ethylenic dibromide is treated with argentic acetate, and thus converted into ethylenic diacetate :—

$$\left\{ \begin{array}{l} CH_2Br \\ CH_2Br \end{array} \right. + 2CMeOAgo = \left\{ \begin{array}{l} CH_2\text{-}O\text{-}CMeO \\ CH_2\text{-}O\text{-}CMeO \end{array} \right. + 2AgBr.$$

Ethylenic   Argentic acetate.   Ethylenic diacetate.   Argentic
dibromide.   (Diacetic glycol.)   bromide.

The ethylenic diacetate is now acted upon by potassic hydrate, when it yields potassic acetate and glycol :—

$$\left\{ \begin{array}{l} CH_2\text{-}O\text{-}CMeO \\ CH_2\text{-}O\text{-}CMeO \end{array} \right. + 2KHo = \left\{ \begin{array}{l} CH_2Ho \\ CH_2Ho \end{array} \right. + 2CMeOKo.$$

Ethylenic diacetate.   Potassic   Glycol.   Potassic acetate.
                           hydrate.

Reactions.—1. Glycol is easily oxidized, the first product of its oxidation being glycollic acid :—

$$\left\{ \begin{array}{l} CH_2Ho \\ CH_2Ho \end{array} \right. + O_2 = \left\{ \begin{array}{l} CH_2Ho \\ COHo \end{array} \right. + OH_2.$$

   Glycol.   Glycollic   Water.
                                 acid.

2. By further oxidation oxalic acid is formed :—

$$\left\{ \begin{array}{l} CH_2Ho \\ CH_2Ho \end{array} \right. + O_4 = \left\{ \begin{array}{l} COHo \\ COHo \end{array} \right. + 2OH_2.$$

   Glycol.   Oxalic   Water.
                           acid.

3. Oxalic acid is also produced by heating glycol and potassic hydrate together to 250° :—

$$\left\{ \begin{array}{l} CH_2Ho \\ CH_2Ho \end{array} \right. + 2KHo = \left\{ \begin{array}{l} COKo \\ COKo \end{array} \right. + H_8.$$

   Glycol.   Potassic   Potassic
                 hydrate.   oxalate.

4. Treated with potassium or sodium, the hydrogen of the hydroxyl in glycol is replaced in two successive stages :—

$$\left\{\begin{array}{l} CH_2Nao \\ CH_2Ho \end{array}\right. ;$$

Monosodic glycol.

$$\left\{\begin{array}{l} CH_2Nao \\ CH_2Nao \end{array}\right.$$

Disodic glycol.

The following list contains some of the principal derivatives of glycol :—

$$\left\{\begin{array}{l} CH_2Ho \\ CH_2Ho \end{array}\right.$$

Glycol.

$$\left\{\begin{array}{l} CH_2Hs \\ CH_2Hs \end{array}\right.$$

Sulphur glycol.

$$\left\{\begin{array}{l} CH_2Ho \\ CH_2Cl \end{array}\right.$$

Chlorhydric glycol.

$$\left\{\begin{array}{l} CH_2Ho \\ CH_2Br \end{array}\right.$$

Bromhydric glycol.

$$\left\{\begin{array}{l} CH_2Eto \\ CH_2Br \end{array}\right.$$

Bromethylic glycol.

$$\left\{\begin{array}{l} CH_2Eto \\ CH_2Ho \end{array}\right.$$

Hydric ethylic glycol.

$$\left\{\begin{array}{l} CH_2Eto \\ CH_2Eto \end{array}\right.$$

Diethylic glycol.

$$\left\{\begin{array}{l} CH_2Br \\ CH_2 \\ O \\ CO \\ CH_3 \end{array}\right. \quad or \left\{\begin{array}{l} CH_2Br \\ CH_2\text{-}O\text{-}CMeO \end{array}\right.$$

Glycollic acetobromide.

$$\left\{\begin{array}{l} CH_2Ho \\ CH_2 \\ O \\ CO \\ CH_3 \end{array}\right. \quad or \left\{\begin{array}{l} CH_2Ho \\ CH_2\text{-}O\text{-}CMeO \end{array}\right.$$

Monaceti glycol.

$$\left\{\begin{array}{l} CH_3 \\ CO \\ O \\ CH_2 \\ CH_2 \\ O \\ CO \\ CH_3 \end{array}\right. \quad or \left\{\begin{array}{l} CH_2\text{-}O\text{-}CMeO \\ CH_2\text{-}O\text{-}CMeO \end{array}\right.$$

Diacetic glycol.

$$\left\{ \begin{array}{l} \mathbf{CH_3} \\ \mathbf{CO} \\ \mathbf{O} \\ \mathbf{CH_2} \\ \mathbf{CH_2} \\ \mathbf{O} \\ \mathbf{CO} \\ \mathbf{C(C_2H_5)H_2} \end{array} \right. \quad \text{or} \quad \left\{ \begin{array}{l} \mathbf{CH_2\text{-}O\text{-}CMeO} \\ \mathbf{CH_2\text{-}O\text{-}CPrO} \end{array} \right. \cdot$$

Acetobutyric glycol.

POLYETHYLENIC GLYCOLS.

Polyethylenic Alcohols.

These bodies are produced by heating ethylenic oxide with glycol in sealed tubes, and by other processes. They may be regarded as formed by the addition of ethylenic oxide to glycol.

Diethylenic glycol ...
$$\left\{ \begin{array}{l} \mathbf{CH_2Ho} \\ \mathbf{CH_2} \\ \mathbf{CH_2} \\ \mathbf{O} \\ \mathbf{CH_2Ho} \end{array} \right. \quad \text{or} \quad \left\{ \begin{array}{l} \mathbf{CH_2Ho} \\ \mathbf{C_2H_4} \\ \mathbf{O} \\ \mathbf{CH_2Ho} \end{array} \right. \cdot$$

Triethylenic glycol ...
$$\left\{ \begin{array}{l} \mathbf{CH_2Ho} \\ \mathbf{CH_2} \\ \mathbf{CH_2} \\ \mathbf{O} \\ \mathbf{CH_2} \\ \mathbf{CH_2} \\ \mathbf{O} \\ \mathbf{CH_2Ho} \end{array} \right. \quad \text{or} \quad \left\{ \begin{array}{l} \mathbf{CH_2Ho} \\ \mathbf{C_2H_4} \\ \mathbf{O} \\ \mathbf{C_2H_4} \\ \mathbf{O} \\ \mathbf{CH_2Ho} \end{array} \right.$$

Tetrethylenic glycol
$$\left\{ \begin{array}{l} \mathbf{CH_2Ho} \\ \mathbf{CH_2} \\ \mathbf{CH_2} \\ \mathbf{O} \\ \mathbf{CH_2} \\ \mathbf{CH_2} \\ \mathbf{O} \\ \mathbf{CH_2} \\ \mathbf{CH_2} \\ \mathbf{O} \\ \mathbf{CH_2Ho} \end{array} \right. \quad \text{or} \quad \left\{ \begin{array}{l} \mathbf{CH_2Ho} \\ \mathbf{C_2H_4} \\ \mathbf{O} \\ \mathbf{C_2H_4} \\ \mathbf{O} \\ \mathbf{C_2H_4} \\ \mathbf{O} \\ \mathbf{CH_2Ho} \end{array} \right.$$

Pentethylenic and hexethylenic glycols have also been formed.

CHAPTER XXXIV.

TRIACID ALCOHOLS.

THESE alcohols contain three atoms of hydroxyl united with three separate atoms of carbon; consequently the lowest term contains three atoms of carbon.

Only two of these alcohols have been obtained :—

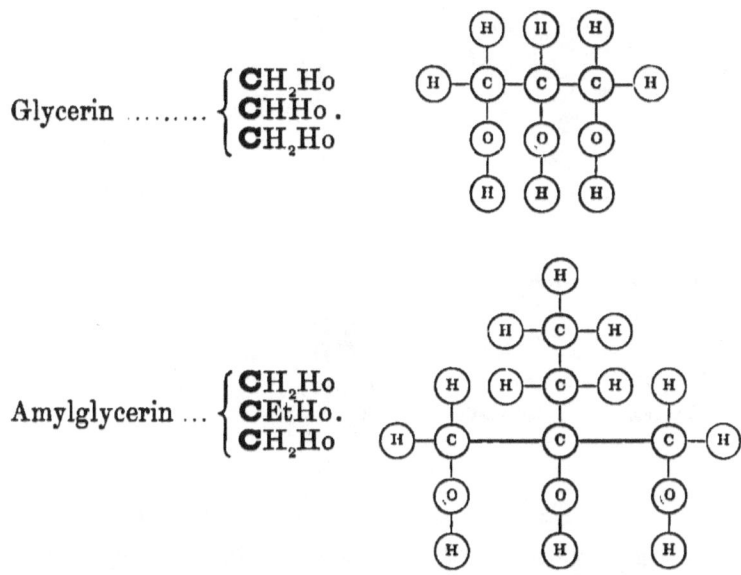

Glycerin .... $\begin{cases} \mathbf{CH_2Ho} \\ \mathbf{CHHo} \\ \mathbf{CH_2Ho} \end{cases}$.

Amylglycerin ... $\begin{cases} \mathbf{CH_2Ho} \\ \mathbf{CEtHo} \\ \mathbf{CH_2Ho} \end{cases}$.

The constitution of amylglycerin is not at present established. Its formula may possibly be

$$\begin{cases} \mathbf{CEtHHo} \\ \mathbf{CHHo} \\ \mathbf{CH_2Ho} \end{cases}.$$

The action of oxidizing agents on amylglycerin will probably throw light upon its internal structure.

GLYCERIN.

$$\begin{cases} CH_2Ho \\ CHHo \\ CH_2Ho \end{cases}.$$

Sources.—Most animal and vegetable fats consist of mixtures of the glycerin ethereal salts of the fatty and of the oleic series of acids. Glycerin is liberated from these by water at high temperatures, or by bases giving salts insoluble in water :—

$$\begin{cases} CH_2\text{-O-}C(C_{17}H_{35})O \\ CH\text{-O-}C(C_{17}H_{35})O \\ CH_2\text{-O-}C(C_{17}H_{35})O \end{cases} + 3OH_2 = \begin{cases} CH_2Ho \\ CHHo \\ CH_2Ho \end{cases} + 3\begin{cases} C(C_{18}H_{33})H_2 \\ COHo \end{cases},$$

Stearin.   Water.   Glycerin.   Stearic acid.

Relation of Glycerin to Isopropylic Alcohol.

By the action of hydriodic acid, glycerin is converted into isopropylic iodide :—

$$\begin{cases} CH_2Ho \\ CHHo \\ CH_2Ho \end{cases} + 5HI = \begin{cases} CH_3 \\ CHI \\ CH_3 \end{cases} + 2I_2 + 3OH_2.$$

Glycerin. Hydriodic acid. Isopropylic iodide. Water.

Relation of Glycerin to Allylic Alcohol.

When diphosphorous tetriodide is brought into contact with glycerin, an energetic reaction ensues, allylic iodide being formed :—

$$'P''_2I_4 + 2\begin{cases} CH_2Ho \\ CHHo \\ CH_2Ho \end{cases} = 2''\begin{cases} CH_2 \\ CH \\ CH_2I \end{cases} + 2POHHo_2 + I_2.$$

Diphosphorous tetriodide. Glycerin. Allylic iodide. Phosphorous acid.

Relations of Glycerin to Propylic Glycol.

The several atoms of hydroxyl in glycerin are capable of being substituted by chlorine, bromine, &c.; thus, by the action of hydrochloric acid on glycerin, one atom of hydroxyl is displaced by chlorine, monochlorhydrin being formed :—

$$\left\{\begin{array}{l} CH_2Ho \\ CHHo \\ CH_2Ho \end{array}\right. + HCl = \left\{\begin{array}{l} CH_2Cl \\ CHHo \\ CH_2Ho \end{array}\right. + OH_2.$$

Glycerin.　　Hydrochloric　　Monochlor-　　Water.
　　　　　　　　acid.　　　　hydrin.

Monochlorhydrin is identical with monochlorinated propylic glycol :—

$$\left\{\begin{array}{l} CH_2Cl \\ CHHo \\ CH_2Ho \end{array}\right. = \left\{\begin{array}{l} C(CH_2Cl)HHo \\ CH_2Ho \end{array}\right. .$$

Monochlorhydrin.　　　Monochlorinated
　　　　　　　　　　　propylic glycol.

By the action of sodium amalgam and water, monochlorinated propylic glycol is readily converted into propylic glycol :—

$$\left\{\begin{array}{l} C(CH_2Cl)HHo \\ CH_2Ho \end{array}\right. + H_2 = \left\{\begin{array}{l} C(CH_3)HHo \\ CH_2Ho \end{array}\right. + HCl.$$

Monochlorinated　　　　　Propylic　　Hydrochloric
propylic glycol.　　　　　glycol.　　　acid.

Relations of Glycerin to the Trihydric Acids—Glyceric Acid, and Tartronic Acid.

By the slow action of nitric acid, glycerin is converted into glyceric acid :—

$$\left\{\begin{array}{l} CH_2Ho \\ CHHo \\ CH_2Ho \end{array}\right. + O_2 = \left\{\begin{array}{l} CH_2Ho \\ CHHo \\ COHo \end{array}\right. + OH_2.$$

Glycerin.　　　　　　Glyceric　　Water.
　　　　　　　　　　acid.

A second atom of oxatyl has not been produced in glycerin, so as to convert the latter into a dibasic acid; but there can be little doubt that tartronic acid, which is formed by the spontaneous decomposition of nitrotartaric acid, is the acid in question, and that it has the following constitution :—

$$\left\{\begin{array}{l} COHo \\ CHHo. \\ COHo \end{array}\right.$$

Tartronic acid.

Relations of Glycerin to Acrylic Acid.

By the action of substances having an attraction for water, such as phosphoric anhydride or sulphuric acid, glycerin is converted into acrolein :—

$$\left\{\begin{array}{l} \mathbf{CH_2Ho} \\ \mathbf{CHHo} \\ \mathbf{CH_2Ho} \end{array}\right. = 2\mathbf{OH_2} + \left\{\begin{array}{l} \mathbf{CMe''H} \\ \mathbf{COH} \end{array}\right. .$$

Glycerin.   Water.   Acrolein.

By the absorption of oxygen, acrolein is transformed into acrylic acid :—

$$\left\{\begin{array}{l} \mathbf{CMe''H} \\ \mathbf{COH} \end{array}\right. + \mathbf{O} = \left\{\begin{array}{l} \mathbf{CMe''H} \\ \mathbf{COHo} \end{array}\right. .$$

Acrolein.   Acrylic acid.

Both these reactions are accomplished simultaneously when glycerin is added to fused potassic hydrate :—

$$\left\{\begin{array}{l} \mathbf{CH_2Ho} \\ \mathbf{CHHo} \\ \mathbf{CH_2Ho} \end{array}\right. + \mathbf{KHo} = \left\{\begin{array}{l} \mathbf{CMe''H} \\ \mathbf{COKo} \end{array}\right. + \mathbf{H_2} + 2\mathbf{OH_2}.$$

Glycerin.   Potassic hydrate.   Potassic acrylate.   Water.

OTHER POLYACID ALCOHOLS.

Erythrite (*Erythroglucin, Erythromannite, Phycite, Pseudorcin*) is a tetracid alcohol, and the acid corresponding to it is tartaric acid. Citric acid may also be considered to be derived from an unknown alcohol of this series. A glance at the formulæ of these alcohols and acids will show their relations :—

$$\left\{\begin{array}{l} \mathbf{CH_2Ho} \\ \mathbf{CHHo} \\ \mathbf{CHHo} \\ \mathbf{CH_2Ho} \end{array}\right. \left\{\begin{array}{l} \mathbf{COHo} \\ \mathbf{CHHo} \\ \mathbf{CHHo} \\ \mathbf{COHo} \end{array}\right. \; \Big| \; \left\{\begin{array}{l} \mathbf{CHHo(CH_2Ho)} \\ \mathbf{CH(CH_2Ho)} \\ \mathbf{CH_2(CH_2Ho)} \end{array}\right. \; \left\{\begin{array}{l} \mathbf{CHHo(COHo)} \\ \mathbf{CH(COHo)} \\ \mathbf{CH_2(COHo)} \end{array}\right. .$$

Erythrite.   Tartaric acid.   Tetracid alcohol (unknown).   Citric acid.

When reduced by hydriodic acid, erythrite yields butylic iodide :—

$$\begin{cases} CH_2Ho \\ CHHo \\ CHHo \\ CH_2Ho \end{cases} + \ 7HI \ = \ 4OH_2 \ + \ \begin{cases} CH_3 \\ CH_2 \\ CH_2 \\ CH_2I \end{cases} + \ 3I_2.$$

Erythrite.   Hydriodic   Water.   Butylic
                  acid.   iodide.

No pentacid alcohol is known; but two acids corresponding to compounds of this class have been obtained; they are apo-sorbic acid and desoxalic acid :—

$$\begin{cases} CH_2Ho \\ CHHo \\ CHHo. \\ CHHo \\ CH_2Ho \end{cases} \quad \begin{cases} COHo \\ CHHo \\ CHHo. \\ CHHo \\ COHo \end{cases} \quad \begin{cases} CH_2Ho \\ CHHo \\ CHo(CH_2Ho). \\ CH_2Ho \end{cases} \quad \begin{cases} COHo \\ CHHo \\ CHo(COHo). \\ COHo \end{cases}$$

Pentacid alcohol   Aposorbic   Unknown   Desoxalic
(unknown).   acid.   alcohol.   acid.

Mannite is a hexacid alcohol. There are two isomeric acids corresponding to this alcohol: these are saccharic and mucic acids :—

$$\begin{cases} CH_2Ho \\ CHHo \\ CHHo \\ CHHo \cdot \\ CHHo \\ CH_2Ho \end{cases} \quad \begin{cases} COHo \\ CHHo \\ CHHo \\ CHHo \cdot \\ CHHo \\ COHo \end{cases}$$

Mannite.   Saccharic
                or mucic acid.

Mannite is closely related to *glucose*, the latter containing two atoms of hydrogen less than the former. Glucose can, in fact, be converted into mannite by the action of nascent hydrogen :—

$$''\begin{cases} CH_2Ho \\ CHHo \\ CHo \\ CHo \\ CHHo \\ CH_2Ho \end{cases} + \ H_2 \ = \ \begin{cases} CH_2Ho \\ CHHo \\ CHHo \\ CHHo \cdot \\ CHHo \\ CH_2Ho \end{cases}$$

Glucose.   Mannite.

CHAPTER XXXV.

THE ETHERS.

THESE compounds are the oxides of the basylous or positive radicals.

Each series of basylous radicals produces a series of ethers; we have thus ethers of the monacid, diacid, and triacid alcohols, of which the following are the general formulæ :—

	Methyl series.	Vinyl series.	Phenyl series.
Ethers of the monacid alcohols	$\left\{ \begin{array}{l} C_nH_{2n+1} \\ O \\ C_nH_{2n+1} \end{array} \right.$	$\left\{ \begin{array}{l} C_nH_{2n-1} \\ O \\ C_nH_{2n-1} \end{array} \right.$	$\left\{ \begin{array}{l} C_nH_{2n-7} \\ O \\ C_nH_{2n-7} \end{array} \right.$.

Ethers of the diacid alcohols... $C_nH_{2n}O.$

Ethers of the triacid alcohols $\left\{ \begin{array}{l} C_nH_{2n}\ \text{-O-}H_{2n}\ C_n \\ C_nH_{2n-1}\text{-O-}H_{2n-1}C_n \\ C_nH_{2n}\ \text{-O-}H_{2n}\ C_n \end{array} \right\}$.

ETHERS OF THE MONACID ALCOHOLS.

THESE bodies are derived from the alcohols by the substitution of the hydrogen of the hydroxyl contained in the latter by a positive monad radical.

METHYL SERIES.

The following list contains some of the ethers of this series :—

				Boiling-points.
Methylic ether	$\left\{ \begin{array}{l} CH_3 \\ O \\ CH_3 \end{array} \right.$		or OMe_2	$-21°.$
Methylic ethylic ether	$\left\{ \begin{array}{l} CH_3 \\ O \\ C_2H_5 \end{array} \right.$ or	$\left\{ \begin{array}{l} CH_3 \\ O \\ CMeH_2 \end{array} \right.$	or $OMeEt$	$+11°.$
Ethylic ether	$\left\{ \begin{array}{l} C_2H_5 \\ O \\ C_2H_5 \end{array} \right.$ or	$\left\{ \begin{array}{l} CMeH_2 \\ O \\ CMeH_2 \end{array} \right.$	or OEt_2	$34°.$
Methylic amylic ether	$\left\{ \begin{array}{l} CH_3 \\ O \\ C_5H \end{array} \right.$ or	$\left\{ \begin{array}{l} CH_3 \\ O \\ CBuH_2 \end{array} \right.$	or $OMeAy$	$92°.$

					Boiling-points.
Ethylic butylic ether.	$\left\{\begin{array}{l}C_2H_5\\O\\C_4H_9\end{array}\right.$	or	$\left\{\begin{array}{l}CMeH_2\\O\\CPrH_2\end{array}\right.$	or $OEtBu$	$80°?$
Ethylic amylic ether.	$\left\{\begin{array}{l}C_2H_5\\O\\C_3H_{11}\end{array}\right.$	or	$\left\{\begin{array}{l}CMeH_2\\O\\CBuH_2\end{array}\right.$	or $OEtAy$	$112°.$
Butylic ether	$\left\{\begin{array}{l}C_4H_9\\O\\C_4H_9\end{array}\right.$	or	$\left\{\begin{array}{l}CPrH_2\\O\\CPrH_2\end{array}\right.$	or OBu_2	$104°.$
Amylic ether	$\left\{\begin{array}{l}C_5H_{11}\\O\\C_5H_{11}\end{array}\right.$	or	$\left\{\begin{array}{l}CBuH_2\\O\\CBuH_2\end{array}\right.$	or OAy_2	$176°.$

Formation.—1. By the action of sulphuric acid upon the $C_nH_{2n+1}Ho$ alcohols. The process may be divided into the two following stages:—

$$\underset{\text{Alcohol.}}{C_nH_{2n+1}Ho} + \underset{\substack{\text{Sulphuric}\\\text{acid.}}}{SO_2Ho_2} = \underset{\text{Sulpho-acid.}}{SO_2Ho(C_nH_{2n+1}O)} + \underset{\text{Water.}}{OH_2}.$$

$$\underset{\text{Sulpho-acid.}}{SO_2Ho(C_nH_{2n+1}O)} + \underset{\text{Alcohol.}}{C_nH_{2n+1}Ho} = \underset{\text{Ether.}}{\left\{\begin{array}{l}C_nH_{2n+1}\\O\\C_nH_{2n+1}\end{array}\right.} + \underset{\substack{\text{Sulphuric}\\\text{acid.}}}{SO_2Ho_2}.$$

2. By converting the $C_nH_{2n+1}Ho$ alcohols into sodium or potassium compounds, and then acting upon the latter with the iodides of the monad alcohol radicals:—

$$\underset{\text{Alcohol.}}{2C_nH_{2n+1}Ho} + Na_2 = \underset{\text{Sodic-alcohol.}}{2C_nH_{2n+1}Nao} + H_2.$$

$$\underset{\text{Sodic alcohol.}}{C_nH_{2n+1}Nao} + \underset{\text{Iodide.}}{C_nH_{2n+1}I} = \underset{\text{Ether.}}{\left\{\begin{array}{l}C_nH_{2n+1}\\O\\C_nH_{2n+1}\end{array}\right.} + \underset{\text{Sodic iodide.}}{NaI}.$$

Reaction.—The ethers can be reconverted into the corresponding alcohols by treating them with sulphuric acid, and then distilling with water the sulpho-acid so produced:—

$$\underset{\text{Ether.}}{\left\{\begin{array}{l}C_nH_{2n+1}\\O\\C_nH_{2n+1}\end{array}\right.} + \underset{\text{Sulphuric acid.}}{2SO_2Ho_2} = \underset{\text{Sulpho-acid.}}{2SO_2Ho(C_nH_{2n+1}O)} + \underset{\text{Water.}}{OH_2}.$$

$$\underset{\text{Sulpho-acid.}}{SO_2Ho(C_nH_{2n+1}O)} + \underset{\text{Water.}}{OH_2} = \underset{\text{Sulphuric acid.}}{SO_2Ho_2} + \underset{\text{Alcohol.}}{C_nH_{2n+1}Ho}.$$

N 5

METHYLIC ETHER, *Methylic Oxide.*

$$\left\{ \begin{array}{l} \mathbf{CH_3} \\ \mathbf{O} \\ \mathbf{CH_3} \end{array} \right. \text{ or } \mathbf{OMe_2}.$$

Molecular weight $= 46$. *Molecular volume* ⊏⊐. 1 *litre of methylic ether vapour weighs* 23 *criths. Boils at* $-21°$.

Preparation.—By heating methylic alcohol with sulphuric acid or boracic anhydride:—

$$\mathbf{CH_3Ho} + \mathbf{SO_2Ho_2} = \mathbf{SO_2Ho(CH_3O)} + \mathbf{OH_2};$$

| Methylic alcohol. | Sulphuric acid. | Sulphomethylic acid. | Water. |

$$\mathbf{SO_2Ho(CH_3O)} + \mathbf{CH_3Ho} = \left\{ \begin{array}{l} \mathbf{CH_3} \\ \mathbf{O} \\ \mathbf{CH_3} \end{array} \right. + \mathbf{SO_2Ho_2}.$$

| Sulphomethylic acid. | Methylic alcohol. | Methylic ether. | Sulphuric acid. |

Reaction.—Methylic ether is acted upon by chlorine under the influence of light, the hydrogen being displaced atom for atom by chlorine. The following compounds are formed:—

$$\left\{ \begin{array}{l} \mathbf{CH_2Cl} \\ \mathbf{O} \\ \mathbf{CH_2Cl} \end{array} \right. ; \quad \left\{ \begin{array}{l} \mathbf{CHCl_2} \\ \mathbf{O} \\ \mathbf{CHCl_2} \end{array} \right. ; \quad \left\{ \begin{array}{l} \mathbf{CCl_3} \\ \mathbf{O} \\ \mathbf{CCl_3} \end{array} \right. .$$

ETHYLIC ETHER, *Ethylic Oxide, Ether, Sulphuric Ether.*

$$\left\{ \begin{array}{l} \mathbf{CMeH_2} \\ \mathbf{O} \\ \mathbf{CMeH_2} \end{array} \right. \text{ or } \mathbf{OEt_2}.$$

Molecular weight $= 74$. *Molecular volume* ⊏⊐. 1 *litre of ether vapour weighs* 37 *criths. Sp. gr.* $= 0\cdot723$. *Fuses at* $-31°$. *Boils at* $35°\cdot6$.

Preparation.—A mixture of equal volumes of sulphuric acid and alcohol is heated to a temperature of from 140° to 145°, and a constant stream of alcohol is allowed to flow into the mixture. Ether and water distil over together. Two reactions take place successively; in the first the alcohol is converted into sulphovinic acid, and in the second the sulphovinic acid is converted by a further quantity of alcohol into sulphuric acid and ether :—

$$\underset{\text{Alcohol.}}{EtHo} + \underset{\text{Sulphuric acid.}}{SO_2Ho_2} = \underset{\text{Sulphovinic acid.}}{SO_2EtoHo} + \underset{\text{Water.}}{OH_2.}$$

$$\underset{\text{Sulphovinic acid.}}{SO_2EtoHo} + \underset{\text{Alcohol.}}{EtHo} = \underset{\text{Ether.}}{OEt_2} + \underset{\text{Sulphuric acid.}}{SO_2Ho_2.}$$

In this manner the same quantity of sulphuric acid can convert an unlimited quantity of alcohol into ether.

The formation of ether is not due to the simple removal of water from two molecules of alcohol by sulphuric acid. This is proved, first, by the sulphuric acid not becoming more dilute, and, secondly, by the fact that if sulphamylic acid be acted upon by ethylic alcohol, the mixed ethylic amylic ether is formed :—

$$\underset{\text{Sulphamylic acid.}}{SO_2AyoHo} + \underset{\substack{\text{Ethylic}\\\text{alcohol.}}}{EtHo} = \underset{\substack{\text{Sulphuric}\\\text{acid.}}}{SO_2Ho_2} + \underset{\substack{\text{Ethylic amylic}\\\text{ether.}}}{OAyEt.}$$

Reactions.—1. Ethylic ether, when mixed with an equal volume of sulphuric acid, produces sulphovinic acid :—

$$\underset{\text{Ethylic ether.}}{OEt_2} + \underset{\text{Sulphuric acid.}}{2SO_2Ho_2} = \underset{\text{Sulphovinic acid.}}{2SO_2EtoHo} + \underset{\text{Water.}}{OH_2.}$$

2. Hot nitric acid converts ethylic ether into carbonic, acetic, and oxalic acids.

3. Exposed to the air, it gradually absorbs oxygen and is transformed into acetic acid :—

$$\underset{\text{Ethylic ether.}}{\left\{\begin{array}{l} CH_3 \\ CH_2 \\ O \\ CH_2 \\ CH_3 \end{array}\right.} + O_4 = 2\underset{\text{Acetic acid.}}{\left\{\begin{array}{l} CH_3 \\ COHo \end{array}\right.} + \underset{\text{Water.}}{OH_2.}$$

ETHYLIC SULPHIDE, *Sulphur Ether*.

$$\left\{ \begin{array}{l} \mathbf{CMeH_2} \\ S \\ \mathbf{CMcH_2} \end{array} \right. \text{ or } \mathbf{SEt_2}.$$

Molecular weight $=90$. *Molecular volume* ☐☐. 1 *litre of ethylic sulphide vapour weighs* 45 *criths.* *Boils at* 73°.

Preparation.—By adding ethylic chloride to potassic sulphide, and distilling :—

$$2\mathbf{CMcH_2Cl} \ + \ \mathbf{SK_2} \ = \ \left\{ \begin{array}{l} \mathbf{CMeH_2} \\ S \\ \mathbf{CMeH_2} \end{array} \right. \ + \ 2\mathrm{KCl}.$$

Ethylic chloride.   Potassic   Ethylic sulphide.   Potassic
          sulphide.   chloride.

Reaction.—Ethylic sulphide combines directly with ethylic iodide, forming

Sulphurous triethylo-iodide .... $\mathbf{SEt_3I}$.

By the action of argentic oxide on this iodide, the corresponding hydrate may be formed :—

$$\mathbf{SEt_3I} \ + \ \mathrm{AgHo} \ = \ \mathbf{SEt_3Ho} \ + \ \mathrm{AgI}.$$

Sulphurous   Argentic   Sulphurous   Argentic
triethylo-   hydrate.   triethylo-   iodide.
iodide.   hydrate.

ETHERS OF THE VINYL AND PHENYL SERIES.

Of the ethers of the vinyl series, allylic ether, $\left\{ \begin{array}{l} \mathbf{C_3H_5} \\ O \\ \mathbf{C_3H_5} \end{array} \right.$, alone is known.

In the phenyl series, phenylic ether, $\left\{ \begin{array}{l} \mathbf{C_6H_5} \\ O \\ \mathbf{C_6H_5} \end{array} \right.$, and toluylic ether, $\left\{ \begin{array}{l} \mathbf{C_7H_7} \\ O \\ \mathbf{C_7H_7} \end{array} \right.$, have been obtained.

ETHERS OF THE DIACID ALCOHOLS.

Of these the three following are known, but the first only has been carefully studied :—

Boiling-points.

Ethylenic oxide, C_2H_4O .... 13°·5.

Propylenic oxide, C_3H_6O .... 35°·0.

Amylenic oxide, $C_5H_{10}O$ .... 95°.

ETHYLENIC OXIDE, *Ethylenic Ether.*

$$C_2H_4O \quad = \quad \left\{ \begin{matrix} CH_2 \\ CH_2 \end{matrix} O. \right.$$

Molecular weight $=44$. *Molecular volume* ⬜. 1 *litre of ethylenic oxide vapour weighs* 22 *criths.* *Boils at* 13°·5.

Preparation.—Ethylenic oxide is obtained from glycol by converting the latter, first into ethylenic chlorhydrate, or chlorhydric glycol, by the action of hydrochloric acid, and subsequently treating the compound thus formed with potassic hydrate :—

$$\left\{ \begin{matrix} CH_2Ho \\ CH_2Ho \end{matrix} \right. + HCl = \left\{ \begin{matrix} CH_2Ho \\ CH_2Cl \end{matrix} \right. + OH_2 :$$

Glycol.   Hydrochloric acid.   Ethylenic chlorhydrate.   Water.

$$\left\{ \begin{matrix} CH_2Ho \\ CH_2Cl \end{matrix} \right. + KHo = \left\{ \begin{matrix} CH_2 \\ CH_2 \end{matrix} O \right. + OH_2 + KCl.$$

Ethylenic chlorhydrate.   Potassic hydrate.   Ethylenic oxide.   · Water.   Potassic chloride.

Isomers.—Ethylenic oxide is isomeric with vinylic alcohol and acetic aldehyde. The nature of this isomerism is seen in the following formulæ :—

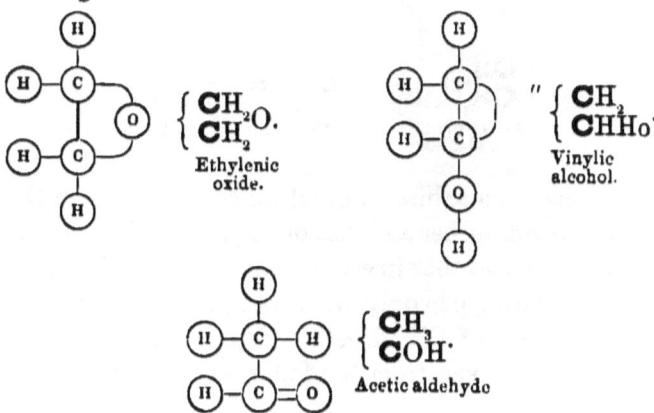

$$\left\{ \begin{matrix} CH_2 \\ CH_2 \end{matrix} O. \right.$$
Ethylenic oxide.

$$" \left\{ \begin{matrix} CH_2 \\ CHHo \end{matrix} \right.$$
Vinylic alcohol.

$$\left\{ \begin{matrix} CH_3 \\ COH \end{matrix} \right.$$
Acetic aldehyde

Reactions.—1. Ethylenic oxide unites with nascent hydrogen, forming alcohol :—

$$\left\{ \begin{array}{l} CH_2 \\ CH_2 \end{array} O \right. + H_2 = \left\{ \begin{array}{l} CH_3 \\ CH_2Ho \end{array} \right.$$

Ethylenic   Alcohol.
oxide.

2. It also unites with oxygen, forming glycollic acid :—

$$\left\{ \begin{array}{l} CH_2 \\ CH_2 \end{array} O \right. + O_2 = \left\{ \begin{array}{l} CH_2Ho \\ COHo \end{array} \right.$$

Ethylenic   Glycollic
oxide.   acid.

3. It is a basic substance, and unites directly with acids :—

$$\left\{ \begin{array}{l} CH_2 \\ CH_2 \end{array} O \right. + HCl = \left\{ \begin{array}{l} CH_2Cl \\ CH_2Ho \end{array} \right.$$

Ethylenic   Hydro-   Ethylenic
oxide.   chloric   chlorhydrate
            acid.   or chlorhydric
                   glycol.

4. Ethylenic oxide precipitates many oxides from solutions of their salts, such as ferric oxide, aluminic oxide, cupric oxide, magnesic oxide, &c. :—

$$2\left\{ \begin{array}{l} CH_2 \\ CH_2 \end{array} O \right. + MgCl_2 + 2OH_2 = 2\left\{ \begin{array}{l} CH_2Cl \\ CH_2Ho \end{array} \right. + MgHo_2.$$

Ethylenic   Magnesic   Water.   Ethylenic   Magnesic
oxide.   chloride.   chlorhydrate. hydrate.

5. It also combines directly with water, reproducing glycol :—

$$\left\{ \begin{array}{l} CH_2 \\ CH_2 \end{array} O \right. + OH_2 = \left\{ \begin{array}{l} CH_2Ho \\ CH_2Ho \end{array} \right.$$

Ethylenic   Water.   Glycol.
oxide.

These reactions exhibit a great difference between the behaviour of ethylenic ether and that of ethylic ether. This difference arises from the fact that in ethylic ether the ethyl atoms are held together by the oxygen only, whereas in ethylenic ether the linking of the atoms of CH_2 does not depend on the oxygen atom alone, as will be seen from the following formulæ :—

Ethylic ether......

$$= \begin{cases} \mathbf{C_2H_5} \\ \mathbf{O} \\ \mathbf{C_2H_5} \end{cases}.$$

Ethylenic ether...

$$= \begin{cases} \mathbf{CH_2} \\ \mathbf{CH_2} \end{cases} O.$$

On account of this peculiarity of constitution, ethylenic oxide can combine directly with many substances without the disruption of its molecule,—a property which obviously cannot be possessed by ethylic ether.

ETHERS OF THE TRIACID ALCOHOLS.

Of these only one is known, viz. :—

GLYCYLIC ETHER, *Glycylic Oxide.*

$$= \begin{cases} \mathbf{CH_2\text{-}O\text{-}H_2C} \\ \mathbf{CH\text{-}O\text{-}H\,C} \\ \mathbf{CH_2\text{-}O\text{-}H_2C} \end{cases}.$$

Preparation.—By the action of potassic hydrate on so-called iodhydrin :—

$$\begin{cases} \mathbf{CH_2} \\ \mathbf{CH} \\ \mathbf{CH_2} \\ \mathbf{O} \\ \mathbf{CH_2} \\ \mathbf{CHI} \\ \mathbf{CH_2Ho} \end{cases}O \; + \; \mathbf{KHo} \; = \; \mathbf{KI} \; + \; \mathbf{OH_2} \; + \; \begin{cases} \mathbf{CH_2\text{-}O\text{-}H_2C} \\ \mathbf{CH\text{-}O\text{-}H\,C} \\ \mathbf{CH_2\text{-}O\text{-}H_2C} \end{cases}.$$

Iodhydrin.   Potassic   Potassic   Water.   Glycylic ether.
              hydrate.   iodide.

CHAPTER XXXVI.

THE HALOID ETHERS.

EACH series of radicals forms its own series of haloid ethers.

These ethers are produced by the substitution of hydroxyl in the alcohols by chlorine, bromine, iodine, fluorine, or cyanogen.

Haloid Ethers of the Monad Positive Radicals.

As these radicals can only unite with one atom of hydroxyl, they can only form one haloid ether. Each series of radicals therefore forms one series of haloid ethers :—

I. Haloid ethers of the form $C_nH_{2n+1}Cl$.

II. „　„　„　„　$C_nH_{2n-1}Cl$.

III. „　„　„　„　$C_nH_{2n-7}Cl$.

The following will serve as examples of the three series :—

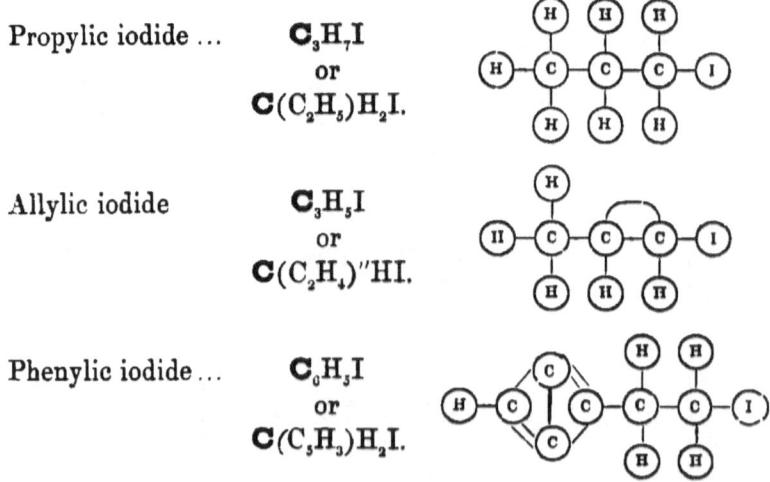

Propylic iodide ...　C_3H_7I
or
$C(C_2H_5)H_2I$.

Allylic iodide　C_3H_5I
or
$C(C_2H_4)''HI$.

Phenylic iodide ...　C_6H_5I
or
$C(C_5H_3)H_2I$.

Haloid Ethers of the Dyad Positive Radicals.

As the diacid alcohols contain two atoms of hydroxyl, it follows that there are two classes of haloid ethers derivable from them. The first is formed by the substitution of one of the hydroxyl atoms by chlorine, bromine, &c., and the second by the like displacement of both atoms of hydroxyl :—

I. Haloid ethers of the form $C_nH_{2n}HoCl$.
II.   ,,   ,,   ,,   ,, $C_nH_{2n}Cl_2$.

The following examples will suffice to illustrate the constitution of both these classes of haloid ethers :—

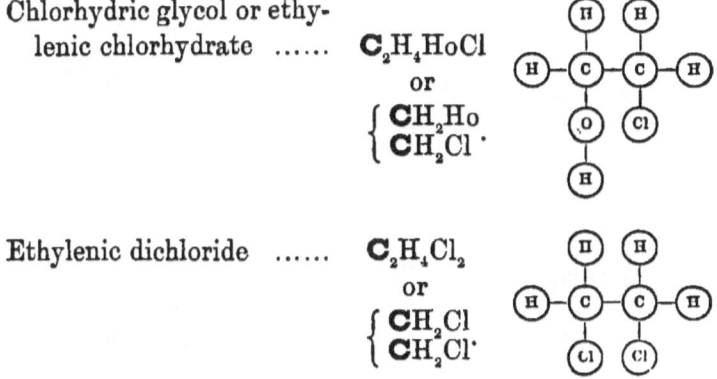

Chlorhydric glycol or ethylenic chlorhydrate ...... C_2H_4HoCl
or
$\begin{cases} CH_2Ho \\ CH_2Cl \end{cases}$

Ethylenic dichloride ...... $C_2H_4Cl_2$
or
$\begin{cases} CH_2Cl \\ CH_2Cl \end{cases}$

Haloid Ethers of the Triad Positive Radicals.

There are three classes of haloid ethers which are derived from the triacid alcohols, by the successive substitution of the three atoms of hydroxyl contained in these alcohols by chlorine, bromine, &c.,—and a fourth class, which stands intermediately between the ethers and the haloid ethers, and which is formed by the substitution of one of the atoms of hydroxyl in the alcohol by a monad negative radical, such as chlorine, bromine, or cyanogen, and the remaining two atoms of hydroxyl by the dyad atom of oxygen:—

    I.   Haloid ethers of the form $C_n^{\cdot}H_{2n-1}Ho_2Cl.$
    II.   „   „   „ $C_nH_{2n-1}HoCl_2.$
    III.   „   „   „ $C_nH_{2n-1}Cl_3.$
    IV.   „   „   „ $C_nH_{2n-1}OCl.$

The following are examples of each of these classes:—

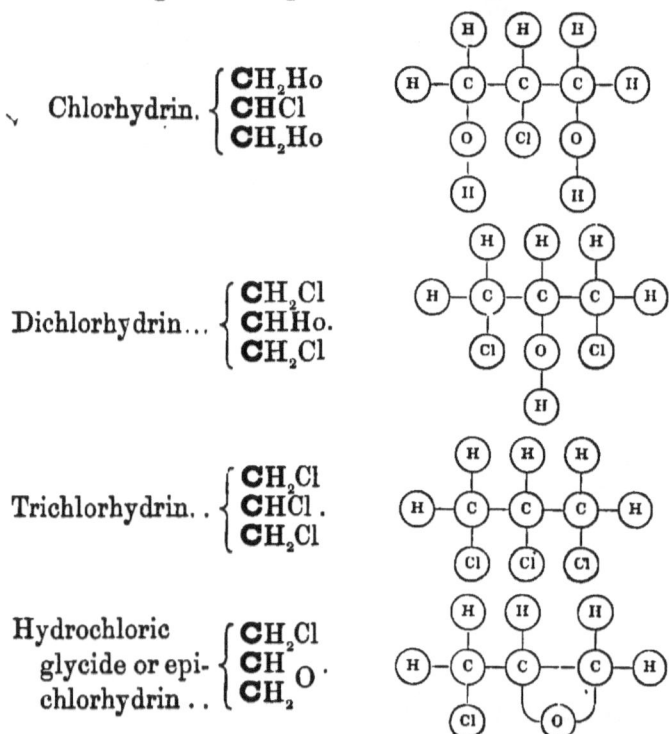

Chlorhydrin. $\begin{cases} CH_2Ho \\ CHCl \\ CH_2Ho \end{cases}$

Dichlorhydrin... $\begin{cases} CH_2Cl \\ CHHo. \\ CH_2Cl \end{cases}$

Trichlorhydrin.. $\begin{cases} CH_2Cl \\ CHCl. \\ CH_2Cl \end{cases}$

Hydrochloric glycide or epi-chlorhydrin .. $\begin{cases} CH_2Cl \\ CH \\ CH_2 \end{cases} O.$

HALOID ETHERS OF THE MONAD POSITIVE RADICALS.

Preparation.—These ethers are produced by the following general reactions:—

1. By the action of the hydracids upon the alcohols:—

$$C_nH_{2n+1}Ho \quad + \quad HCl \quad = \quad C_nH_{2n+1}Cl \quad + \quad OH_2.$$

Alcohol.   Hydrochloric acid.   Haloid ether.   Water.

2. By the action of phosphorous trichloride on the alcohols :—

$$3C_nH_{2n+1}Ho \quad + \quad PCl_3 \quad = \quad 3C_nH_{2n+1}Cl \quad + \quad POHHo_2.$$

Alcohol.   Phosphorous trichloride.   Haloid ether.   Phosphorous acid.

3. By the action of chlorine on the hydrides of the radicals :—

$$C_nH_{2n+1}H \quad + \quad Cl_2 \quad = \quad C_nH_{2n+1}Cl \quad + \quad HCl.$$

Hydride.   Haloid ether.   Hydrochloric acid.

It is obvious that in these reactions bromine and iodine may be used instead of chlorine.

These reactions apply equally to the C_nH_{2n-1} and C_nH_{2n-7} series.

For the preparation of the cyanides of the radicals, two special reactions are employed.

1. The distillation in the dry state of a mixture of the potassic sulphate of the radical with potassic cyanide :—

$$SO_2Ko(C_nH_{2n+1}O) \quad + \quad KCy \quad = \quad SO_2Ko_2 \quad + \quad C_nH_{2n+1}Cy.$$

Potassic sulphate of the radical.   Potassic cyanide.   Potassic sulphate.   Cyanide.

2. The fatty acids are converted into ammonium salts and distilled with phosphoric anhydride, when the cyanides of the positive radicals which they contain are produced :—

$$\left\{ \begin{matrix} C_nH_{2n+1} \\ CO(N'H_4O) \end{matrix} \right. \quad + \quad 2P_2O_5 \quad = \quad \left\{ \begin{matrix} C_nH_{2n+1} \\ CN''' \end{matrix} \right. \quad + \quad 4PO_2Ho.$$

Ammonium salt.   Phosphoric anhydride.   Cyanide.   Metaphosphoric acid.

Reactions.—1. Treated with alcoholic solution of potash, all the haloid ethers of the C_nH_{2n+1} series, except the cyanides, are reconverted into alcohols :—

$$C_nH_{2n+1}Cl \quad + \quad KHo \quad = \quad C_nH_{2n+1}Ho \quad + \quad KCl.$$

Haloid ether.   Potassic hydrate.   Alcohol.   Potassic chloride.

2. The cyanides under similar circumstances are converted into potassic salts of the acids which contain the positive radical of the cyanide:—

$$\left\{ \begin{array}{l} C_nH_{2n+1} \\ \mathbf{CN'''} \end{array} \right. + \quad \mathbf{KHo} \quad + \quad \mathbf{OH_2} \quad = \quad \left\{ \begin{array}{l} C_nH_{2n+1} \\ \mathbf{COKo} \end{array} \right. + \quad \mathbf{NH_3}.$$

Cyanide.   Potassic   Water.   Potassic   Ammonia.
hydrate.   salt.

3. When the iodides are digested with zinc or magnesium, the radicals are either liberated or unite with the metal:—

$$2C_nH_{2n+} I \quad + \quad 2Zn \quad = \quad \mathbf{Zn}(C_nH_{2n+1})_2 \quad + \quad \mathbf{ZnI_2},$$

Iodide.   Organo-zinc   Zincic
compound.   iodide.

or

$$2C_nH_{2n+1}I \quad + \quad Zn \quad = \quad \left\{ \begin{array}{l} C_nH_{2n+1} \\ C_nH_{2n+1} \end{array} \right. + \quad \mathbf{ZnI_2}.$$

Iodide.   Free   Zincic
radical.   iodide.

4. When the iodides are submitted to the action of sodic ethylate, a mixed ether (or a simple ether if $n=2$) is formed:—

$$\mathbf{C_2H_5Nao} \quad + \quad C_nH_{2n+1}I \quad = \quad \left\{ \begin{array}{l} C_nH_{2n+1} \\ O \\ \mathbf{C_2H_5} \end{array} \right. + \quad \mathbf{NaI}.$$

Sodic   Iodide.   Ether.
ethylate.

5. The haloid ethers are the representatives of the hydracids of mineral chemistry, and unite directly with ammonia, producing salts which, when treated with potassic hydrate, yield compound ammonias containing the basylous radical of the haloid ether in the place of one atom of hydrogen:—

$$\mathbf{NH_3} \quad + \quad \mathbf{EtI} \quad = \quad \mathbf{NH_3EtI}.$$

Ammonia.   Ethylic   Ethylammonic
iodide.   iodide.

$$\mathbf{NH_3EtI} \quad + \quad \mathbf{KHo} \quad = \quad \mathbf{NEtH_2} \quad + \quad \mathbf{KI} \quad + \quad \mathbf{OH_2}.$$

Ethylammonic   Potassic   Ethylamine.   Potassic   Water.
iodide.   hydrate.   iodide.

METHYLIC CHLORIDE.

CH_3Cl or MeCl.

Molecular weight $=50.5$. *Molecular volume* ☐☐. 1 *litre of methylic chloride vapour weighs* 25.25 *criths. Boils at* $-21°$.

Preparation.—By heating together sodic chloride, methylic alcohol, and sulphuric acid :—

$$SO_2Ho_2 \quad + \quad MeHo \quad = \quad SO_2MeoHo \quad + \quad OH_2.$$

Sulphuric acid.   Methylic alcohol.   Sulphomethylic acid.   Water.

$$SO_2MeoHo \quad + \quad NaCl \quad = \quad MeCl \quad + \quad SO_2HoNao.$$

Sulphomethylic acid.   Sodic chloride.   Methylic chloride.   Hydric sodic sulphate.

Reaction.—By the action of chlorine, methylic chloride produces three substitution derivatives :—

			Boiling-points.
Monochlorinated methylic chloride, CH_2Cl_2.			31°
Dichlorinated ,, ,, $CHCl_3$.			60°·8
Trichlorinated ,, ,, CCl_4.			78°

CHLOROFORM, *Dichlorinated Methylic Chloride.*

$CHCl_3$.

Molecular weight $=119.5$. *Molecular volume* ☐☐. 1 *litre of chloroform vapour weighs* 59.75 *criths. Sp. gr.* 1.48. *Boils at* $60°·8$.

Preparation.—This compound is manufactured in large quantities by heating alcohol with a solution of calcic chlorohypochlorite (*chloride of lime*). It may also be made by treating methylic alcohol in the same manner. For the reaction see p. 250.

Reactions.—1. Chloroform is transformed into potassic formate by boiling with alcoholic potash :—

$$CHCl_3 \quad + \quad 4KHo \quad = \quad CHOKo \quad + \quad 3KCl \quad + \quad 2OH_2.$$

Chloroform.   Potassic hydrate.   Potassic formate.   Potassic chloride.   Water.

2. When acted upon by chlorine in the presence of sunlight, the hydrogen is displaced by chlorine, and carbonic tetrachloride (CCl_4) formed.

ETHYLIC CHLORIDE.

C_2H_5Cl or EtCl.

Molecular weight $=64\cdot5$. *Molecular volume* ☐. 1 *litre of ethylic chloride vapour weighs* $32\cdot25$ *criths.* *Sp. gr.* $0\cdot874$. *Boils at* $11°\cdot5$.

Preparation.—Ethylic alcohol is saturated with hydrochloric acid, and digested in sealed tubes at 100° for one or two hours, when the mixture separates into two layers, the upper one being the ethylic chloride :—

$$\text{EtHo} + \text{HCl} = \text{EtCl} + \text{OH}_2.$$

Alcohol.　　　Hydrochloric　　Ethylic　　Water.
　　　　　　　　acid.　　　　chloride.

ETHYLIC IODIDE.

C_2H_5I or EtI.

Molecular weight $=156$. *Molecular volume* ☐. 1 *litre of ethylic iodide vapour weighs* 78 *criths.* *Sp. gr.* $1\cdot9464$. *Boils at* $72°\cdot2$.

Preparation.—By placing in a retort two parts by weight of alcohol and one of amorphous phosphorus, and then introducing five parts of iodine and distilling in a water-bath :—

$$3C_2H_5Ho + P + I_3 = 3C_2H_5I + POHHo_2.$$

Alcohol.　　　　　　　　　　Ethylic　　Phosphorous
　　　　　　　　　　　　　iodide.　　　acid.

Reaction.—Ethylic iodide, when heated with water in a sealed tube, produces ether and hydriodic acid :—

$$2C_2H_5I \ + \ OH_2 \ = \ \begin{cases} C_2H_5 \\ O \\ C_2H_5 \end{cases} + \ 2HI.$$

Ethylic   Water.   Ether.   Hydriodic
iodide.   acid.

The methylic and amylic iodides are similar liquids, and obtained by analogous processes ; the methylic iodide, CH_3I, has a sp. gr. 2·237, and boils at 42° C.   Amylic iodide, $C_5H_{11}I$, has a sp. gr. 1·511, and boils at 146°.

The haloid compounds of the allylic and phenylic series are of comparatively little importance.

HALOID ETHERS OF THE DYAD POSITIVE RADICALS.

I. *Haloid ethers of the form* $C_nH_{2n}HoCl$.

Preparation.—These ethers are prepared by the action of the hydracids on the glycols.   The following will serve as examples of this class :—

Ethylenic chlorhydrate or chlorhydric glycol $\begin{cases} CH_2Ho \\ CH_2Cl \end{cases}$.

Ethylenic iodhydrate or iodhydric glycol ... $\begin{cases} CH_2Ho \\ CH_2I \end{cases}$.

Treated with potassic hydrate, both these bodies give ethylenic oxide, as previously described (p. 277).

II. *Haloid ethers of the form* $C_nH_{2n}Cl_2$.

Preparation.—These ethers are generally formed by the direct union of the dyad radicals with the chlorous elements.

The following list comprises the chief members of this class:—

		Boiling-points.
Methylenic chloride.........	CH_2Cl_2	40°
„ iodide	CH_2I_2	181°
Ethylenic chloride	$C_2H_4Cl_2$	85°

			Boiling-points.
Ethylenic bromide	....	$C_2H_4Br_2$	...... 129°
„ iodide	....	$C_2H_4I_2$	...... ——
Propylenic chloride	....	$C_3H_6Cl_2$	...... 103°
„ bromide	....	$C_3H_6Br_2$	...... 144°
„ iodide	....	$C_3H_6I_2$	...... ——
Butylenic chloride	....	$C_4H_8Cl_2$	...... 127°
„ bromide	....	$C_4H_8Br_2$	...... 160°
Amylenic chloride	....	$C_5H_{10}Cl_2$	...... ——
„ bromide	....	$C_5H_{10}Br_2$	...... 175°

By the action of potassium, sodium, or zinc, the radicals are again liberated, except in the case of the methylene compounds. The bromides are the most important members of the series.

ETHYLENIC BROMIDE.

$$C_2H_4Br_2 \quad \text{or} \quad \begin{cases} CH_2Br \\ CH_2Br \end{cases} \quad \text{or} \quad Et''Br_2.$$

Molecular weight =188. *Molecular volume* ☐☐. *1 litre of ethylenic bromide vapour weighs 94 criths. Sp. gr. 2·16. Fuses at 9°. Boils at 129°.*

Preparation.—By agitating bromine and water with ethylene.

Reactions.—1. Boiled with alcoholic potash it yields bromethylene or vinylic bromide :—

$$C_2H_4Br_2 \;+\; KHo \;=\; C_2H_3Br \;+\; KBr \;+\; OH_2.$$

Ethylenic bromide.   Potassic hydrate.   Vinylic bromide or bromethylene.   Potassic bromide.   Water.

2. Heated with an alcoholic solution of potassic acetate, it yields monacetic glycol.

$$\begin{cases} CH_2Br \\ CH_2Br \end{cases} + \; 2CMeOKo \;+\; OH_2 \;=\; \begin{cases} CH_2\text{-}O\text{-}CMeO \\ CH_2Ho \end{cases}$$

Ethylenic bromide.   Potassic acetate.   Water.   Monacetic glycol.

$$+ \; CMeOHo \;+\; 2KBr.$$

Acetic acid.   Potassic bromide.

ETHYLENIC CYANIDE.

$$\begin{cases} \mathbf{CH_2Cy} \\ \mathbf{CH_2Cy} \end{cases} = \mathbf{C_2H_4Cy_2}.$$

Fuses at 37°.

Preparation.—By heating ethylenic bromide with potassic cyanide to 100° for sixteen hours :—

$$\begin{cases} \mathbf{CH_2Br} \\ \mathbf{CH_2Br} \end{cases} + 2\mathbf{CN'''K} = \begin{cases} \mathbf{CH_2(CN''')} \\ \mathbf{CH_2(CN''')} \end{cases} + 2\mathbf{KBr}.$$

| Ethylenic bromide. | Potassic cyanide. | Ethylenic cyanide. | Potassic bromide. |

Reaction.—When boiled with alcoholic potash, ethylenic cyanide yields potassic succinate :—

$$\begin{cases} \mathbf{CH_2(CN''')} \\ \mathbf{CH_2(CN''')} \end{cases} + 2\mathbf{KHo} + 2\mathbf{OH_2} = \begin{cases} \mathbf{CH_2(COKo)} \\ \mathbf{CH_2(COKo)} \end{cases} + 2\mathbf{NH_3}.$$

| Ethylenic cyanide. | Potassic hydrate. | Water. | Potassic succinate. | Ammonia. |

HALOID ETHERS OF THE TRIAD POSITIVE RADICALS.

I. of the form $C_nH_{2n-1}Ho_2Cl$.

Boiling-point.

Chlorhydrin ... $\begin{cases} \mathbf{CH_2Ho} \\ \mathbf{CHCl} \\ \mathbf{CH_2Ho} \end{cases}$ .... 227°.

Bromhydrin ... $\begin{cases} \mathbf{CH_2Ho} \\ \mathbf{CHBr} \\ \mathbf{CH_2Ho} \end{cases}$ .... 180° *in vacuo.*

II. Of the form $C_nH_{2n-1}HoCl_2$.

Dichlorhydrin... $\begin{cases} \mathbf{CH_2Cl} \\ \mathbf{CHHo} \\ \mathbf{CH_2Cl} \end{cases}$ .... 180°.

III. Of the form $C_nH_{2n-1}Cl_3$.

Trichlorhydrin.. $\begin{cases} \mathbf{CH_2Cl} \\ \mathbf{CHCl} \\ \mathbf{CH_2Cl} \end{cases}$ .... 155°.

o

IV. Of the form $C_nH_{2n-1}OCl$.

Boiling-
point.

Hydrochloric
glycide or epi-
chlorhydrin ...
$\left\{\begin{array}{l} CH_2O \\ CH^2O \\ CH_2Cl \end{array}\right.$ .... 118°.

Preparation.—The ethers of the first three forms are obtained by the action of the hydracids upon glycerin; whilst those of the fourth are produced by the action of alkalies upon the second form of compounds.

CHAPTER XXXVII.

THE ALDEHYDES.

THESE compounds are intermediate between the alcohols and the acids. They are formed from alcohols by the abstraction of hydrogen; hence the name, which is an abbreviation of *alcohol dehydrogenatum*.

Three series of aldehydes are known, corresponding to the three series of monacid alcohols, viz. :—

A. Aldehydes derived from $C_nH_{2n+1}Ho$ alcohols.
B.   „   „   „ $C_nH_{2n-1}Ho$ „
C.   „   „   „ $C_nH_{2n-7}Ho$ „

Preparation.—1. The aldehydes are formed by the oxidation of the alcohols; ethylic alcohol, for instance, yields acetic aldehyde :—

$$\left\{\begin{array}{l} CH_3 \\ CH_2Ho \end{array}\right. + O = \left\{\begin{array}{l} CH_3 \\ COH \end{array}\right. + OH_2.$$

Ethylic   Acetic   Water.
alcohol.   aldehyde*.

2. Aldehydes are also formed by distilling a mixture of equiva-

---

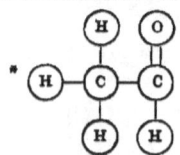

lent quantities of the potassic salt of a fatty acid and of potassic formate:—

$$\left\{\begin{array}{l} CH_3 \\ COKo \end{array}\right. + \left\{\begin{array}{l} H \\ COKo \end{array}\right. = \left\{\begin{array}{l} CH_3 \\ COH \end{array}\right. + COKo_2.$$

Potassic acetate.   Potassic formate.   Acetic aldehyde.   Potassic carbonate.

This is an important reaction, as by its means the series of fatty acids can be ascended; for the aldehyde may next be transformed into an alcohol by nascent hydrogen, then the alcohol converted into a cyanide, which by treatment with potassic hydrate gives the potassic salt of the next higher acid. Thus:—

$$\left\{\begin{array}{l} CH_3 \\ COH \end{array}\right. + H_2 = \left\{\begin{array}{l} CH_3 \\ CH_2Ho \end{array}\right. ;$$

Acetic aldehyde.   Ethylic alcohol.

$$\left\{\begin{array}{l} CH_3 \\ CH_2Ho \end{array}\right. + SO_2Ho_2 = SO_2HoEto + OH_2 ;$$

Ethylic alcohol.   Sulphuric acid.   Sulphovinic acid.   Water.

$$SO_2HoEto + CN'''K = SO_2KoHo + \left\{\begin{array}{l} CMeH_2 \\ CN''' \end{array}\right. ;$$

Sulphovinic acid.   Potassic cyanide.   Hydric potassic sulphate.   Ethylic cyanide.

$$\left\{\begin{array}{l} CMeH_2 \\ CN''' \end{array}\right. + KHo + OH_2 = \left\{\begin{array}{l} CMeH_2 \\ COKo_2 \end{array}\right. + NH_3.$$

Ethylic cyanide.   Potassic hydrate.   Water.   Potassic propionate.   Ammonia.

Starting again with potassic propionate, instead of potassic acetate, the same series of reactions can be performed, resulting in potassic butyrate, and so on.

Reactions.—1. By direct absorption of oxygen, the aldehydes are transformed into the corresponding acids:—

$$\left\{\begin{array}{l} C_nH_{2n+1} \\ COH \end{array}\right. + O = \left\{\begin{array}{l} C_nH_{2n+1} \\ COHo \end{array}\right. .$$

Aldehyde.   Acid.

2. Also heated with ammoniacal solution of argentic oxide, the aldehydes are converted into acids, metallic silver being deposited:—

$$\left\{\begin{array}{l} C_nH_{2n+1} \\ COH \end{array}\right. + OAg_2 = \left\{\begin{array}{l} C_nH_{2n+1} \\ COHo \end{array}\right. + Ag_2.$$

Aldehyde.   Argentic oxide.   Acid.

3. When heated with potassic hydrate, the aldehydes yield the potassic salts of the corresponding acids, with evolution of hydrogen:—

$$\left\{ \begin{array}{l} C_nH_{2n+1} \\ \mathbf{C}OH \end{array} \right. + \mathbf{K}Ho = \left\{ \begin{array}{l} C_nH_{2n+1} \\ \mathbf{C}OKo \end{array} \right. + \mathbf{H_2}.$$

Aldehyde.   Potassic hydrate.   Potassic salt.

4. Treated with nascent hydrogen, they are converted into the corresponding alcohols :—

$$\left\{ \begin{array}{l} C_nH_{2n+1} \\ \mathbf{C}OH \end{array} \right. + \mathbf{H_2} = \left\{ \begin{array}{l} C_nH_{2n+1} \\ \mathbf{C}H_2Ho \end{array} \right. .$$

Aldehyde.   Alcohol.

5. Most aldehydes combine directly with ammonia, forming crystalline compounds :—

$$\left\{ \begin{array}{l} C_nH_{2n+1} \\ \mathbf{C}OH \end{array} \right. + \mathbf{NH_3} = \left\{ \begin{array}{l} C_nH_{2n+1} \\ \mathbf{C}O(N'H_4) \end{array} \right. .$$

Aldehyde.   Ammonia.   Ammonium compound.

6. Aldehydes also combine with the alkaline hydric sulphites, producing crystalline compounds :—

$$\left\{ \begin{array}{l} C_nH_{2n+1} \\ \mathbf{C}OH \end{array} \right. + \mathbf{SOKoHo} = \mathbf{SOKoHo} , \left\{ \begin{array}{l} C_nH_{2n+1} \\ \mathbf{C}OH \end{array} \right. .$$

Aldehyde.   Hydric potassic sulphite.

A. ALDEHYDES DERIVED FROM THE $C_nH_{2n+1}Ho$ SERIES OF ALCOHOLS.

The following are known :—

			Fusing-point.	Boiling-point.
Acetic aldehyde	$\left\{ \begin{array}{l} CH_3 \\ \mathbf{C}OH \end{array} \right.$		—	21°.
Propionic aldehyde	$\left\{ \begin{array}{l} CMeH_2 \\ \mathbf{C}OH \end{array} \right.$ or $\left\{ \begin{array}{l} C(CH_3)H_2 \\ \mathbf{C}OH \end{array} \right.$		—	(55°–65°).
Butyric aldehyde	$\left\{ \begin{array}{l} CEtH_2 \\ \mathbf{C}OH \end{array} \right.$ or $\left\{ \begin{array}{l} C(C_2H_5)H_2 \\ \mathbf{C}OH \end{array} \right.$		—	(68°–75°).
Valeric aldehyde	$\left\{ \begin{array}{l} CPrH_2 \\ \mathbf{C}OH \end{array} \right.$ or $\left\{ \begin{array}{l} C(C_3H_7)H_2 \\ \mathbf{C}OH \end{array} \right.$		—	93°.
Œnanthic aldehyde	$\left\{ \begin{array}{l} CAyH_2 \\ \mathbf{C}OH \end{array} \right.$ or $\left\{ \begin{array}{l} C(C_5H_{11})H_2 \\ \mathbf{C}OH \end{array} \right.$	below −12°.	152°.
Capric aldehyde	$\left\{ \begin{array}{l} C(C_8H_{17})H_2 \\ \mathbf{C}OH \end{array} \right.$		− 2°.	228° ?
Euodic aldehyde	$\left\{ \begin{array}{l} C(C_8H_{16})H_2 \\ \mathbf{C}OH \end{array} \right.$		+ 7°.	213°.
Lauric aldehyde	$\left\{ \begin{array}{l} C(C_{10}H_{21})H_2 \\ \mathbf{C}OH \end{array} \right.$		—	232°.
Palmitic aldehyde	$\left\{ \begin{array}{l} C(C_{15}H_{29})H_2 \\ \mathbf{C}OH \end{array} \right.$		52°.	

ACETIC ALDEHYDE, *Aldehyde.*

$$\left\{ \begin{array}{l} CH_3 \\ COH \end{array} \right.$$

Molecular weight $=44$. *Molecular volume* ☐. 1 *litre of aldehyde vapour weighs* 22 *criths. Sp. gr.* $=0.79$. *Boils at* $21°.8$.

Preparation.—1. By oxidizing alcohol with chromic acid, chlorine water, or manganic oxide and sulphuric acid:—

$$\left\{ \begin{array}{l} CH_3 \\ CH_2Ho \end{array} \right. + \ O \ = \ \left\{ \begin{array}{l} CH_3 \\ COH \end{array} \right. + \ OH_2.$$

Ethylic alcohol.   Acetic aldehyde.   Water.

2. By oxidation, casein, fibrin, and albumen also yield aldehyde.

3. Aldehyde is formed when the vapour of alcohol or ether is passed through a tube heated to dull redness.

Reactions.—1. It gradually absorbs oxygen from the air, forming acetic acid, into which it is also readily converted by oxidizing agents:—

$$\left\{ \begin{array}{l} CH_3 \\ COH \end{array} \right. + \ .O \ = \ \left\{ \begin{array}{l} CH_3 \\ COHo \end{array} \right.$$

Acetic aldehyde.   Acetic acid.

2. It reduces silver salts, depositing lustrous metallic silver on the sides of the vessel.

3. When submitted to the action of potassium, one atom of hydrogen is substituted by an atom of the metal, the compound

$$\left\{ \begin{array}{l} CH_3 \\ COK \end{array} \right.$$

being formed.

4. Hydrocyanic acid transforms aldehyde into alanin:—

$$\left\{ \begin{array}{l} CH_3 \\ COH \end{array} \right. + \ CN'''H \ + \ OH_2 \ = \ \left\{ \begin{array}{l} CMeHHo \\ CO(N'''H_2) \end{array} \right.$$

Acetic aldehyde.   Hydrocyanic acid.   Water.   Alanin.

By the action of nitrous anhydride, alanin is converted into lactic acid :—

$$2\left\{\begin{array}{l}\mathbf{CMeHHo}\\\mathbf{CO(N'''H_2)}\end{array}\right. + \mathbf{N_2O_3} = 2\left\{\begin{array}{l}\mathbf{CMeHHo}\\\mathbf{COHo}\end{array}\right. + 2\mathbf{N_2} + \mathbf{OH_2}.$$

Alanin.   Nitrous anhydride.   Lactic acid.   Water.

There are three isomeric modifications of aldehyde :—

Metaldehyde, crystalline, subliming at 120°.

Paraldehyde, liquid, boiling at 125°.

Elaldehyde, crystalline, fusing at 2°, boiling at 94°.

B. *ALDEHYDES DERIVED FROM THE* $C_nH_{2n-1}Ho$ *ALCOHOLS.*

ACROLEIN.   *Acrylic Aldehyde.*

$$\left\{\begin{array}{l}\mathbf{CMe''H}\\\mathbf{COH}\end{array}\right. \cdot$$

Molecular weight =56. *Molecular volume* ⊡. 1 *litre of acrolein vapour weighs* 28 *criths.*   *Boils at* 52°·4.

Preparation.—1. By the action of phosphoric anhydride or of sulphuric acid on glycerin :—

$$\left\{\begin{array}{l}\mathbf{CH_2Ho}\\\mathbf{CHHo}\\\mathbf{CH_2Ho}\end{array}\right. = 2\mathbf{OH_2} + \left\{\begin{array}{l}\mathbf{CMe''H}\\\mathbf{COH}\end{array}\right. \cdot$$

Glycerin.   Water.   Acrolein.

2. By the oxidation of allylic alcohol :—

$$\left\{\begin{array}{l}\mathbf{CMe''H}\\\mathbf{CH_2Ho}\end{array}\right. + \mathbf{O} = \left\{\begin{array}{l}\mathbf{CMe''H}\\\mathbf{COH}\end{array}\right. + \mathbf{OH_2}.$$

Allylic alcohol.   Acrolein.   Water.

3. By the action of heat on the product of the union of acetone with bromine :—

$$\left\{\begin{array}{l}\mathbf{COMe}\\\mathbf{CH_3}\end{array}\right. + \mathbf{Br_2} = \left\{\begin{array}{l}\mathbf{CMeBr_2}\\\mathbf{CH_2Ho}\end{array}\right. :$$

Acetone.

$$\left\{\begin{array}{l}\mathbf{CMeBr_2}\\\mathbf{CH_2Ho}\end{array}\right. = \left\{\begin{array}{l}\mathbf{CMe''H}\\\mathbf{COH}\end{array}\right. + 2\mathbf{HBr}.$$

Acrolein.   Hydrobromic acid.

Reaction.—By oxidation, acrolein yields acrylic acid:—

$$\left\{ \begin{array}{l} \mathbf{C}\mathrm{Me''H} \\ \mathbf{C}\mathrm{OH} \end{array} \right. + \mathrm{O} = \left\{ \begin{array}{l} \mathbf{C}\mathrm{Me''H} \\ \mathbf{C}\mathrm{OHo} \end{array} \right.$$
$$\quad\text{Acrolein.} \qquad\qquad\qquad \text{Acrylic acid.}$$

C. *ALDEHYDES DERIVED FROM THE* $C_nH_{2n-7}Ho$ *ALCOHOLS.*

Benzoic aldehyde......
$$\left\{ \begin{array}{l} \mathbf{C}(C_5H_3)H_2 \\ \mathbf{C}\mathrm{OH} \end{array} \right. \overset{\text{Boiling-}}{\underset{\text{point.}}{\dots 180°.}}$$

Cuminic aldehyde ...
$$\left\{ \begin{array}{l} \mathbf{C}(C_nH_9)H_2 \\ \mathbf{C}\mathrm{OH} \end{array} \right. \dots 229°·4.$$

BENZOIC ALDEHYDE, *Oil of Bitter Almonds, Hydride of Benzoyl.*

$$\left\{ \begin{array}{l} \mathbf{C}(C_5H_3)H_2 \\ \mathbf{C}\mathrm{OH} \end{array} \right.$$

Molecular weight $=106$. *Molecular volume* $\square$. 1 *litre of benzoic aldehyde vapour weighs* 53 *criths. Sp. gr.* 1·043. *Boils at* 180°.

Preparation.—1. By the oxidation of amygdalin by nitric acid, and by the action of a mixture of manganic oxide and sulphuric acid on albumen, fibrin, casein, and gelatin.

2. By digesting bitter almonds with water for five or six hours at 30°–40°. The synaptase present acts as a ferment on the amygdalin, converting it into glucose, benzoic aldehyde, and hydrocyanic acid:—

$$C_{20}H_{27}NO_{11}+2OH_2 = \left\{ \begin{array}{l} \mathbf{C}(C_5H_3)H_2 \\ \mathbf{C}\mathrm{OH} \end{array} \right. + \mathbf{C}N'''H + 2C_6H_{12}O_6.$$
$$\text{Amygdalin.}\quad \text{Water.} \qquad \text{Benzoic} \qquad \text{Hydrocyanic}\quad \text{Glucose.}$$
$$\qquad\qquad\qquad\qquad \text{aldehyde.} \qquad \text{acid.}$$

Reactions.—1. When exposed to the air, benzoic aldehyde absorbs oxygen and is converted into benzoic acid:—

$$\left\{ \begin{array}{l} \mathbf{C}(C_5H_3)H_2 \\ \mathbf{C}\mathrm{OH} \end{array} \right. + \mathrm{O} = \left\{ \begin{array}{l} \mathbf{C}(C_5H_3)H_2 \\ \mathbf{C}\mathrm{OHo} \end{array} \right.$$
$$\text{Benzoic aldehyde.} \qquad\qquad\qquad \text{Benzoic acid.}$$

2. Heated with solid potassic hydrate, it gives hydrogen and potassic benzoate:—

$$\left\{ \begin{array}{l} \mathbf{C}(C_5H_3)H_2 \\ \mathbf{C}\mathrm{OH} \end{array} \right. + \mathrm{KHo} = \left\{ \begin{array}{l} \mathbf{C}(C_5H_3)H_2 \\ \mathbf{C}\mathrm{OKo} \end{array} \right. + H_2.$$
$$\text{Benzoic}\qquad\qquad \text{Potassic}\qquad\quad \text{Potassic}$$
$$\text{aldehyde.}\qquad\qquad \text{hydrate.}\qquad\quad \text{benzoate.}$$

CHAPTER XXXVIII.

THE ACIDS.

THE acids form the most numerous family of organic compounds.

Many of them are contained in plants in the free state, or in combination as metallic or ethereal salts.

Others are produced by the action of chemical agents on organic matters.

Some are formed in the animal organism, as, for instance, formic, paralactic, oleic, and stearic acids.

The organic acids are divided into three great classes, according to their basicity.

    1. Monobasic acids.

    2. Dibasic acids.

    3. Tribasic acids.

The basicity of organic acids is determined by the following simple law:—*an organic acid containing* n *atoms of oxatyl is* n-*basic.*

MONOBASIC ACIDS.

The monobasic acids, which always contain a single atom of oxatyl (COHo), include the six following series:—

General formulæ.

1. Acetic or fatty series ...... $\begin{cases} C_n H_{2n+1} \\ COHo \end{cases}$.

2. Acrylic or oleic series ...... $\begin{cases} C(C_n H_{2n})''(C_m H_{2m+1}) \\ COHo \end{cases}$.

3. Lactic series .... $\begin{cases} C(C_n H_{2n+1})(C_m H_{2m+1})Ho \\ COHo \end{cases}$.

4. Pyruvic series .... $\begin{cases} CO(C_n H_{2n+1}) \\ COHo \end{cases}$.

5. Glyoxylic series.............. $\begin{cases} C(C_n H_{2n+1})Ho_2 \\ COHo \end{cases}$.

6. Benzoic or aromatic series $\begin{cases} C(C_n H_{2n-7})H_2 \\ COHo \end{cases}$.

The 1st, 2nd, 3rd, 5th, and 6th of these series may be regarded as the derivatives from corresponding series of alcohols.

1. The Acetic series from the Methyl series of alcohols.
2.   ,,   Acrylic   ,,   ,,   Allyl   ,,   ,,
3.   ,,   Lactic   ,,   ,,   Ethylene ,,   ,,
5.   ,,   Glyoxylic ,,   ,,   Glycerin ,,   ,,
6.   ,,   Benzoic ,,   ,,   Benzoic ,,   ,,

The acids of the first, second, fourth, and sixth series are termed *monohydric* as well as monobasic; whilst the acids of the third series are termed *dihydric* and monobasic, indicating their origin from the diacid alcohols, and that they contain two atoms of hydroxyl, one of which is in the oxatyl, and the other in the positive part of the compound. The hydrogen of the latter hydroxyl may be displaced by very positive metals, in the same manner as the hydrogen of the hydroxyl in alcohols; but it cannot be displaced by double decomposition with bases in the same manner as the hydrogen in the oxatyl may be substituted.

The acids of the fifth series are termed *trihydric* and monobasic, indicating that they are derived from the triacid alcohols, and that they contain, besides the hydroxyl in the oxatyl, two other atoms of hydroxyl in the positive part of the compound.

1. *ACETIC OR FATTY SERIES OF ACIDS.*

$$\text{General formula} \dots \left\{ \begin{array}{l} (C_nH_{2n+1}) \\ COHo \end{array} \right.$$

These acids may be conveniently arranged under three divisions, viz. :—

A. Normal acids.

$$\text{General formula} \dots \left\{ \begin{array}{l} C(C_nH_{2n+1})H_2 \\ COHo \end{array} \right.$$

B. Secondary acids.

$$\text{General formula} \dots \left\{ \begin{array}{l} C(C_nH_{2n+1})_2H \\ COHo \end{array} \right.$$

C. Tertiary acids.

General formula ... $\left\{ \begin{array}{l} \mathbf{C}(C_nH_{2n+1})_3 \\ \mathbf{C}OHo \end{array} \right.$

A. *NORMAL ACIDS OF THE ACETIC OR FATTY SERIES.*

General formula ... $\left\{ \begin{array}{l} \mathbf{C}(C_nH_{2n+1})H_2 \\ \mathbf{C}OHo \end{array} \right.$

In formic acid, which is generally considered to be the first term of this division, the radical $\mathbf{C}(C_nH_{2n+1})H_2$ is replaced by H ; and in acetic acid the value of $n=0$. The following is a list of the normal fatty acids :—

	Fusing-point.	Boiling-point.
Formic acid $\left\{ \begin{array}{l} H \\ \mathbf{C}OHo \end{array} \right.$	$+1°.$	$100°.$
Acetic acid $\left\{ \begin{array}{l} Me \\ \mathbf{C}OHo \end{array} \right.$ or $\left\{ \begin{array}{l} \mathbf{C}H_3 \\ \mathbf{C}OHo \end{array} \right.$	$+17°.$	$117°.$
Propionic acid $\left\{ \begin{array}{l} \mathbf{C}MeH_2 \\ \mathbf{C}OHo \end{array} \right.$ or $\left\{ \begin{array}{l} \mathbf{C}(CH_3)H_2 \\ \mathbf{C}OHo \end{array} \right.$	——	$141°.$
Butyric acid $\left\{ \begin{array}{l} \mathbf{C}EtH_2 \\ \mathbf{C}OHo \end{array} \right.$ or $\left\{ \begin{array}{l} \mathbf{C}(C_2H_5)H_2 \\ \mathbf{C}OHo \end{array} \right.$	below $-20°.$	$161°.$
Valeric acid $\left\{ \begin{array}{l} \mathbf{C}PrH_2 \\ \mathbf{C}OHo \end{array} \right.$ or $\left\{ \begin{array}{l} \mathbf{C}(C_3H_7)H_2 \\ \mathbf{C}OHo \end{array} \right.$	——	$175°.$
Caproic acid $\left\{ \begin{array}{l} \mathbf{C}BuH_2 \\ \mathbf{C}OHo \end{array} \right.$ or $\left\{ \begin{array}{l} \mathbf{C}(C_4H_9)H_2 \\ \mathbf{C}OIIo \end{array} \right.$	$+5°.$	$198°.$
Œnanthylic acid $\left\{ \begin{array}{l} \mathbf{C}AyH_2 \\ \mathbf{C}OHo \end{array} \right.$ or $\left\{ \begin{array}{l} \mathbf{C}(C_5H_{11})H_2 \\ \mathbf{C}OHo \end{array} \right.$	——	$212°.$
Caprylic acid $\left\{ \begin{array}{l} \mathbf{C}CpH_2 \\ \mathbf{C}OHo \end{array} \right.$ or $\left\{ \begin{array}{l} \mathbf{C}(C_6H_{13})H_2 \\ \mathbf{C}OHo \end{array} \right.$	$+14°.$	$236°.$
Pelargonic acid $\left\{ \begin{array}{l} \mathbf{C}(C_7H_{15})H_2 \\ \mathbf{C}OHo \end{array} \right.$	$+18°?$	$260°.$

		Fusing-point.
Capric acid	$\mathbf{C}(C_5H_{17})H_2$ $\mathbf{C}OHo$	27°·2(30°).
Lauric acid	$\mathbf{C}(C_{10}H_{21})H_2$ $\mathbf{C}OHo$	43°·6.
Myristic acid	$\mathbf{C}(C_{12}H_{25})H_2$ $\mathbf{C}OHo$	53°·8.
Palmitic acid	$\mathbf{C}(C_{14}H_{29})H_2$ $\mathbf{C}OHo$	62°.
Margaric acid	$\mathbf{C}(C_{15}H_{31})H_2$ $\mathbf{C}OHo$	59°·9 ?
Stearic acid	$\mathbf{C}(C_{16}H_{33})H_2$ $\mathbf{C}OHo$	69°·2.
Arachidic acid	$\mathbf{C}(C_{18}H_{37})H_2$ $\mathbf{C}OHo$	75°.
Behenic acid	$\mathbf{C}(C_{20}H_{41})H_2$ $\mathbf{C}OHo$	76°.
Cerotic acid	$\mathbf{C}(C_{25}H_{51})H_2$ $\mathbf{C}OHo$	78°.
Melissic acid	$\mathbf{C}(C_{29}H_{57})H_2$ $\mathbf{C}OHo$	88°.

Occurrence.—The greater number of the acids of this series are met with ready formed in nature, some in the free state, as formic acid in ants and nettles, valeric acid in the valerian root, pelargonic acid in the essential oil of the *Pelargonium roseum*, and cerotic acid in bees-wax.

Others are met with as the ethereal salts of monacid alcohols. Thus spermaceti is cetylic palmitate, and chinese wax cerylic cerotate.

A large number exist as natural fats in the form of the ethereal salts of glycerin: this is the case with butyric, palmitic, and stearic acids, which, united with glycerin, form respectively butyrin, palmitin, and stearin.

Formation.—1. By the oxidation of the normal alcohols of the methyl series, as in the conversion of alcohol into acetic acid by heating it with a solution of chromic acid :—

$$\begin{Bmatrix} CH_3 \\ CH_2Ho \end{Bmatrix} + O_2 = \begin{Bmatrix} CH_3 \\ COHo \end{Bmatrix} + OH_2.$$

$$\qquad\text{Alcohol.}\qquad\qquad\qquad\quad\text{Acetic}\qquad\quad\text{Water.}$$
$$\qquad\qquad\qquad\qquad\qquad\qquad\text{acid.}$$

2. By the action of alkalies or acids upon the cyanides of the C_nH_{2n+1} series of radicals :—

$$\begin{Bmatrix} C_nH_{2n+1} \\ CN''' \end{Bmatrix} + KHo + OH_2 = \begin{Bmatrix} C_nH_{2n+1} \\ COKo \end{Bmatrix} + NH_3 ;$$

$$\quad\text{Cyanide.}\qquad\text{Potassic}\qquad\text{Water.}\qquad\text{Potassic}\qquad\text{Ammonia.}$$
$$\qquad\qquad\text{hydrate.}\qquad\qquad\qquad\text{salt.}$$

and

$$\begin{Bmatrix} C_nH_{2n+1} \\ CN''' \end{Bmatrix} + HCl + 2OH_2 = \begin{Bmatrix} C_nH_{2n+1} \\ COHo \end{Bmatrix} + NH_4Cl.$$

$$\quad\text{Cyanide.}\qquad\text{Hydrochloric}\quad\text{Water.}\qquad\text{Acid.}\qquad\text{Ammonic}$$
$$\qquad\qquad\text{acid.}\qquad\qquad\qquad\qquad\qquad\text{chloride.}$$

Instances of these reactions are seen in the treatment of ethylic cyanide by a boiling solution of potassic hydrate, when it is converted into potassic propionate, ammonia being evolved, thus—

$$\left\{ \begin{matrix} CMeH_2 \\ CN''' \end{matrix} \right. + KHo + OH_2 = \left\{ \begin{matrix} CMeH_2 \\ COKo \end{matrix} \right. + NH_3 ;$$

Ethylic cyanide. Potassic hydrate. Water. Potassic propionate. Ammonia.

and in the conversion of ethylic cyanide, by the action of hydrochloric acid, into ammonic chloride and propionic acid—

$$\left\{ \begin{matrix} CMeH_2 \\ CN''' \end{matrix} \right. + HCl + 2OH_2 = \left\{ \begin{matrix} CMeH_2 \\ COHo \end{matrix} \right. + NH_4Cl.$$

Ethylic cyanide. Hydrochloric acid. Water. Propionic acid. Ammonic chloride.

3. By the action of the potassium or sodium compound of the C_nH_{2n+1} radicals upon carbonic anhydride—

$$CO_2 + C_nH_{2n+1}Na = \left\{ \begin{matrix} C_nH_{2n+1} \\ CONao' \end{matrix} \right.$$

Carbonic anhydride. Sodium compound. Sodic salt.

as, for example, in the formation of sodic propionate by the absorption of carbonic anhydride by sodic ethide :—

$$CO_2 + CMeH_2Na = \left\{ \begin{matrix} CMeH_2 \\ CONao' \end{matrix} \right.$$

Carbonic anhydride. Sodic ethide. Sodic propionate.

4. By the oxidation of aldehydes—

$$\left\{ \begin{matrix} C_nH_{2n+1} \\ COH \end{matrix} \right. + O = \left\{ \begin{matrix} C_nH_{2n+1} \\ COHo \end{matrix} \right.'$$

Aldehyde. Acid.

as in the conversion of acetic aldehyde into acetic acid by the absorption of atmospheric oxygen :—

$$\left\{ \begin{matrix} CH_3 \\ COH \end{matrix} \right. + O = \left\{ \begin{matrix} CH_3 \\ COHo \end{matrix} \right.'$$

Acetic aldehyde. Acetic acid.

Besides these reactions of general application, there are numerous special methods for the production of certain members of this series. In most of these methods, however, the reactions cannot be clearly traced.

Thus, by the oxidation of albumen, fibrin, casein, and other similar substances, there are produced formic, acetic, propionic, butyric, valeric, and caproic acids.

Propionic and butyric acids are produced in some kinds of fermentation; and acetic acid is obtained by the destructive distillation of wood and other similar substances.

Relations of the Normal Fatty Acids to the C_nH_{2n+1} Series of Radicals.

1. When these acids are submitted to the action of nascent oxygen evolved by electrolysis, the negative radical oxatyl is converted into carbonic anhydride and water, the positive radical being set at liberty:—

$$2\left\{\begin{matrix}C_nH_{2n+1}\\COHo\end{matrix}\right. + O = \left\{\begin{matrix}C_nH_{2n+1}\\C_nH_{2n+1}\end{matrix}\right. + 2CO_2 + OH_2.$$

Acid.   Positive radical.   Carbonic anhydride.   Water.

On electrolyzing a solution of potassic valerate, hydric potassic carbonate and the normal radical butyl are formed:—

$$2\left\{\begin{matrix}C_4H_9\\COKo\end{matrix}\right. + OH_2 + O = \left\{\begin{matrix}C_4H_9\\C_4H_9\end{matrix}\right. + 2COHoKo.$$

Potassic valerate.   Water.   Butyl.   Hydric potassic carbonate.

2. When the ammonic salts of these acids are heated with phosphoric anhydride, they are converted into cyanides of the radicals of the C_nH_{2n+1} series—

$$\left\{\begin{matrix}C_nH_{2n+1}\\CO(N^vH_4O)\end{matrix}\right. + 2P_2O_5 = \left\{\begin{matrix}C_nH_{2n+1}\\CN'''\end{matrix}\right. + 4PO_2Ho,$$

Ammonic salt.   Phosphoric anhydride.   Cyanide.   Metaphosphoric acid.

as in the transformation of ammonic acetate into methylic cyanide by distillation with phosphoric anhydride:—

$$\left\{\begin{matrix}CH_3\\CO(N^vH_4O)\end{matrix}\right. + 2P_2O_5 = \left\{\begin{matrix}CH_3\\CN'''\end{matrix}\right. + 4PO_2Ho.$$

Ammonic acetate.   Phosphoric anhydride.   Methylic cyanide.   Metaphosphoric acid.

These cyanides are converted into normal monacid alcohols by Mendius's reaction (see p. 248).

From the alcohols so obtained, the C_nH_{2n+1} radicals are isolated as described at page 246.

Relations of the Normal Fatty Acids to the $C_nH_{2n+1}Ho$ Series of Alcohols.

1. By oxidation the normal alcohols yield these acids, as above shown.

2. Conversely, the normal fatty acids can be converted into the normal $C_nH_{2n+1}Ho$ alcohols,—

1st, by Mendius's reaction (see p. 248);

2nd, by Piria and Wurtz's reactions, viz. :—

Distillation of the potassic salt of the fatty acid with an equivalent quantity of potassic formate, by which the acid is converted into the aldehyde—

$$\left\{ \begin{matrix} C_nH_{2n+1} \\ COKo \end{matrix} \right. \ + \ \left\{ \begin{matrix} H \\ COKo \end{matrix} \right. \ = \ \left\{ \begin{matrix} C_nH_{2n+1} \\ COH \end{matrix} \right. \ + \ COKo_2,$$

Potassic   Potassic   Aldehyde.   Potassic
salt.   formate.   carbonate.

and subsequent treatment of the aldehyde by nascent hydrogen—

$$\left\{ \begin{matrix} C_nH_{2n+1} \\ COH \end{matrix} \right. \ + \ H_2 \ = \ \left\{ \begin{matrix} C_nH_{2n+1} \\ CH_2Ho \end{matrix} \right. .$$

Aldehyde.   Normal
alcohol.

Relations of the Normal Fatty Acids to each other. Ascent of the Series.

If the hydrogen constituting the positive part of formic acid were substituted successively by methyl, ethyl, &c., the whole series of normal fatty acids would be obtained:—

Formic acid......... $\left\{ \begin{matrix} H \\ COHo \end{matrix} \right.$

Acetic acid .... $\left\{ \begin{matrix} CH_3 \\ COHo \end{matrix} \right.$

Propionic acid ... $\left\{ \begin{matrix} C_2H_5 \\ COHo \end{matrix} \right.$

Butyric acid ...... $\left\{ \begin{matrix} C_3H_7 \\ COHo \end{matrix} \right.$

This substitution has not yet been accomplished; but an analogous series of reactions has been effected with acetic acid.

By the action of sodium on acetic ether, one of the hydrogen atoms in the positive part of the compound becomes substituted by sodium, producing

$$\text{Monosodacetic ether} \ldots\ldots \begin{cases} \text{CH}_2\text{Na} \\ \text{COEto} \end{cases}$$

By acting on this body with the iodides of the C_nH_{2n+1} radicals, ethylic salts of the higher acids are produced.

On submitting monosodacetic ether to the action of methylic iodide, for instance, propionic ether is formed—

$$\begin{cases} \text{CH}_2\text{Na} \\ \text{COEto} \end{cases} + \text{CH}_3\text{I} = \begin{cases} \text{C(CH}_3)\text{H}_2 \\ \text{COEto} \end{cases} + \text{NaI;}$$

| Monosodacetic] ether. | Methylic iodide. | Propionic ether. | Sodic iodide. |

and by the action of ethylic iodide, butyric ether is produced—

$$\begin{cases} \text{CH}_2\text{Na} \\ \text{COEto} \end{cases} + \text{C}_2\text{H}_5\text{I} = \begin{cases} \text{C(C}_2\text{H}_5)\text{H}_2 \\ \text{COEto} \end{cases} + \text{NaI.}$$

| Monosodacetic ether. | Ethylic iodide. | Butyric ether. | Sodic iodide. |

FORMIC ACID.

$$\begin{cases} \text{H} \\ \text{COHo} \end{cases} \text{ or } \text{CHOHo.}$$

Molecular weight $=46$. *Molecular volume* ⬜. *1 litre of formic acid vapour weighs* 23 *criths.* *Sp.gr.* 1·2353. *Fuses at* 1° C. *Boils at* 100° C.

Occurrence.—If red ants be made to pass over blue litmus-paper and be at the same time irritated, they leave a red streak behind them, produced by the formic acid which they eject. By placing the hand on an ant-hill, a tingling sensation is felt from the same cause, and the hand acquires the powerful and pleasant odour of formic acid.

Formic acid also occurs in the hairs of certain caterpillars, and in the sting of nettles.

Formation.—1. Formic acid is produced in a very large number of chemical reactions, as in the oxidation of many organic matters (such as starch, woody fibre, or tartaric acid) by a mixture of sulphuric acid and manganic oxide, or by potassic hydrate or chromic acid.

2. By the action of potassic hydrate on chloroform, potassic formate is generated :—

$$\text{CHCl}_3 \quad + \quad 4\text{KHo} \quad = \quad \text{CHOKo} \quad + \quad 3\text{KCl} \quad + \quad 2\text{OH}_2.$$

Chloroform.   Potassic   Potassic   Potassic   Water.
hydrate.   formate.   chloride.

3. By the oxidation of methylic alcohol :—

$$\left\{ \begin{array}{l} \text{H} \\ \text{CH}_2\text{Ho} \end{array} \right. + \text{O}_2 = \left\{ \begin{array}{l} \text{H} \\ \text{COHo} \end{array} \right. + \text{OH}_2.$$

Methylic   Formic   Water.
alcohol.   acid.

4. By heating equal weights of dry oxalic acid and glycerin together to 75° C., when the oxalic acid splits into formic acid and carbonic anhydride :—

$$\left\{ \begin{array}{l} \text{COHo} \\ \text{COHo} \end{array} \right. = \left\{ \begin{array}{l} \text{H} \\ \text{COHo} \end{array} \right. + \text{CO}_2.$$

Oxalic   Formic   Carbonic
acid.   acid.   anhydride.

5. By digesting together at 100°, for forty-eight hours, potassic hydrate and carbonic oxide :—

$$\text{KHo} \quad + \quad \text{CO} \quad = \quad \left\{ \begin{array}{l} \text{H} \\ \text{COKo} \end{array} \right.$$

Potassic   Carbonic   Potassic
hydrate.   oxide.   formate.

Formic acid from any of these sources is obtained in the concentrated state by decomposing plumbic formate with sulphuretted hydrogen, and afterwards rectifying the acid over plumbic formate :—

$$\left. \begin{array}{l} \text{CHO} \\ \text{CHO} \end{array} \right\} \text{Pbo}'' + \text{SH}_2 = 2 \left\{ \begin{array}{l} \text{H} \\ \text{COHo} \end{array} \right. + \text{PbS}''.$$

Plumbic   Sulphuretted   Formic   Plumbic
formate.   hydrogen.   acid.   sulphide.

Character.—When heated with concentrated sulphuric acid, formic acid splits into water and carbonic oxide:—

$$\left\{ \begin{array}{l} H \\ \mathbf{C}OHo \end{array} \right. = \mathbf{C}O + OH_2.$$

Formic   Carbonic   Water.
acid.   oxide.

Chlorine converts formic acid into hydrochloric acid and carbonic anhydride:—

$$\left\{ \begin{array}{l} H \\ \mathbf{C}OHo \end{array} \right. + Cl_2 = 2HCl + \mathbf{C}O_2.$$

Formic   Hydrochloric   Carbonic
acid.   acid.   anhydride.

When heated with excess of mercuric oxide, it is converted into carbonic anhydride and water, the mercury being reduced to the metallic state:—

$$\left\{ \begin{array}{l} H \\ \mathbf{C}OHo \end{array} \right. + \mathbf{H}gO = \mathbf{C}O_2 + Hg + OH_2.$$

Formic   Mercuric   Carbonic   Water.
acid.   oxide.   anhydride.

ACETIC ACID.

$$\left\{ \begin{array}{l} \mathbf{C}H_3 \\ \mathbf{C}OHo \end{array} \right.$$

Molecular weight $= 60.$ *Molecular volume* □□.   1 *litre of acetic acid vapour weighs* 30 *criths.*   *Sp. gr.* 1·064.   *Fuses at* $+17°.$ *Boils at* $117°.$

Occurrence.—Found in small quantities in the juices of plants and in animal fluids.

Manufacture.—1. By the destructive distillation of wood, a liquid is obtained which contains acetic acid; the acid is purified by being converted first into a calcic, and then into a sodic salt, the latter being afterwards decomposed by sulphuric acid.

2. By the oxidation of ethylic alcohol:—

$$\left\{ \begin{array}{l} \mathbf{C}H_3 \\ \mathbf{C}H_2Ho \end{array} \right. + O_2 = \left\{ \begin{array}{l} \mathbf{C}H_3 \\ \mathbf{C}OHo \end{array} \right. + OH_2.$$

Ethylic   Acetic   Water.
alcohol.   acid.

Preparation.—Pure acetic acid may be obtained by distilling potassic diacetate :—

$$\left\{ \begin{array}{l} CH_3 \\ COKo' \end{array} \right. \left\{ \begin{array}{l} CH_3 \\ COHo \end{array} \right. = \left\{ \begin{array}{l} CH_3 \\ COKo \end{array} \right. + \left\{ \begin{array}{l} CH_3 \\ COHo \end{array} \right.$$

Potassic diacetate.   Potassic acetate.   Acetic acid.

Character.—Chlorine acts on acetic acid in sunlight, producing three chlorinated acids, in which chlorine is substituted for hydrogen :—

$$\left\{ \begin{array}{l} CH_3 \\ COHo \end{array} \right. + Cl_2 = \left\{ \begin{array}{l} CH_2Cl \\ COHo \end{array} \right. + HCl.$$

Acetic acid.   Monochloracetic acid.   Hydrochloric acid.

$$\left\{ \begin{array}{l} CH_2Cl \\ COHo \end{array} \right. + Cl_2 = \left\{ \begin{array}{l} CHCl_2 \\ COHo \end{array} \right. + HCl.$$

Monochloracetic acid.   Dichloracetic acid.   Hydrochloric acid.

$$\left\{ \begin{array}{l} CHCl_2 \\ COHo \end{array} \right. + Cl_2 = \left\{ \begin{array}{l} CCl_3 \\ COHo \end{array} \right. + HCl.$$

Dichloracetic acid.   Trichloracetic acid.   Hydrochloric acid.

The salts of acetic acid in which the hydrogen of the oxatyl is replaced by monad metals have the general formula

$$\left\{ \begin{array}{l} CH_3 \\ COMo \end{array} \right.$$

The acetates of the dyad metals have the constitution represented by the following general formula :—

$$\left\{ \begin{array}{l} CH_3 \\ CO\text{-}O \\ CO\text{-}O \\ CH_3 \end{array} M'' \text{ or } \left\{ \begin{array}{l} CMeO \\ Mo'' \\ CMeO \end{array} \right. \text{ or } \begin{array}{l} CMeO\text{-}O \\ CMeO\text{-}O \end{array} M''. \right.$$

By the action of phosphorous trichloride, acetic acid yields acetylic chloride :—

$$3 \left\{ \begin{array}{l} CH_3 \\ COHo \end{array} \right. + PCl_3 = 3 \left\{ \begin{array}{l} CH_3 \\ COCl \end{array} \right. + POHHo_2.$$

Acetic acid.   Phosphorous trichloride.   Acetylic chloride.   Phosphorous acid.

PROPIONIC ACID, *Methacetic Acid.*

$$\begin{cases} \mathbf{C}\text{MeH}_2 \\ \mathbf{C}\text{OHo} \end{cases}$$

Molecular weight = 74. *Molecular volume* ☐ 1 *litre of propionic acid vapour weighs* 37 *criths.* *Boils at* 141°.

Preparation.—1. By the oxidation of metacetone—a liquid obtained by the distillation of a mixture of sugar and lime.

2. By the action of concentrated solution of potassic hydrate on sugar.

3. By the fermentation of glycerin, and also of sugar, by means of putrid cheese in the presence of calcic carbonate.

4. By the action of potassic hydrate or hydrochloric acid on ethylic cyanide (see p. 300).

5. By the action of carbonic anhydride on sodic ethide (see p. 301).

6. By the action of hydriodic acid on lactic acid:—

$$\begin{cases} \mathbf{C}\text{MeHHo} \\ \mathbf{C}\text{OHo} \end{cases} + 2\text{HI} = \begin{cases} \mathbf{C}\text{MeH}_2 \\ \mathbf{C}\text{OHo} \end{cases} + \text{OH}_2 + \text{I}_2.$$

Lactic acid.   Hydriodic acid.   Propionic acid.   Water.

7. By the action of methylic iodide on sodacetic ether (see page 304).

BUTYRIC ACID, *Ethacetic Acid.*

$$\begin{cases} \mathbf{C}\text{EtH}_2 \\ \mathbf{C}\text{OHo} \end{cases}$$

Molecular weight = 88. *Molecular volume* ☐. 1 *litre of butyric acid vapour weighs* 44 *criths.* *Sp. gr.* 0·9886. *Fuses below* −20°. *Boils at* 161° C.

Occurrence.—In butter, juice of flesh, perspiration, and many animal secretions.

Preparation.—1. By the fermentation of sugar with putrid cheese.

2. By the action of ethylic iodide on sodacetic ether (for reaction, see page 304).

VALERIC ACID, *Valerianic Acid.*

$$\begin{cases} \mathbf{C}PrH_2 \\ \mathbf{C}OHo \end{cases}$$

Molecular weight $= 102$. *Molecular volume* ☐☐. 1 *litre of valeric acid vapour weighs* 51 *criths. Sp. gr.* 0·937. *Boils at* 175°.

Occurrence.—In many plants, as in the roots of valerian and angelica.

Preparation.—By the oxidation of amylic alcohol with a mixture of sulphuric acid and dipotassic dichromate :—

$$\begin{cases} \mathbf{C}(C_3H_7)H_2 \\ \mathbf{C}H_2Ho \end{cases} + O_2 = \begin{cases} \mathbf{C}(C_3H_7)H_2 \\ \mathbf{C}OHo \end{cases} + OH_2.$$

Amylic alcohol.   Valeric acid.   Water.

Isomeric forms.—There are four possible isomers of valeric acid :—

Normal valeric acid or propylacetic acid ...
$$\begin{cases} \mathbf{C}PrH_2 \\ \mathbf{C}OHo \end{cases} \text{ or } \begin{cases} \mathbf{C}H_2(CH_2[CH_2(CH_3)]) \\ \mathbf{C}OHo \end{cases}.$$

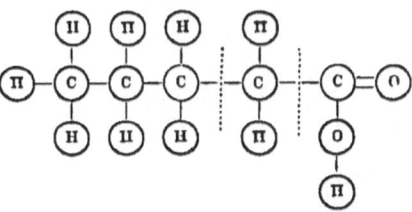

β Propylacetic acid ...
$$\begin{cases} \mathbf{C}Pr\beta H_2 \\ \mathbf{C}OHo \end{cases} \text{ or } \begin{cases} \mathbf{C}H_2[CH(CH_3)_2] \\ \mathbf{C}OHo \end{cases}.$$

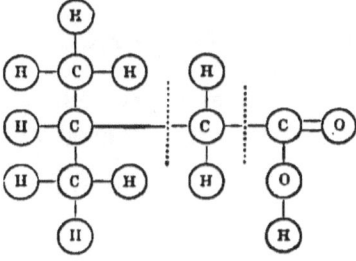

Methethacetic acid   ...$\begin{cases} \mathbf{CEtMeH} \\ \mathbf{COHo} \end{cases}$ or $\begin{cases} \mathbf{CH(CH_3)[CH_2(CH_3)]} \\ \mathbf{COHo} \end{cases}$

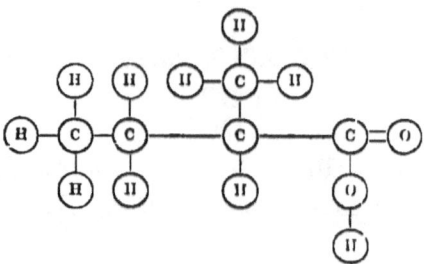

Trimethacetic acid ......$\begin{cases} \mathbf{CMe_3} \\ \mathbf{COHo} \end{cases}$ or $\begin{cases} \mathbf{C(CH_3)_3} \\ \mathbf{COHo} \end{cases}$.

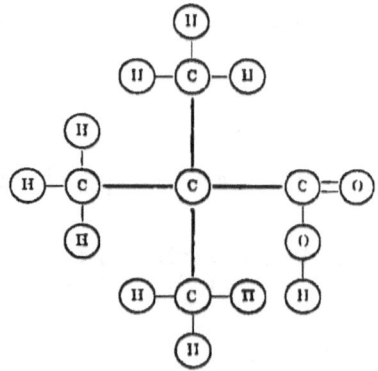

Of these, the normal, the β propylacetic, and the trimethacetic acid are known.

B. *SECONDARY FATTY ACIDS.*

General formula... $\begin{cases} \mathbf{C(C_nH_{2n+1})_2H} \\ \mathbf{COHo} \end{cases}$.

Formation.—By the action of the iodides of the C_nH_{2n+1} radicals on disodacetic ether, the ethereal salts of these acids are produced.

DIMETHACETIC ACID, *or Isobutyric Acid.*

$$\left\{ \begin{array}{l} \mathbf{C}Me_2H \\ \mathbf{C}OHo \end{array} \right. \cdot$$

Molecular weight =88. *Molecular volume* ☐☐. 1 *litre of dimethacetic acid vapour weighs* 44 *criths. Boils at* 152°. *Isomeric with butyric or ethacetic acid.*

Preparation.—As a potassic salt, by the action of methylic iodide on disodacetic ether and the subsequent decomposition, by alcoholic solution of potash, of the ethereal salt so formed :—

$$\left\{ \begin{array}{l} \mathbf{C}Na_2H \\ \mathbf{C}OEto \end{array} \right. + 2MeI = \left\{ \begin{array}{l} \mathbf{C}Me_2H \\ \mathbf{C}OEto \end{array} \right. + 2NaI.$$

Disodacetic ether.　　Methylic iodide.　　Dimethacetic ether.　　Sodic iodide.

$$\left\{ \begin{array}{l} \mathbf{C}Me_2H \\ \mathbf{C}OEto \end{array} \right. + KHo = \left\{ \begin{array}{l} \mathbf{C}Me_2H \\ \mathbf{C}OKo \end{array} \right. + EtHo.$$

Dimethacetic ether.　　Potassic hydrate.　　Potassic dimethacetate.　　Alcohol.

Diethacetic acid, $\left\{ \begin{array}{l} \mathbf{C}Et_2H \\ \mathbf{C}OHo \end{array} \right.$, isomeric with caproic acid, and

diamylacetic acid, $\left\{ \begin{array}{l} \mathbf{C}Ay_2H \\ \mathbf{C}OHo \end{array} \right.$, isomeric with lauric acid, have been prepared by the substitution of ethylic and amylic iodides in the above reaction.

C. *TERTIARY FATTY ACIDS.*

General formula... $\left\{ \begin{array}{l} \mathbf{C}(C_nH_{2n+1})_3 \\ \mathbf{C}OHo \end{array} \right.$.

Formation.—As ethereal salts, by the action of the iodides of the C_nH_{2n+1} radicals on trisodacetic ether.

TRIMETHACETIC ACID.

$$\begin{cases} \mathbf{C}\text{Me}_3 \\ \mathbf{C}\text{OHo} \end{cases}$$

Molecular weight $=102$. *Molecular volume* ⬜. 1 *litre of trimethacetic acid vapour weighs* 51 *criths. Isomeric with valerianic acid.*

For the graphic formula of trimethacetic acid see p. 310.

Preparation.—As an ethereal salt, by the action of methylic iodide on trisodacetic ether :—

$$\begin{cases} \mathbf{C}\text{Na}_3 \\ \mathbf{C}\text{OEto} \end{cases} + 3\text{MeI} = \begin{cases} \mathbf{C}\text{Me}_3 \\ \mathbf{C}\text{OEto} \end{cases} + 3\text{NaI}.$$

Trisodacetic ether.   Methylic iodide.   Trimethacetic ether.   Sodic iodide.

CHAPTER XXXIX.

THE ACIDS.

2. *ACRYLIC OR OLEIC SERIES OF ACIDS.*

General formula of normal and secondary acids .... $\begin{cases} \mathbf{C}(C_nH_{2n})''(C_mH_{2m+1}) \\ \mathbf{C}\text{OHo} \end{cases}$.

This series is divided into normal, secondary, and olefine acids. In the normal acid $m=0$; in the secondary it must be a positive integer.

Most of the normal acids exist as glycerin ethers in natural fats and oils.

The following is a list of the acrylic series of acids :—

A. *NORMAL ACIDS.*

Acrylic acid .... $\begin{cases} \mathbf{C}\text{Me}''\text{H} \\ \mathbf{C}\text{OHo} \end{cases}$ or $\begin{cases} \mathbf{C}(CH_2)''\text{H} \\ \mathbf{C}\text{OHo} \end{cases}$.

Crotonic acid .... $\begin{cases} \mathbf{C}\text{Et}''\text{H} \\ \mathbf{C}\text{OHo} \end{cases}$ or $\begin{cases} \mathbf{C}(C_2H_4)''\text{H} \\ \mathbf{C}\text{OHo} \end{cases}$.

Angelic acid .... $\begin{cases} \mathbf{C}\text{Pr}''\text{H} \\ \mathbf{C}\text{OHo} \end{cases}$ or $\begin{cases} \mathbf{C}(C_3H_6)''\text{H} \\ \mathbf{C}\text{OHo} \end{cases}$.

Pyroterebic acid............ $\begin{cases} \mathbf{C}Bu''H \\ \mathbf{C}OHo \end{cases}$ or $\begin{cases} \mathbf{C}(C_4H_9)''H \\ \mathbf{C}OHo \end{cases}$.

Damaluric acid .... $C_7H_{12}O_2$.

Damolic acid .... $C_{13}H_{24}O_2$.

Moringic acid............ $\left.\right\}$
Cinicic acid .... $\left.\right\}$ .... $C_{15}H_{24}O_2$.

Physetoleic acid .... $\left.\right\}$
Hypogœic acid .... $\left.\right\}$ .... $C_{16}H_{30}O_2$.
Gaïdic acid .... $\left.\right\}$

Oleic acid .... $\begin{cases} \mathbf{C}(C_{16}H_{32})''H \\ \mathbf{C}OHo \end{cases}$.

Elaïdic acid..................................... $C_{18}H_{34}O_2$.

Doeglic acid .... $C_{19}H_{36}O_2$.

Brassic acid. (*Erucic* $\left.\right\}$
acid.) .... $\left.\right\}$ .... $C_{22}H_{42}O_2$.

B. *SECONDARY ACIDS.*

Methacrylic acid............ $\begin{cases} \mathbf{C}Me''Me \\ \mathbf{C}OHo \end{cases}$ or $\begin{cases} \mathbf{C}(CH_3)''(CH_3) \\ \mathbf{C}OHo \end{cases}$.

Methylcrotonic acid ...... $\begin{cases} \mathbf{C}Et''Me \\ \mathbf{C}OHo \end{cases}$ or $\begin{cases} \mathbf{C}(C_2H_4)''(CH_3) \\ \mathbf{C}OHo \end{cases}$.

Ethylcrotonic acid .... $\begin{cases} \mathbf{C}Et''Et \\ \mathbf{C}OHo \end{cases}$ or $\begin{cases} \mathbf{C}(C_2H_4)''(C_2H_5) \\ \mathbf{C}OHo \end{cases}$.

C. *OLEFINE ACIDS.*

β Crotonic acid .... $\begin{cases} \mathbf{C}Me''H \\ \mathbf{C}H_2 \\ \mathbf{C}OHo \end{cases}$ or $\begin{cases} \mathbf{C}(CH_2)''H \\ \mathbf{C}H_2 \\ \mathbf{C}OHo \end{cases}$.

Formation of Normal Acids.—1. By the oxidation of the alcohols of the vinyl or $C_nH_{2n-1}Ho$ series :—

$\begin{cases} \mathbf{C}(C_nH_{2n})''H \\ \mathbf{C}H_2Ho \end{cases}$ + O_2 = $\begin{cases} \mathbf{C}(C_nH_{2n})''H \\ \mathbf{C}OHo \end{cases}$ + OH_2.
$\quad$ Alcohol. $\qquad\qquad\qquad\qquad\quad$ Acid. $\qquad\qquad$ Water.

2. By the oxidation of the aldehydes of the acrolein or $\begin{cases} \mathbf{C}(C_nH_{2n})''H \\ \mathbf{C}OH \end{cases}$ series :—

$\begin{cases} \mathbf{C}(C_nH_{2n})''H \\ \mathbf{C}OH \end{cases}$ + O = $\begin{cases} \mathbf{C}(C_nH_{2n})''H \\ \mathbf{C}OHo \end{cases}$.
$\qquad$ Aldehyde. $\qquad\qquad\qquad\qquad\qquad$ Acid.

P

Formation of Secondary Acids.—By the action of the phosphorus chlorides or of phosphoric anhydride upon the ethereal salts of secondary acids of the lactic series, the elements of water are removed and the ethereal salts of the acrylic secondary division of acids produced :—

$$3 \left\{ \begin{array}{l} \mathbf{C}(C_nH_{2n+1})_2Ho \\ \mathbf{CO}Eto \end{array} \right. + \mathbf{PCl_3} = 3 \left\{ \begin{array}{l} \mathbf{C}(C_nH_{2n})''(C_nH_{2n+1}) \\ \mathbf{CO}Eto \end{array} \right.$$

Ethereal salt of   Phosphorous   Ethereal salt of
the lactic series.   trichloride.   the acrylic series.

$$+ \quad \mathbf{POHHo_2} + \quad 3HCl.$$

Phosphorous   Hydrochloric acid.
acid.   ric acid.

Formation of Olefine Acids.—By the action of potassic hydrate upon the cyanides of the C_nH_{2n-1} family of radicals :—

$$\left\{ \begin{array}{l} \mathbf{C}Me''H \\ \mathbf{CH_2} \\ \mathbf{CN}''' \end{array} \right. + KHo + OH_2 = \left\{ \begin{array}{l} \mathbf{C}Me''H \\ \mathbf{CH_2} \\ \mathbf{CO}Ko \end{array} \right. + \mathbf{NH_3}.$$

Allylic cyanide.   Potassic   Water.   Potassic   Ammonia.
hydrate.   β crotonate.

Relations of the Acrylic to the Acetic Series of Acids.

The normal and secondary acids of the acrylic series, when treated with fused potassic hydrate, yield the potassic salts of two normal acids of the acetic series :—

$$\left\{ \begin{array}{l} \mathbf{C}(C_nH_{2n})''(C_mH_{2m+1}) \\ \mathbf{CO}Ho \end{array} \right. + 2KHo = \left\{ \begin{array}{l} \mathbf{C}(C_mH_{2m+1})H_2 \\ \mathbf{CO}Ko \end{array} \right.$$

Acid of the acrylic series.   Potassic   Potassic salt of acid
hydrate.   of the acetic series.

$$+ \left\{ \begin{array}{l} \mathbf{C}(C_{n-2}H_{2n-3})H_2 \\ \mathbf{CO}Ko \end{array} \right. + H_2.$$

Potassic salt of acid
of the acetic series.

All the members of the acrylic series found in nature give acetic acid as one of the acids produced in this reaction. From this and other considerations it is believed that their basylous radicals all contain one atom of hydrogen and a dyad radical. They are normal acids; and by the action of fused potassic hydrate the dyad radical becomes substituted by two atoms of hydrogen. Thus:—

$$\begin{Bmatrix} \mathbf{CMe''H} \\ \mathbf{COHo} \end{Bmatrix} + 2KHo = \begin{Bmatrix} \mathbf{CH_3} \\ \mathbf{COKo} \end{Bmatrix} + \begin{Bmatrix} \mathbf{H} \\ \mathbf{COKo} \end{Bmatrix} + H_2.$$

Acrylic acid.   Potassic hydrate.   Potassic acetate.   Potassic formate.

$$\begin{Bmatrix} \mathbf{CEt''H} \\ \mathbf{COHo} \end{Bmatrix} + 2KHo = \begin{Bmatrix} \mathbf{CH_3} \\ \mathbf{COKo} \end{Bmatrix} + \begin{Bmatrix} \mathbf{CH_3} \\ \mathbf{COKo} \end{Bmatrix} + H_2$$

Crotonic acid.   Potassic hydrate.   Potassic acetate.   Potassic acetate.

$$\begin{Bmatrix} \mathbf{CPr''H} \\ \mathbf{COHo} \end{Bmatrix} + 2KHo = \begin{Bmatrix} \mathbf{CH_3} \\ \mathbf{COKo} \end{Bmatrix} + \begin{Bmatrix} \mathbf{CMeH_2} \\ \mathbf{COKo} \end{Bmatrix} + H_2.$$

Angelic acid.   Potassic hydrate.   Potassic acetate.   Potassic propionate.

Some of the secondary acids also give acetic acid when treated with fused potassic hydrate; but this can only happen when the dyad radical is ethylene, thus:—

$$\begin{Bmatrix} \mathbf{CMe''Me} \\ \mathbf{COHo} \end{Bmatrix} + 2KHo = \begin{Bmatrix} \mathbf{CMeH_2} \\ \mathbf{COKo} \end{Bmatrix} + \begin{Bmatrix} \mathbf{H} \\ \mathbf{COKo} \end{Bmatrix} + H_2.$$

Methacrylic acid.   Potassic hydrate.   Potassic propionate.   Potassic formate.

$$\begin{Bmatrix} \mathbf{CEt''Me} \\ \mathbf{COHo} \end{Bmatrix} + 2KHo = \begin{Bmatrix} \mathbf{CMeH_2} \\ \mathbf{COKo} \end{Bmatrix} + \begin{Bmatrix} \mathbf{CH_3} \\ \mathbf{COKo} \end{Bmatrix} + H_2.$$

Methylcrotonic acid.   Potassic hydrate.   Potassic propionate.   Potassic acetate.

$$\begin{Bmatrix} \mathbf{CEt''Et} \\ \mathbf{COHo} \end{Bmatrix} + 2KHo = \begin{Bmatrix} \mathbf{CEtH_2} \\ \mathbf{COKo} \end{Bmatrix} + \begin{Bmatrix} \mathbf{CH_3} \\ \mathbf{COKo} \end{Bmatrix} + H_2.$$

Ethylcrotonic acid.   Potassic hydrate.   Potassic butyrate.   Potassic acetate.

ACRYLIC ACID.

$$\begin{Bmatrix} \mathbf{CMe''H} \\ \mathbf{COHo} \end{Bmatrix}$$

Molecular weight $=72$.   *Boils at about* $100°$.

Preparation.—By the oxidation of acrolein with argentic oxide:—

$$\begin{Bmatrix} \mathbf{CMe''H} \\ \mathbf{COH} \end{Bmatrix} + \mathbf{OAg_2} = \begin{Bmatrix} \mathbf{CMe''H} \\ \mathbf{COHo} \end{Bmatrix} + Ag_2.$$

Acrolein.   Argentic oxide.   Acrylic acid.

Reactions.—1. Acrylic acid, under the influence of nascent hydrogen, produces propionic acid :—

$$\left\{ \begin{array}{l} \mathbf{C}Me''H \\ \mathbf{C}OHo \end{array} \right. + H_2 = \left\{ \begin{array}{l} \mathbf{C}MeH_2 \\ \mathbf{C}OHo \end{array} \right. .$$

Acrylic acid.   Propionic acid.

2. Acrylic acid also combines directly with bromine, producing dibromopropionic acid.

OLEIC ACID.

$$\left\{ \begin{array}{l} \mathbf{C}(C_{16}H_{32})''H \\ \mathbf{C}OHo \end{array} \right. .$$

Preparation.—Obtained in the purification of stearic acid.

Reaction.—Heated with potassic hydrate, it gives potassic acetate and palmitate :—

$$\left\{ \begin{array}{l} \mathbf{C}(C_{16}H_{32})''H \\ \mathbf{C}OHo \end{array} \right. + 2KHo = \left\{ \begin{array}{l} \mathbf{C}H_3 \\ \mathbf{C}OKo \end{array} \right. + \left\{ \begin{array}{l} \mathbf{C}(C_{14}H_{29})H_2 \\ \mathbf{C}OKo \end{array} \right. + H_2 .$$

Oleic acid.   Potassic hydrate.   Potassic acetate.   Potassic palmitate.

CHAPTER XL.

THE ACIDS.

3. *LACTIC SERIES OF ACIDS.*

General formula of normal and secondary acids :—

$$\left\{ \begin{array}{l} \mathbf{C}(C_nH_{2n+1})(C_mH_{2m+1})Ho \\ \mathbf{C}OHo \end{array} \right. .$$

In the normal acids m in this formula $=0$; but in the secondary acids it must be a positive integer.

The members of the lactic series may be defined as acids containing one atom of oxatyl, the fourth bond of the carbon of which is united with the carbon of a basylous group, containing one atom, and one only, of hydroxyl, or of the peroxide of a monad radical, either alcoholic or acid. The following examples will serve to illustrate this definition :—

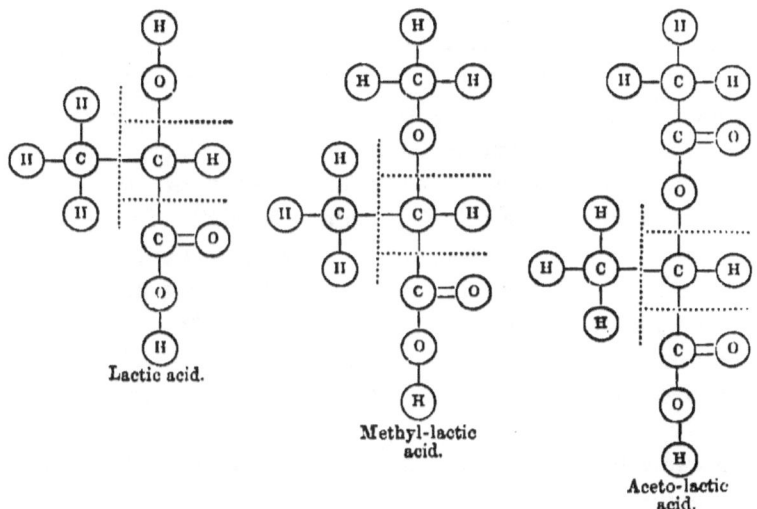

Lactic acid.

Methyl-lactic acid.

Aceto-lactic acid.

The acids of this series at present known, or which could be obtained by obvious processes, are classified into the following eight divisions :—

1. Normal Acids.
2. Etheric Normal Acids.
3. Secondary Acids.
4. Etheric Secondary Acids.
5. Normal Olefine Acids.
6. Etheric Normal Olefine Acids.
7. Secondary Olefine Acids.
8. Etheric Secondary Olefine Acids.

1st. *Normal Acids.*—A normal acid of the lactic series may be defined as one in which an atom of carbon is united with oxatyl, hydroxyl, and at least one atom of hydrogen. The general formula of these acids is therefore

$$\left\{ \begin{array}{l} \mathbf{C\overset{+}{R}HHo} \\ \mathbf{COHo} \end{array} \right. \cdot$$

In this formula $\overset{+}{R}$ may be either hydrogen or any monad alcohol radical; and the number of acids possessing the same atomic weight, and belonging to this division, is determined by the number of isomeric modifications of which the alcohol radical is susceptible. Thus, of the acids containing two, three, or four atoms of carbon, there can be only one of each belonging to this division, because these acids cannot contain an alcohol radical higher in the series than ethyl, and this radical is not susceptible of isomeric modification; but a normal acid containing propyl can have one isomer in this division, the two acids containing respectively propyl ($\mathbf{CEtH_2}$) and isopropyl ($\mathbf{CMe_2H}$). For acids of this division containing normal alcohol radicals only, the following general graphic formula may be given :—

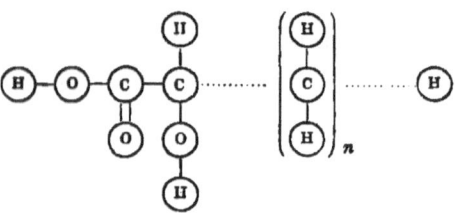

In the case of glycollic acid $n = 0$.

The following are the acids at present known belonging to this division :—

Glycollic acid	$\left\{ \begin{array}{l} \mathbf{CH_2Ho} \\ \mathbf{COHo} \end{array} \right. \cdot$
Lactic acid	$\left\{ \begin{array}{l} \mathbf{CMeHHo} \\ \mathbf{COHo} \end{array} \right. \cdot$
Oxybutyric acid	$\left\{ \begin{array}{l} \mathbf{CEtHHo} \\ \mathbf{COHo} \end{array} \right. \cdot$

Valerolactic acid .... $\left\{\begin{array}{l} \mathbf{CPrHHo} \\ \mathbf{COHo} \end{array}\right.$.

Leucic acid.................. $\left\{\begin{array}{l} \mathbf{CBuHHo} \\ \mathbf{COHo} \end{array}\right.$.

2nd. *Etheric Normal Acids.*—An etheric normal acid of the lactic series is constituted like a normal acid, but contains a monad organic radical, chlorous or basylous, in the place of the hydrogen of the *non-oxatylic* hydroxyl. The following is therefore the general formula of these acids: in the graphic formula *n*, as before, may $=0$.

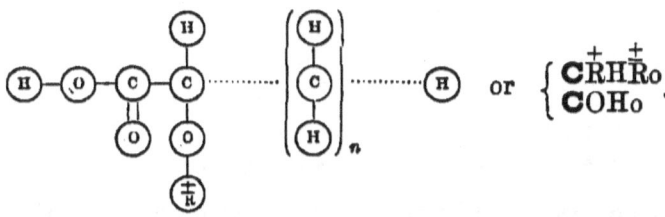 or $\left\{\begin{array}{l} \mathbf{C\overset{+}{R}H\overset{+}{R}o} \\ \mathbf{COHo} \end{array}\right.$.

The number of possible isomers belonging to this division is very great; for, in addition to those of which the normal acids containing $\overset{+}{R}$ of the same value are susceptible, a host of others must result from the complementary variation of $\overset{+}{R}$ and $\overset{+}{R}$. The lowest member of the division, methylglycollic acid (isomeric with lactic acid), is the only one incapable of isomeric modification.

The following examples will serve to illustrate the constitution of the acids belonging to this division:—

Methylglycollic acid........... $\left\{\begin{array}{l} \mathbf{CH_2Meo} \\ \mathbf{COHo} \end{array}\right.$.

Ethyl-lactic acid .... $\left\{\begin{array}{l} \mathbf{CMeHEto} \\ \mathbf{COHo} \end{array}\right.$.

Aceto-lactic acid .... $\left\{\begin{array}{l} \mathbf{CMeHAco*} \\ \mathbf{COHo} \end{array}\right.$.

3rd. *Secondary Acids.*—A secondary acid of the lactic series

* Aco $=$ peroxide of acetyl, $C_2H_3O_2$.

is one in which an atom of carbon is united with oxatyl, hydroxyl, and *two atoms* of a monad alcohol radical. The general formula of these acids is

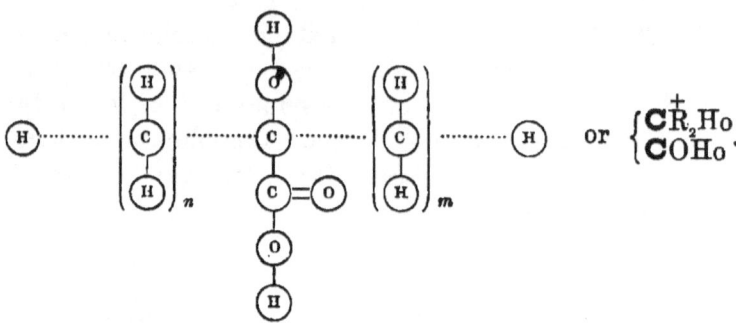

In the graphic expression, the values of n and m may differ; but both are positive integers, and neither may $=0$. In the symbolic formula $\overset{+}{R}$ must be a monad alcohol radical. The following examples will serve to illustrate their constitution :—

Dimethoxalic acid .... $\left\{ \begin{array}{l} \mathbf{CMe_2Ho} \\ \mathbf{COHo} \end{array} \right.$.

Ethomethoxalic acid .... $\left\{ \begin{array}{l} \mathbf{CEtMeHo} \\ \mathbf{COHo} \end{array} \right.$.

Diethoxalic acid .... $\left\{ \begin{array}{l} \mathbf{CEt_2Ho} \\ \mathbf{COHo} \end{array} \right.$.

The number of acids possessing the same atomic weight, and belonging to this division, is determined, first, by the complementary variation of the two alcohol radicals, and, secondly, by the number of possible isomers of these radicals. The two lowest terms of the series are alone incapable of isomeric modification by either of the causes mentioned.

4th. *Etheric Secondary Acids.*—These acids stand in the same relation to the secondary as the etheric normal to the normal acids; they consequently contain a monad organic radical in the place of the hydrogen of the non-oxatylic hydroxyl. The following is therefore the general formula of these acids :—

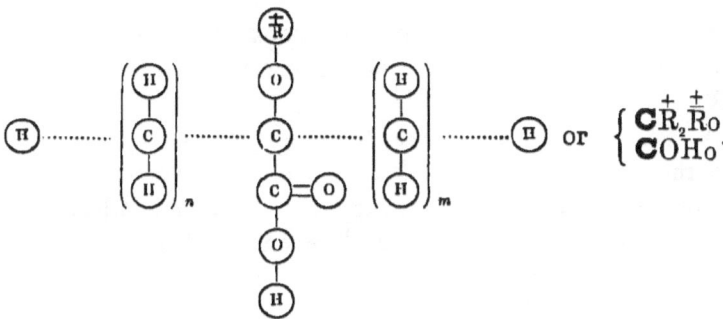

$$\left\{ \begin{array}{l} C\overset{+}{R}_2\overset{+}{R}o \\ COHo \end{array} \right.$$

5th. *Normal Olefine Acids.*—A normal olefine acid belonging to the lactic series is one in which the atom of carbon united with oxatyl is *not* combined with hydroxyl, and in which the atom of carbon united with hydroxyl is combined with not less than one atom of hydrogen. The following are the general graphic and symbolic formulæ of the acids belonging to this division :—

$$\left\{ \begin{array}{l} C\overset{+}{R}HHo \\ (CH_2)_n \\ COHo \end{array} \right.$$

In both these formulæ n must be a positive integer and cannot $=0$, but $\overset{+}{R}$ may be either hydrogen or a monad alcohol radical. The olefines of these acids may belong either to the ethylene or ethylidene series.

The following are the only acids at present known belonging to this division :—

Paralactic acid $\quad\dots\dots\dots\dots\quad \left\{ \begin{array}{l} CH_2Ho \\ CH_2 \\ COHo \end{array} \right.$

Paraleucic acid $\dots\dots\dots\dots\quad \left\{ \begin{array}{l} CH_2Ho \\ (C_4H_9)'' \\ COHo \end{array} \right.$

P 5

The number of isomers in this division will obviously depend, first, upon the complementary variations of $\overset{+}{R}$ and $(CH_2)_n$; secondly, upon the isomeric modifications of which $\overset{+}{R}$ is susceptible; and thirdly, upon the isomeric modifications of $(CH_2)_n$.

6th. *Etheric Normal Olefine Acids.*—These acids only differ from the normal olefine acids in having the hydrogen of the non-oxatylic hydroxyl replaced by an organic monad radical, positive or negative; their general formula is therefore,

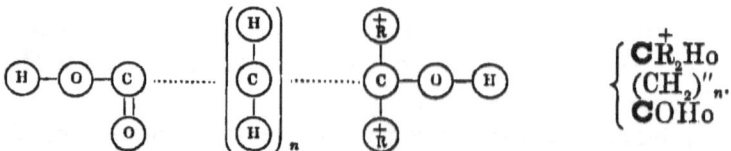

As in the fifth division, n must be a positive integer and cannot $=0$, whilst $\overset{+}{R}$ may be either hydrogen or a monad alcohol radical; but $\overset{+}{R}$ must be a monad compound radical, either acid or alcoholic.

7th. *Secondary Olefine Acids.*—A secondary olefine acid of this series is one in which the atom of carbon united with oxatyl is *not* combined with hydroxyl, and in which the atom of carbon united with hydroxyl is also combined with two monad alcohol radicals, as shown in the following formulæ:—

In both of these formulæ n must be a positive integer and cannot $=0$, and $\overset{+}{R}$ must be a monad alcohol radical.

8th. *Etheric Secondary Olefine Acids.*—These acids are related to the secondary olefine acids in the same way as the sixth division to the fifth. No member of the seventh or eighth division has yet been formed.

Formation of the Normal Acids.—1. By the oxidation of the glycols, or diacid alcohols.

$$\left\{ \begin{array}{l} \mathbf{CH_2Ho} \\ \mathbf{CH_2Ho} \end{array} \right. + \ \mathbf{O_2} \ = \ \left\{ \begin{array}{l} \mathbf{CH_2Ho} \\ \mathbf{COHo} \end{array} \right. + \ \mathbf{OH_2.}$$

Glycol.   Glycollic acid.   Water.

2. By the oxidation of the $C_nH_{2n+1}Ho$ alcohols :—

$$\left\{ \begin{array}{l} \mathbf{CH_3} \\ \mathbf{CH_2Ho} \end{array} \right. + \ \mathbf{O_2} \ = \ \left\{ \begin{array}{l} \mathbf{CH_2Ho} \\ \mathbf{COHo} \end{array} \right. + \ \mathbf{OH_2.}$$

Ethylic alcohol.   Glycollic acid.   Water.

3. From the fatty acids, by converting them first into chloro-substitution acids, and then acting upon these compounds with potassic hydrate :—

$$\left\{ \begin{array}{l} \mathbf{C(C_nH_{2n+1})H_2} \\ \mathbf{COHo} \end{array} \right. + \ \mathbf{Cl_2} \ = \ \left\{ \begin{array}{l} \mathbf{C(C_nH_{2n+1})HCl} \\ \mathbf{COHo} \end{array} \right. + \ \mathbf{HCl} ;$$

Fatty acid.   Chlorofatty acid.   Hydrochloric acid.

$$\left\{ \begin{array}{l} \mathbf{C(C_nH_{2n+1})HCl} \\ \mathbf{COHo} \end{array} \right. + \ \mathbf{KHo} \ = \ \left\{ \begin{array}{l} \mathbf{C(C_nH_{2n+1})HHo} \\ \mathbf{COHo} \end{array} \right. + \ \mathbf{KCl.}$$

Chlorofatty acid.   Potassic hydrate.   Normal acid of the lactic series.   Potassic chloride.

Formation of Secondary Acids.—By the action of the zinc compounds of the monad radicals upon ethylic oxalate, and the subsequent addition of water :—

$$\left\{ \begin{array}{l} \mathbf{COEto} \\ \mathbf{COEto} \end{array} \right. + \ \mathbf{2Zn(C_nH_{2n+1})_2} = \left\{ \begin{array}{l} \mathbf{C(C_nH_{2n+1})_2(Zn''C_nH_{2n+1}O)} \\ \mathbf{COEto} \end{array} \right.$$

Ethylic oxalate.   Zinc compound of monad radical.

$$+ \ \mathbf{Zn(C_nH_{2n+1})Eto} ;$$

$$\left\{ \begin{array}{l} \mathbf{C(C_nH_{2n+1})_2(Zn''C_nH_{2n+1}O)} \\ \mathbf{COEto} \end{array} \right. + \ \mathbf{2OH_2} \ = \ \left\{ \begin{array}{l} \mathbf{C(C_nH_{2n+1})_2Ho} \\ \mathbf{COEto} \end{array} \right.$$

Water.   Secondary acid.

$$+ \ \left\{ \begin{array}{l} \mathbf{C_nH_{2n+1}} \\ \mathbf{H} \end{array} \right. + \ \mathbf{ZnHo_2.}$$

Hydride of radical.   Zincic hydrate.

Formation of Olefine Acids.—By uniting a dyad radical with carbonic oxydichloride (*phosgene gas*) under the influence of sunlight, and subsequently acting upon the product with potassic hydrate :—

$$'' \begin{Bmatrix} CH_2 \\ CH_2 \end{Bmatrix} + COCl_2 = \begin{Bmatrix} CH_2Cl \\ CH_2(COCl) \end{Bmatrix};$$

Ethylene.　Carbonic oxydichloride.　β Chlorpropionylic (Phosgene gas.)　chloride.

$$\begin{Bmatrix} CH_2Cl \\ CH_2(COCl) \end{Bmatrix} + 3KHo = \begin{Bmatrix} CH_2Ho \\ CH_2(COKo) \end{Bmatrix} + 2KCl + OH_2.$$

β Chlorpropionylic　Potassic　Potassic　Potassic　Water.
chloride.　hydrate.　paralactate.　chloride.

Relations of the Lactic to the Acetic Series of Acids.

1. The transformation of the acetic or fatty into the normal lactic series of acids has been mentioned above (p. 323).

2. The converse operation is effected with the normal and secondary acids of the lactic series by means of hydriodic acid :—

$$\begin{Bmatrix} C(C_nH_{2n+1})(C_mH_{2m+1})Ho \\ COHo \end{Bmatrix} + 2HI = \begin{Bmatrix} C(C_nH_{2n+1})(C_mH_{2m+1})H \\ COHo \end{Bmatrix}$$

Acid of lactic series.　Hydriodic acid.　Acid of acetic series.

$$+ \ OH_2 \ + \ I_2.$$

Water.

If m does not $=0$ the fatty acid will be a secondary one, like the member of the lactic series from which it is derived.

Relations of the Lactic to the Acrylic Series of Acids.

If the ethereal salts of the secondary acids of the lactic series be treated with phosphorous trichloride, the ethereal salts of the secondary acids of the acrylic series are produced :—

$$3\begin{Bmatrix} C(C_nH_{2n+1})(C_mH_{2m+1})Ho \\ COEto \end{Bmatrix} + PCl_3 = 3\begin{Bmatrix} C(C_nH_{2n+1})(C_mH_{2m})'' \\ COEto \end{Bmatrix}$$

Ether of lactic series.　Phosphorous trichloride.　Ether of acrylic series.

$$+ \ POHHo_2 \ + \ 3HCl.$$

Phosphorous acid.　Hydrochloric acid.

This reaction has not yet been produced in the case of the normal acids of the lactic series.

A secondary lactic acid *minus* $OH_2 =$ an acrylic acid. The reverse of this operation has not been performed.

LACTIC ACID.

$$\begin{cases} \mathbf{CMeHHo} \\ \mathbf{COHo} \end{cases}$$

Sp. gr. 1·215.

Occurrence.—In sour milk, *Sauerkraut,* fluids of muscular tissue, gastric juice, saliva of diabetic patients. In the acid liquor of starch-factories, in blood, urine, tears, bile, &c. It is also a general product of putrefactive fermentation. The acid contained in animal fluids is *paralactic acid* (see p. 327).

Preparation.—By fermenting sugar with putrid cheese.

For other processes, see pages 329 and 330.

Its salts have the following general formulæ :—

$$\begin{cases} \mathbf{CMcHHo} \\ \mathbf{COMo} \end{cases}$$

Salts of monad metals.

$$\begin{cases} \mathbf{CMeHHo} \\ \mathbf{CO\text{-}O} \\ \mathbf{CO\text{-}O} \\ \mathbf{CMeHHo} \end{cases} M''.$$

Salts of dyad metals.

Isomerism in the Lactic Series.

The synthetical study of the acids of this series affords an insight into numerous and interesting cases of isomerism. Commencing with the lowest member of the series, we have for glycollic acid the formula

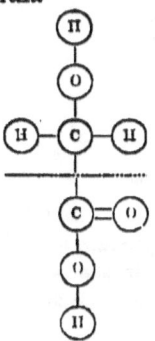

An inspection of this formula shows that glycollic acid admits of no isomeric modification, except with a total change of type. The part of the formula below the dotted line represents oxatyl, which cannot be altered without sacrificing the acid character of the compound; there remains therefore only the part of the formula above the dotted line, which admits of the following modification :—

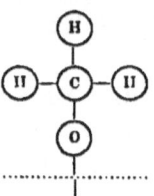

The acid represented by the formula so modified no longer comes within the definition of the lactic series. It is carbomethylic acid, and differs essentially from glycollic acid and the lactic series in general, inasmuch as the carbon of its chlorous radical, oxatyl, is linked to the carbon of the basylous radical by oxygen[*].

* Bearing this constitution of carbomethylic acid in mind, we have only to go one step further in order to perceive the constitution of carbonic acid itself, and the anomalous basicity of this acid; for if, in the above graphic formula for carbomethylic acid, we replace the methyl by hydrogen, we have

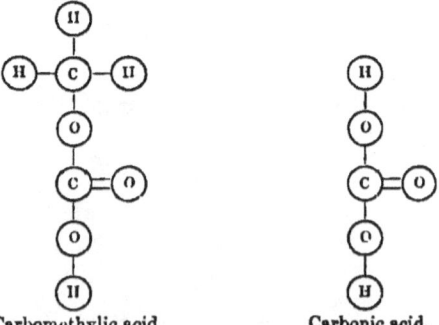

Carbomethylic acid.   Carbonic acid.

It is thus evident that the radical oxatyl, when united with hydroxyl, has sufficient chlorous power to produce a feebly dibasic acid; but inasmuch as carbonic acid is not included in the category of organic acids, it forms no exception to the law that an organic acid containing n atoms of oxatyl is n-basic.

There being no decisive evidence that *homolactic acid* differs from glycollic acid, experiment and theory both agree in asserting that the formula $C_2H_4O_3$ represents only one acid in the lactic series.

Proceeding now one step higher in this series, we have in the formula of lactic acid an expression capable of the following three variations without quitting the lactic type :—

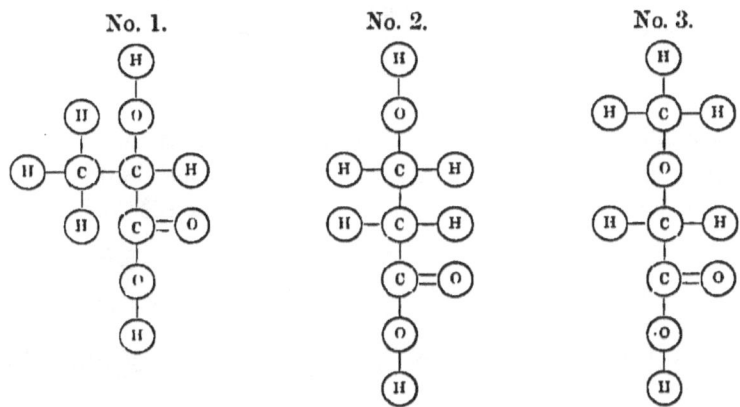

Or, expressed symbolically :—

No. 1.

$$\begin{cases} \mathbf{CMeHHo} \\ \mathbf{COHo} \end{cases}.$$

No. 2.

$$\begin{cases} \mathbf{CH_2Ho} \\ \mathbf{CH_2(COHo)} \end{cases} \text{ or } \begin{cases} \mathbf{CH_2Ho} \\ \mathbf{CH_2} \\ \mathbf{COHo} \end{cases}.$$

No. 3.

$$\begin{cases} \mathbf{CH_2Meo} \\ \mathbf{COHo} \end{cases}.$$

All the acids represented by the above formulæ are known. The first expresses the constitution of lactic acid, which belongs to the normal division $\left(\begin{cases} \mathbf{C\overset{+}{R}HHo} \\ \mathbf{COHo} \end{cases} \right)$ of the series, described at page 317; the second shows the atomic arrangement of paralactic acid; whilst the third represents methylglycollic acid. The proof that the first two of these acids are so constituted is afforded by the synthetic processes sometimes employed to produce them; for ethylidene cyanhydrate is converted by ebullition with potash into a salt of lactic acid, whilst ethylene cyanhydrate is transformed under similar circumstances into paralactic acid. It has also been mentioned above that by the action of phosgene gas upon ethylene, para-lactic acid is produced. Now the formation of ethylidene, or rather of its compounds, scarcely leaves a doubt that this body, if isolated, would have the following atomic constitution:—

$$\text{or} \quad \left\{ \begin{array}{l} \mathbf{CH_3} \\ \mathbf{''CH} \end{array} \right. :$$

it would consist of an atom of methyl and an atom of hydrogen, both united with an atom of carbon, two of whose bonds satisfy each other. Thus the formation of ethylidene dichloride from aldehyde and phosphoric chloride takes place as follows:—

$$\left\{ \begin{array}{l} \mathbf{CH_3} \\ \mathbf{CHO} \end{array} \right. + \mathbf{PCl_5} = \left\{ \begin{array}{l} \mathbf{CH_3} \\ \mathbf{CHCl_2} \end{array} \right. + \mathbf{POCl_3},$$

| Aldehyde. | Phosphoric chloride. | Ethylidene dichloride. | Phosphoric oxytrichloride. |

the oxygen in the aldehyde being simply replaced by chlorine. There now only remains one possible formula for ethylene, viz.

$$\text{or} \ '' \left\{ \begin{array}{l} \mathbf{CH_2} \\ \mathbf{CH_2} \end{array} \right. .$$

Such, then, being the constitution of ethylidene and ethylene, it follows that the former ought to give rise to an acid of the constitution shown in formula No. 1, whilst ethylene should produce an acid agreeing with formula No. 2. The acids actually produced from these sources are lactic and paralactic acids; hence No. 1 is the constitutional formula of lactic acid, and No. 2 that of paralactic acid,—a conclusion which harmonizes perfectly with all the reactions in which the production of these acids can be traced. Thus in the formation of lactic acid by the oxidation of propylic glycol, we have

$$\begin{Bmatrix} CMeHHo \\ CH_2Ho \end{Bmatrix} + O_2 = \begin{Bmatrix} CMeHHo \\ COHo \end{Bmatrix} + OH_2.$$

Propylic glycol.   Lactic acid.   Water.

Again, in the production of this acid from ethylidene cyanhydrate,

$$\begin{Bmatrix} CH_3 \\ CHHo(CN''') \end{Bmatrix} + KHo + OH_2 = \begin{Bmatrix} CH_3 \\ CHHo(COKo) \end{Bmatrix} + NH_3.$$

Ethylidene   Potassic   Water.   Potassic lactate.   Ammonia.
cyanhydrate.   hydrate.

The formula given for potassic lactate in this equation is only apparently different in type from that previously used for lactic acid, since

$$\begin{Bmatrix} CH_3 \\ CHHo(COKo) \end{Bmatrix} = CMeHHo(COKo) = \begin{Bmatrix} CMeHHo \\ COKo \end{Bmatrix}.$$

In the reaction by which chloropropionic acid is transformed into lactic acid we have the following change :—

$$\begin{Bmatrix} CMeHCl \\ COHo \end{Bmatrix} + 2KHo = \begin{Bmatrix} CMeHHo \\ COKo \end{Bmatrix} + KCl + OH_2.$$

Chloropropionic   Potassic   Potassic lactate.   Potassic   Water.
acid.   hydrate.   chloride.

The production of lactamic acid (*alanin*), and that of lactic acid from the latter by the action of nitrous acid, are also clearly confirmatory of the above view.

$$\begin{Bmatrix} CH_3 \\ CO(N^vH_4) \end{Bmatrix} + C{}^{N'''}_H + OH_2 + HCl = \begin{Bmatrix} CMeH(N'''H_2) \\ COHo \end{Bmatrix} + NH_4Cl :$$

Ammonic   Hydrocyanic   Water.   Hydro-   Lactamic acid   Ammonic
aldehyde   acid.   chloric   (alanin).   chloride.
                                   acid.

$$\begin{Bmatrix} CMeH(N'''H_2) \\ COHo \end{Bmatrix} + NOHo = \begin{Bmatrix} CMeHHo \\ COHo \end{Bmatrix} + OH_2 + N_2.$$

Lactamic acid   Nitrous   Lactic acid.   Water.
(alanin).   acid.

Not the least interesting reaction illustrative of the constitution of lactic acid is the formation of this acid by the action of nascent hydrogen upon pyruvic acid :—

$$\left\{ \begin{array}{l} \mathbf{COMe} \\ \mathbf{COHo} \end{array} \right. + \mathbf{H_2} = \left\{ \begin{array}{l} \mathbf{CMeHHo} \\ \mathbf{COHo} \end{array} \right. .$$

Pyruvic acid.   Lactic acid.

By an analogous reaction, glyoxalic acid, which is the next lower homologue of pyruvic acid, has been transformed into glycollic acid.

$$\left\{ \begin{array}{l} \mathbf{COH} \\ \mathbf{COHo} \end{array} \right. + \mathbf{H_2} = \left\{ \begin{array}{l} \mathbf{CH_2Ho} \\ \mathbf{COHo} \end{array} \right. .$$

Glyoxalic acid.   Glycollic acid.

In a similar manner it can be demonstrated that the above formula No. 2 expresses the constitution of paralactic acid, which belongs to the fifth or olefine division of these acids,

$$\left\{ \begin{array}{l} \mathbf{C\overset{+}{R}HHo} \\ \mathbf{(CH_2)''_n(COHo)} \end{array} \right. \quad \text{or} \quad \left\{ \begin{array}{l} \mathbf{C\overset{+}{R}HHo} \\ \mathbf{(CH_2)''_n} \\ \mathbf{COHo} \end{array} \right. .$$

That paralactic acid possesses this constitution is proved, first, by its production from cyanhydric glycol—

$$\left\{ \begin{array}{l} \mathbf{CH_2Ho} \\ \mathbf{CH_2(CN''')} \end{array} \right. + \mathbf{KHo} + \mathbf{OH_2} = \left\{ \begin{array}{l} \mathbf{CH_2Ho} \\ \mathbf{CH_2} \\ \mathbf{COKo} \end{array} \right. + \mathbf{NH_3} ;$$

Cyanhydric   Potassic   Water.   Potassic   Ammonia.
glycol.   hydrate.   paralactate.

and secondly, by its formation from phosgene gas and ethylene (see p. 324).

By the action of water upon the chloride of β chlorpropionyl, a body of the composition of chloropropionic acid is obtained ; but inasmuch as this body yields paralactic acid by ebullition with potash, whilst chloropropionic acid gives under the same circumstances lactic acid, it follows that the former chloro-acid must be isomeric, and not identical, with the latter. Now, although the formula of propionic acid does not admit of any isomer, yet that of chloropropionic acid does, as is seen in the following graphic formulæ :—

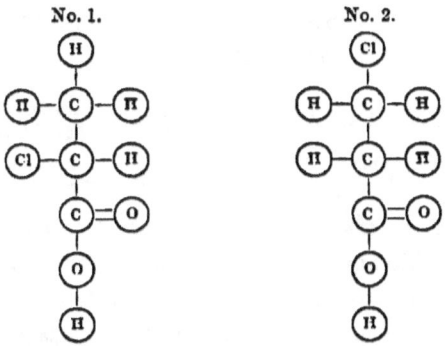

A comparison of these formulæ with those of lactic and para-lactic acids (p. 327) shows at a glance that No. 1 is the chloropropionic acid which yields lactic acid, whilst No. 2 is iso-chloropropionic acid, which, by the substitution of its chlorine by hydroxyl, must yield paralactic acid. By the action of nascent hydrogen, both isomeric chlorides will obviously produce the same propionic acid.

The cause of the isomerism of methyl-glycollic acid (No. 3, p. 327) is so obvious as to require no further explanation. Proceeding to the next higher stage in the series, such is the rapid increase of isomerism, that we now encounter no less than eight possible isomers, all within the lactic family.

Normal. No. 1.	Secondary. No. 2.	Etheric normal.	
		No. 3.	No. 4.
$\begin{cases} \mathbf{CEtHHo} \\ \mathbf{COHo} \end{cases}$.	$\begin{cases} \mathbf{CMe_2Ho} \\ \mathbf{COHo} \end{cases}$.	$\begin{cases} \mathbf{CH_2Eto} \\ \mathbf{COHo} \end{cases}$.	$\begin{cases} \mathbf{CMeHMeo} \\ \mathbf{COHo} \end{cases}$.

Normal olefine.			Etheric normal olefine.
No. 5.	No. 6.	No. 7.	No. 8.
$\begin{cases} \mathbf{CH_2Ho} \\ \mathbf{CH_2} \\ \mathbf{CH_2} \\ \mathbf{COHo} \end{cases}$.	$\begin{cases} \mathbf{CH_2Ho} \\ \mathbf{CMeH} \\ \mathbf{COHo} \end{cases}$.	$\begin{cases} \mathbf{CMeHHo} \\ \mathbf{CH_2} \\ \mathbf{COHo} \end{cases}$.	$\begin{cases} \mathbf{CH_2Meo} \\ \mathbf{CH_2} \\ \mathbf{COHo} \end{cases}$.

Of these acids, Nos. 1, 2, and 3 are known. No. 1 is oxy-butyric acid; No. 2 is dimethoxalic acid, which is probably identical with acetonic acid. On this assumption, the formation of the latter by the action of hydrocyanic and hydrochloric acids upon acetone is easily intelligible :—

$$\left\{\begin{matrix}\mathbf{CH_3}\\\mathbf{COMe}\end{matrix}\right. + \mathbf{CN'''H} + \mathbf{2OH_2} + \mathbf{HCl} = \left\{\begin{matrix}\mathbf{CMe_2Ho}\\\mathbf{COHo}\end{matrix}\right. + \mathbf{NH_4Cl.}$$

Acetone.　　Hydrocyanic　　Water.　　Hydro-　　Acetonic or
　　　　　　acid.　　　　　　　chloric acid. dimethoxalic acid.

The third of the above formulæ is that of ethyl-glycollic acid.

Of the possible acids containing five atoms of carbon, only two, viz. ethomethoxalic acid and valerolactic acid, are known. The cause of the isomerism of these two acids is seen at once from an inspection of their constitutional formulæ :—

$$\text{Ethomethoxalic acid} \ldots\ldots \left\{\begin{matrix}\mathbf{CEtMeHo}\\\mathbf{COHo}\end{matrix}\right. .$$

$$\text{Valerolactic acid} \ldots\ldots\ldots \left\{\begin{matrix}\mathbf{CPrHHo}\\\mathbf{COHo}\end{matrix}\right. .$$

Of acids containing six atoms of carbon, the following three are known :—

$$\text{Leucic acid} \ldots\ldots\ldots\ldots \left\{\begin{matrix}\mathbf{CBuHHo}\\\mathbf{COHo}\end{matrix}\right. .$$

$$\text{Diethoxalic acid} \ldots\ldots\ldots \left\{\begin{matrix}\mathbf{CEt_2Ho}\\\mathbf{COHo}\end{matrix}\right. .$$

$$\text{Paraleucic acid} \ldots\ldots\ldots \left\{\begin{matrix}\mathbf{CH_2Ho}\\\mathbf{(C_4H_4)''}\\\mathbf{COHo}\end{matrix}\right.$$

The above formula for leucic acid is founded upon a reaction for the synthetical production of this acid from valeric aldehyde and hydrocyanic acid. Valeric acid contains butyl; consequently valeric aldehyde has the constitution expressed by the formula $\left\{\begin{matrix}\mathbf{Bu}\\\mathbf{COH}\end{matrix}\right.$, and the reaction in question may therefore be explained by the following equation :—

$$\left\{\begin{matrix}\mathbf{Bu}\\\mathbf{CO(N'H_4)}\end{matrix}\right. + \mathbf{C}{\overset{N'''}{\underset{}{\mathbf{H}}}} + \mathbf{OH_2} + \mathbf{HCl} = \left\{\begin{matrix}\mathbf{CBuH(N'''H_2)}\\\mathbf{COHo}\end{matrix}\right. + \mathbf{NH_4Cl.}$$

Ammonic　　　Hydro-　　Water.　　Hydro-　　　Leucin.　　　Ammonic
valeric aldehyde. cyanic　　　　　　chloric　　　　　　　　　　chloride.
　　　　　　　acid.　　　　　　acid.

Such being the rational formula of leucin, its transformation into leucic acid by nitrous acid determines the constitution of leucic acid:—

$$\left\{ \begin{array}{l} \mathbf{C}BuH(N'''H_2) \\ \mathbf{C}OHo \end{array} \right. + \mathbf{N}OHo = \left\{ \begin{array}{l} \mathbf{C}BuHHo \\ \mathbf{C}OHo \end{array} \right. + \mathbf{O}H_2 + N_2.$$

Leucin. Nitrous acid. Leucic acid. Water.

CHAPTER XLI.

THE ACIDS.

4. *PYRUVIC SERIES.*

General formula...... $\left\{ \begin{array}{l} \mathbf{C}O(C_nH_{2n+1}). \\ \mathbf{C}OHo \end{array} \right.$

In this formula n may $= 0$.

The following list contains all the known members of this series :—

Glyoxalic acid............ $\left\{ \begin{array}{l} \mathbf{C}OH \\ \mathbf{C}OHo \end{array} \right.$

Pyruvic acid .... $\left\{ \begin{array}{l} \mathbf{C}OMe \\ \mathbf{C}OHo \end{array} \right.$ Boils at 165°.

Convolvulinoleic acid... $\left\{ \begin{array}{l} \mathbf{C}O(C_{11}H_{23}). \\ \mathbf{C}OHo \end{array} \right.$ Fuses at 42°.

Jalapinoleic acid......... $\left\{ \begin{array}{l} \mathbf{C}O(C_{14}H_{29}). \\ \mathbf{C}OHo \end{array} \right.$,, 65°.

Ricinoleic acid .... $\left\{ \begin{array}{l} \mathbf{C}O(C_{16}H_{33}). \\ \mathbf{C}OHo \end{array} \right.$,, 0°.

The first two members only of this series are well known.

These acids are the semialdehydes and semiketones of oxalic acid, and they stand in much the same relation to this acid as that which acetic aldehyde occupies with regard to acetic acid :—

$\left\{ \begin{array}{l} \mathbf{C}H_3 \\ \mathbf{C}OHo \end{array} \right.$ $\left\{ \begin{array}{l} \mathbf{C}H_3 \\ \mathbf{C}OH \end{array} \right.$

Acetic acid. Acetic aldehyde.

$$\left\{ \begin{array}{l} \textbf{CO}\text{Ho} \\ \textbf{CO}\text{Ho} \end{array} \right. :$$
Oxalic acid.

$$\left\{ \begin{array}{l} \textbf{CO}\text{H} \\ \textbf{CO}\text{Ho}^{\cdot} \end{array} \right.$$
Glyoxalic acid.

The same chemical change, when repeated upon the other half of oxalic acid, converts this acid into a true aldehyde, viz.

$$\text{Glyoxal} \quad \dots \left\{ \begin{array}{l} \textbf{CO}\text{H} \\ \textbf{CO}\text{H} \end{array} \right. .$$

Both glyoxal and glyoxalic acid are produced by the oxidation of ethylic alcohol by nitric acid :—

$$\left\{ \begin{array}{l} \textbf{CH}_3 \\ \textbf{CH}_2\text{Ho} \end{array} \right. + \ O_3 \ = \ \left\{ \begin{array}{l} \textbf{CO}\text{H} \\ \textbf{CO}\text{H} \end{array} \right. + \ 2\textbf{OH}_2 ;$$

Ethylic alcohol.   Glyoxal.   Water.

$$\left\{ \begin{array}{l} \textbf{CO}\text{H} \\ \textbf{CO}\text{H} \end{array} \right. + \ O \ = \ \left\{ \begin{array}{l} \textbf{CO}\text{H} \\ \textbf{CO}\text{Ho} \end{array} \right. .$$

Glyoxal.   Glyoxalic acid.

Glyoxalic acid reduces silver salts like an aldehyde, and is transformed into oxalic acid :—

$$\left\{ \begin{array}{l} \textbf{CO}\text{H} \\ \textbf{CO}\text{Ho} \end{array} \right. + \ O \ = \ \left\{ \begin{array}{l} \textbf{CO}\text{Ho} \\ \textbf{CO}\text{Ho} \end{array} \right. .$$

Glyoxalic acid.   Oxalic acid.

These reactions show clearly the relations of the pyruvic series to oxalic acid.

The pyruvic series is also closely related to the lactic series —the first two members of the former absorbing hydrogen and being converted into glycollic and normal lactic acids respectively :—

$$\left\{ \begin{array}{l} \textbf{CO}\text{H} \\ \textbf{CO}\text{Ho} \end{array} \right. + \ \text{H}_2 \ = \ \left\{ \begin{array}{l} \textbf{CH}_2\text{Ho} \\ \textbf{CO}\text{Ho} \end{array} \right. .$$

Glyoxalic acid.   Glycollic acid.

$$\left\{ \begin{array}{l} \textbf{CO}\text{Me} \\ \textbf{CO}\text{Ho} \end{array} \right. + \ \text{H}_2 \ = \ \left\{ \begin{array}{l} \textbf{CM}\text{cHHo} \\ \textbf{CO}\text{Ho} \end{array} \right. .$$

Pyruvic acid.   Lactic acid.

5. *THE GLYOXYLIC SERIES OF ACIDS.*

General formula ...
$$\begin{cases} \mathbf{C}(C_nH_{2n+1})\,Ho_2 \\ \mathbf{CO}Ho \end{cases} \text{or} \begin{cases} C_nH_{2n}Ho \\ C_mH_{2m-1}Ho. \\ \mathbf{CO}Ho \end{cases}$$

In the second formula n may $=0$.

The two following acids of this series are known:—

	Formula.	Physical condition.
Glyoxylic acid ...	$\begin{cases} \mathbf{C}HHo_2 \\ \mathbf{CO}Ho. \end{cases}$	Syrupy, crystalline hydrate.
Glyceric acid	$\begin{cases} \mathbf{C}H_2Ho \\ \mathbf{C}HHo. \\ \mathbf{CO}Ho \end{cases}$	Syrupy.

These acids are dihydric, but monobasic, and are related to the glycerin series of alcohols in the same way in which the members of the lactic series are related to the glycols:—

$$\begin{cases} \mathbf{C}H_2Ho \\ \mathbf{C}H_2Ho. \end{cases} \qquad \begin{cases} \mathbf{C}H_2Ho \\ \mathbf{CO}Ho. \end{cases}$$

Glycol.   Glycollic acid.

$$\begin{cases} \mathbf{C}H_2Ho \\ \mathbf{C}HHo. \\ \mathbf{C}H_2Ho \end{cases} \qquad \begin{cases} \mathbf{C}H_2Ho \\ \mathbf{C}HHo. \\ \mathbf{CO}Ho \end{cases}$$

Glycerin.   Glyceric acid.

Glyceric acid has hitherto been but little investigated.

CHAPTER XLII.

THE ACIDS.

6. *THE BENZOIC OR AROMATIC SERIES OF ACIDS.*

General formula ...
$$\begin{cases} \mathbf{C}(C_nH_{2n-7})H_2 \\ \mathbf{CO}Ho \end{cases}.$$

The following terms of this series are known :—

$$\left.\begin{array}{l}\text{Phenoic acid}\\\text{Collinic acid}\end{array}\right\} \cdots \left\{\begin{array}{l}\mathbf{C_5H_3}\\\mathbf{COHo\cdot}\end{array}\right. \qquad \begin{array}{l}\text{Fuses at } 60°.\\\text{Fuses at } 100°.\end{array}$$

$$\text{Benzoic acid} \ \ldots \ldots \left\{\begin{array}{l}\mathbf{C(C_5H_3)H_2}\\\mathbf{COHo}\end{array}\right. \qquad \text{Fuses at } 121°\cdot4.$$

$$\text{Toluylic acid} \ \ldots \ldots \left\{\begin{array}{l}\mathbf{C(C_6H_5)H_2}\\\mathbf{COHo}\end{array}\right.$$

$$\text{Cuminic acid} \ \ldots \ldots \left\{\begin{array}{l}\mathbf{C(C_4H_9)H_2}\\\mathbf{COHo}\end{array}\right. \qquad \text{Fuses at } 92°.$$

These acids have the same constitution as those of the acetic series, but contain the C_nH_{2n-7} radicals.

They have been much less studied than the acetic series, and further investigation will probably bring to light other series holding towards them the same relation as the acrylic, glycollic, pyruvic, and glyoxylic series bear to the acetic series. Already an acryloid acid of this section is known corresponding to an unknown homologue of benzoic acid :—

$$\left\{\begin{array}{l}\mathbf{C(C_7H_7)H_2}\\\mathbf{COHo}\end{array}\right. \qquad\qquad \left\{\begin{array}{l}\mathbf{C(C_7H_6)''H}\\\mathbf{COHo}\end{array}\right.$$
$$\qquad \text{Unknown acid.} \qquad\qquad\qquad \text{Cinnamic acid.}$$

Cinnamic acid is decomposed, like the acids of the acrylic series, when treated with fused potassic hydrate ; it gives, under these circumstances, potassic acetate and benzoate. For the analogous reaction in the acrylic series see p. 314.

Salicylic acid is the lactic acid of benzoic acid :—

$$\left\{\begin{array}{l}\mathbf{C(C_5H_3)H_2}\\\mathbf{COHo}\end{array}\right. \qquad\qquad \left\{\begin{array}{l}\mathbf{C(C_5H_3)HHo}\\\mathbf{COHo}\end{array}\right. ;$$
$$\qquad \text{Benzoic acid.} \qquad\qquad\qquad \text{Salicylic acid.}$$

and the oil of meadow-sweet (*Spiræa ulmaria*) is generally regarded as the aldehyde of salicylic acid :—

$$\left\{\begin{array}{l}\mathbf{C(C_5H_3)HHo}\\\mathbf{COH}\end{array}\right.$$

BENZOIC ACID.

$$\left\{ \begin{array}{l} \mathbf{C}(C_5H_3)H_2 \\ \mathbf{C}OHo \end{array} \right.$$

Molecular weight $=122$. *Molecular volume* $\boxed{}$. 1 *litre of benzoic acid vapour weighs* **61** *criths. Fuses at* $121^{\circ}\cdot4$. *Boils at* 239°.

Occurrence.—In many balsams and gums. In putrid urine.

Preparation.—1. By the oxidation of oil of bitter almonds (p. 295).

2. By the action of fused potassic hydrate on cinnamic acid:—

$$\left\{ \begin{array}{l} \mathbf{C}(C_7H_6)''H \\ \mathbf{C}OHo \end{array} \right. + 2KHo = \left\{ \begin{array}{l} \mathbf{C}H_3 \\ \mathbf{C}OKo \end{array} \right. + \left\{ \begin{array}{l} \mathbf{C}(C_5H_3)H_2 \\ \mathbf{C}OKo \end{array} \right. + H_2.$$

| Cinnamic acid. | Potassic hydrate. | Potassic acetate. | Potassic benzoate. |

3. By boiling hippuric acid with hydrochloric acid:—

$$C_9H_9NO_3 \ + \ OH_2 \ = \ C_2H_5NO_2 \ + \ \left\{ \begin{array}{l} \mathbf{C}(C_5H_3)H_2 \\ \mathbf{C}OHo \end{array} \right. .$$

| Hippuric acid. | Water. | Glycocin. | Benzoic acid. |

4. By the action of oxidizing agents on casein or gelatin.

5. From gum benzoin, by sublimation, or by extraction with potassic hydrate and subsequent precipitation of the acid by hydrochloric acid.

CHAPTER XLIII.

THE ACIDS.

DIBASIC ACIDS.

General formula... $\mathbf{A}(COHo)_2$ or $\left\{ \begin{array}{l} \mathbf{A}(COHo) \\ \mathbf{B}(COHo) \end{array} \right.$

A and B being dyad radicals containing $C_nH_mO_l$.

These acids all contain two atoms of oxatyl; and if in the

Q

general formula n, m, and $l = 0$, oxalic acid will be the first term of the series.

Formation.—Many of the dibasic acids are produced by the oxidation of substances richer in carbon, such as oils and fats. Others are found ready formed in nature.

Reactions.—1. By the action of dehydrating substances, and even sometimes by heat alone, these acids lose water, forming anhydrides :—

$$\left\{ \begin{matrix} \mathbf{A}(COHo) \\ \mathbf{B}(COHo) \end{matrix} \right. = OH_2 + \left\{ \begin{matrix} \mathbf{A}(CO \\ \mathbf{B}(CO \end{matrix} O \right).$$

Acid.   Water.   Anhydride.

2. If the anhydride be submitted to the action of phosphoric chloride, an atom of oxygen is replaced by two of chlorine :—

$$\left\{ \begin{matrix} \mathbf{A}(CO \\ \mathbf{B}(CO \end{matrix} O \right) + PCl_5 = \left\{ \begin{matrix} \mathbf{A}(COCl) \\ \mathbf{B}(COCl) \end{matrix} \right. + POCl_3.$$

Anhydride.   Phosphoric   Chloride.   Phosphoric
         chloride.   oxytrichloride.

3. Both the anhydrides and the chlorides are reconverted into the acids by the action of water :—

$$\left\{ \begin{matrix} \mathbf{A}(COCl) \\ \mathbf{B}(COCl) \end{matrix} \right. + 2OH_2 = \left\{ \begin{matrix} \mathbf{A}(COHo) \\ \mathbf{B}(COHo) \end{matrix} \right. + 2HCl.$$

Chloride.   Water.   Acid.   Hydrochloric
                                             acid.

The dibasic acids may be divided into the four following series :—

    1. Succinic or acetoid series .... $\left\{ \begin{matrix} C_nH_{2n}(COHo) \\ C_mH_{2m}(COHo) \end{matrix} \right.$

In the first member of the series $m = 0$.

    2. Fumaric or acryloid series ......" $\left\{ \begin{matrix} C_nH_{2n-1}(COHo) \\ C_nH_{2n-1}(COHo) \end{matrix} \right.$

    3. Malic or lactoid series .... $\left\{ \begin{matrix} C_nH_{2n-1}Ho(COHo) \\ C_nH_{2n}(COHo) \end{matrix} \right.$

    4. Tartaric or glyoxyloid series ... $\left\{ \begin{matrix} C_nH_{2n-1}Ho(COHo) \\ C_nH_{2n-1}Ho(COHo) \end{matrix} \right.$

The first and second series are dibasic and *dihydric*, the third, dibasic and *trihydric*, and the fourth, dibasic and *tetrahydric*.

1. *THE SUCCINIC OR ACETOID SERIES.*

General formula ...
$\begin{cases} COHo \\ C_nH_{2n} \\ C_mH_{2m} \\ COHo \end{cases}$ or $\begin{cases} COHo \\ C_nH_{2n} \\ COHo \end{cases}$.

The following members of this series are known :—

Malonic acid .... $\begin{cases} COHo \\ CH_2 \\ COHo \end{cases}$. Fuses at 140°.

Succinic acid .... $\begin{cases} COHo \\ C_2H_4 \\ COHo \end{cases}$. Fuses at 180°. Boils at 235°.

Pyrotartaric acid ... $\begin{cases} COHo \\ C_3H_6 \\ COHo \end{cases}$. Fuses at 112°. Boils at 200°.

Adipic acid .... $\begin{cases} COHo \\ C_4H_8 \\ COHo \end{cases}$. Fuses at 140°.

Pimelic acid .... $\begin{cases} COHo \\ C_5H_{10} \\ COHo \end{cases}$. Fuses at 134°.

Suberic acid .... $\begin{cases} COHo \\ C_6H_{12} \\ COHo \end{cases}$. Fuses at 125°.

Anchoic acid .... $\begin{cases} COHo \\ C_7H_{14} \\ COHo \end{cases}$. Fuses at 116°.

Sebacic acid........... $\begin{cases} COHo \\ C_8H_{16} \\ COHo \end{cases}$. Fuses at 127°.

Roccellic acid .... $\begin{cases} COHo \\ C_{15}H_{30} \\ COHo \end{cases}$. Fuses at 132°. Boils at 200°.

It is obvious that there may be several modifications of each of these acids. Thus there may be two succinic acids, one containing ethylene, and the other ethylidene (see p. 346) :—

$\begin{cases} CH_2(COHo) \\ CH_2(COHo) \end{cases}$ and $\begin{cases} CH_3 \\ CH(COHo)_2 \end{cases}$.

Q 2

1. *Relations of the Succinic to the Lactic Series of Acids and to the Glycols.*

These acids are related to the lactic series and to the glycols in the same way as the fatty acids are related to the monacid alcohols :—

$$\begin{cases} CH_2Ho \\ CH_2Ho \end{cases} \qquad \begin{cases} CH_2Ho \\ COHo \end{cases} \qquad \begin{cases} COHo \\ COHo \end{cases}$$

$$\quad\text{Glycol.} \qquad\qquad \text{Glycollic acid.} \qquad\quad \text{Oxalic acid.}$$

This relation, however, does not strictly extend beyond the first member, although it may be partially traced in the relations of malonic and adipic acids to paralactic and paraleucic acids :—

$$\begin{cases} CH_2(CH_2Ho) \\ CH_2Ho \end{cases} \qquad \begin{cases} CH_2(CH_2Ho) \\ COHo \end{cases} \qquad \begin{cases} CH_2(COHo) \\ COHo \end{cases}$$

$$\quad\text{Isopropylic glycol.} \qquad\quad \text{Paralactic acid.} \qquad\quad \text{Malonic acid.}$$
$$\quad\text{(Unknown.)}$$

$$\begin{cases} C_4H_8(CH_2Ho) \\ CH_2Ho \end{cases} \qquad \begin{cases} C_4H_8(CH_2Ho) \\ COHo \end{cases} \qquad \begin{cases} C_4H_8(COHo) \\ COHo \end{cases}$$

$$\quad\text{Unknown glycol.} \qquad\quad \text{Paraleucic acid.} \qquad\quad \text{Adipic acid.}$$

2. *Relations of the Succinic Series to the Dyad Radicals.*

1. The succinic series is intimately related to the dyad radicals, the cyanides of which are readily converted into dibasic acids by ebullition with potassic hydrate or hydrochloric acid :—

$$\begin{cases} C_nH_{2n}(CN''') \\ C_nH_{2n}(CN''') \end{cases} + 2KHo + 2OH_2 = \begin{cases} C_nH_{2n}(COKo) \\ C_nH_{2n}(COKo) \end{cases} + 2NH_3.$$

$$\begin{array}{llll} \text{Cyanide of the} & \text{Potassic} & \text{Water.} & \text{Potassic salt of the} & \text{Ammo-} \\ \text{dyad radical.} & \text{hydrate.} & & \text{dibasic acid.} & \text{nia.} \end{array}$$

2. Some of these acids, when heated with excess of caustic baryta, give up two atoms of carbonic anhydride, yielding the hydrides of the dyad radicals :—

$$\begin{cases} COHo \\ C_6H_{12} \\ COHo \end{cases} = 2CO_2 + (C_6H_{12})''H_2.$$

$$\quad\text{Suberic acid.} \qquad \text{Carbonic} \qquad$$
$$\qquad\qquad\qquad \text{anhydride.}$$

$$\begin{cases} COHo \\ C_8H_{16} \\ COHo \end{cases} = 2CO_2 + (C_8H_{16})''H_2.$$

$$\quad\text{Sebacic acid.} \qquad \text{Carbonic} \qquad$$
$$\qquad\qquad\qquad \text{anhydride.}$$

These reactions are the analogues, in the dyad series, of the one by which marsh-gas is obtained from acetic acid. The hydrides of the dyad radicals so obtained are isomeric with those of the corresponding monad radicals.

The elimination of carbonic anhydride from a monobasic acid can only take place once, while from a dibasic acid it takes place in two successive stages :—

In the case of a monobasic acid,

$$\left\{ \begin{matrix} C_nH_{2n+1} \\ COHo \end{matrix} \right. \quad - \quad CO_2 \quad = \quad C_nH_{2n+1}H.$$

In the case of a dibasic acid,

$$\text{1st stage} \ldots \left\{ \begin{matrix} C_nH_{2n}(COHo) \\ C_nH_{2n}(COHo) \end{matrix} \right. \quad - \quad CO_2 \quad = \quad \left\{ \begin{matrix} C_nH_{2n+1} \\ C_nH_{2n} \\ COHo \end{matrix} \right. :$$

$$\text{2nd stage} \ldots \left\{ \begin{matrix} C_nH_{2n+1} \\ C_nH_{2n} \\ COHo \end{matrix} \right. \quad - \quad CO_2 \quad = \quad \left\{ \begin{matrix} C_nH_{2n+1} \\ C_nH_{2n+1} \end{matrix} \right. .$$

3. *Relations of the Succinic to the Acetic Series of Acids.*

1. By the loss of the elements of carbonic anhydride, the first three members of the succinic series are converted into members of the acetic series, containing one atom of carbon less :—

$$\left\{ \begin{matrix} COHo \\ COHo \end{matrix} \right. = CO_2 + \left\{ \begin{matrix} H \\ COHo \end{matrix} \right. .$$

Oxalic acid.   Carbonic anhydride.   Formic acid.

$$\left\{ \begin{matrix} COHo \\ CH_2 \\ COHo \end{matrix} \right. = CO_2 + \left\{ \begin{matrix} CH_3 \\ COHo \end{matrix} \right. .$$

Malonic acid.   Carbonic anhydride.   Acetic acid.

$$\left\{ \begin{matrix} COHo \\ C_2H_4 \\ COHo \end{matrix} \right. = CO_2 + \left\{ \begin{matrix} CMeH_2 \\ COHo \end{matrix} \right. .$$

Succinic acid.   Carbonic anhydride.   Propionic acid.

In the first two cases the action of heat alone is sufficient to effect the transformation; but in the third the attraction of lime for carbonic anhydride must be superadded.

2. Conversely, the members of the acetic series may be converted into those of the succinic containing one atom of carbon more, by replacing one atom of the methylic hydrogen in acetic acid by cyanogen, and then boiling with potassic hydrate :—

$$\left\{ \begin{matrix} \mathbf{CH_2(CN''')} \\ \mathbf{COHo} \end{matrix} \right. + \quad 2KHo \quad = \quad \left\{ \begin{matrix} \mathbf{CH_2(COKo)} \\ \mathbf{COKo} \end{matrix} \right. + \quad \mathbf{NH_3}.$$

<table>
<tr><td>Cyanacetic acid.</td><td>Potassic hydrate.</td><td>Potassic malonate.</td><td>Ammonia.</td></tr>
</table>

The conversion of formic acid into oxalic acid, by heating with potassic hydrate, also belongs to this class of reactions :—

$$2\left\{ \begin{matrix} \mathbf{H} \\ \mathbf{COHo} \end{matrix} \right. + \quad 2KHo \quad = \quad \left\{ \begin{matrix} \mathbf{COKo} \\ \mathbf{COKo} \end{matrix} \right. + \quad 2\mathbf{OH_2} \quad + \quad \mathbf{H_2}.$$

<table>
<tr><td>Formic acid.</td><td>Potassic hydrate.</td><td>Potassic oxalate.</td><td>Water.</td></tr>
</table>

SUCCINIC ACID.

$$\left\{ \begin{matrix} \mathbf{COHo} \\ \mathbf{C_2H_4} \\ \mathbf{COHo} \end{matrix} \right. .$$

Fuses at 180°. *Boils at* 235°.

Occurrence.—In amber; in some kinds of lignite; in the resin of some kinds of pine; also in many other vegetable and animal substances.

Formation.—1. By the action of potassic hydrate upon ethylenic cyanide (p. 289): this reaction proves that succinic acid contains ethylene, and that its constitutional formula is as above given; but an isosuccinic acid must be capable of formation, though hitherto never obtained. This acid will obviously contain ethylidene in the place of ethylene; and its

formula will be $\left\{ \begin{matrix} \mathbf{COHo} \\ \mathbf{CMeH} \\ \mathbf{COHo} \end{matrix} \right.$ (See p. 346.)

2. By the oxidation of butyric acid by nitric acid :—

$$\left\{\begin{array}{l}CEtH_2 \\ COHo\end{array}\right. + O_3 = \left\{\begin{array}{l}COHo \\ C_2H_4 \\ COHo\end{array}\right. + OH_2.$$

Butyric acid.   Succinic acid.   Water.

The nature of this reaction is more clearly seen with fully developed formulæ, thus :—

$$\left\{\begin{array}{l}CH_3 \\ CH_2 \\ CH_2 \\ COHo\end{array}\right. + O_3 = \left\{\begin{array}{l}COHo \\ CH_2 \\ CH_2 \\ COHo\end{array}\right. + OH_2.$$

Butyric acid.   Succinic acid.   Water.

3. By the reduction of malic acid by fermentation, or by hydriodic acid :—

$$\left\{\begin{array}{l}COHo \\ CHHo \\ CH_2 \\ COHo\end{array}\right. + H_2 = \left\{\begin{array}{l}COHo \\ CH_2 \\ CH_2 \\ COHo\end{array}\right. + OH_2.$$

Malic acid.   Succinic acid.   Water.

4. By the reduction of tartaric acid by hydriodic acid :—

$$\left\{\begin{array}{l}COHo \\ CHHo \\ CHHo \\ COHo\end{array}\right. + 4HI = \left\{\begin{array}{l}COHo \\ CH_2 \\ CH_2 \\ COHo\end{array}\right. + 2OH_2 + 2I_2.$$

Tartaric   Hydriodic   Succinic   Water.
acid.   acid.   acid.

It is evident that this reaction is perfectly analogous to that by which lactic acid is transformed into propionic acid (p. 308).

5. The two isomeric acids, fumaric and maleic acids, are converted by nascent hydrogen into succinic acid :—

$$\left.\begin{array}{l}\{COHo \\ \{CH \\ \{CH \\ \{COHo\end{array}\right. + H_2 = \left\{\begin{array}{l}COHo \\ CH_2 \\ CH_2 \\ COHo\end{array}\right. .$$

Succinic acid.

$$\left\{\begin{array}{l}COHo \\ CH_2 \\ ''C \\ COHo\end{array}\right. + H_2 = \left\{\begin{array}{l}COHo \\ CH_2 \\ CH_2 \\ COHo\end{array}\right. .$$

Succinic acid.

The two processes by which succinic acid is always prepared are, the distillation of amber and the fermentation of calcic malate.

Reactions.—1. By distillation, succinic acid splits almost entirely into succinic anhydride and water:—

$$\left\{\begin{array}{l} \mathbf{COH_o} \\ \mathbf{C_2H_4} \\ \mathbf{COH_o} \end{array}\right. = \left\{\begin{array}{l} \mathbf{CO} \\ \mathbf{C_2H_4}\ \ \mathbf{O} \\ \mathbf{CO} \end{array}\right\rfloor + \mathbf{OH_2}.$$

Succinic   Succinic   Water.
acid.   anhydride.

2. Under the action of nascent oxygen produced by electrolysis, succinic acid yields ethylene, carbonic anhydride, and water:—

$$\left\{\begin{array}{l} \mathbf{COH_o} \\ \mathbf{C_2H_4} \\ \mathbf{COH_o} \end{array}\right. + \mathbf{O} = \mathbf{C_2H_4} + 2\mathbf{CO_2} + \mathbf{OH_2}.$$

Succinic   Ethylene.   Carbonic   Water.
acid.   anhydride.

3. Succinic acid may be boiled for hours with concentrated nitric acid without suffering any change; neither is it affected by a mixture of potassic chlorate and hydrochloric acid; but it produces acetic acid when distilled with sulphuric acid and manganic oxide.

4. Succinic acid forms three kinds of salts, viz.:—

Neutral.   Acid.   Superacid.

$$\left\{\begin{array}{l} \mathbf{COM_o} \\ \mathbf{C_2H_4} \\ \mathbf{COM_o} \end{array}\right. \quad \left\{\begin{array}{l} \mathbf{COH_o} \\ \mathbf{C_2H_4} \\ \mathbf{COM_o} \end{array}\right. \quad \left\{\begin{array}{l} \mathbf{COH_o} \\ \mathbf{C_2H_4} \\ \mathbf{COM_o} \end{array}\right. \left\{\begin{array}{l} \mathbf{COH_o} \\ \mathbf{C_2H_4} \\ \mathbf{COH_o} \end{array}\right.$$

$$\left\{\begin{array}{l} \mathbf{CO} \\ \mathbf{C_2H_4}\ \ \mathbf{Mo''}. \\ \mathbf{CO} \end{array}\right\rfloor$$

CHAPTER XLIV.

THE ACIDS.

2. *FUMARIC OR ACRYLOID SERIES.*

General formula...$''$ $\left\{\begin{array}{l} C_nH_{2n-1}(COHo) \\ C_nH_{2n-1}(COHo) \end{array}\right.$ or $C_nH_{2n-2}(COHo)_2$.

In this series there are three isomeric acids containing four atoms of carbon, viz.

Formula.

$$\left.\begin{array}{l}\text{Fumaric acid}\\\text{Maleic acid}\\\text{Isomaleic acid}\end{array}\right\}\ldots\ldots''\mathbf{C}''_2H_2(COHo)_2;$$

and three other isomeric acids containing five atoms of carbon. viz.

$$\left.\begin{array}{l}\text{Itaconic acid}\\\text{Citraconic acid}\\\text{Mesaconic acid}\end{array}\right\}\ldots\ldots^{vi}(\mathbf{C}_3)^{vi}H_4(COHo)_2.$$

Rational notation predicts the existence of a fourth acid belonging to the four-carbon group. The following are the four possible formulæ for these acids:—

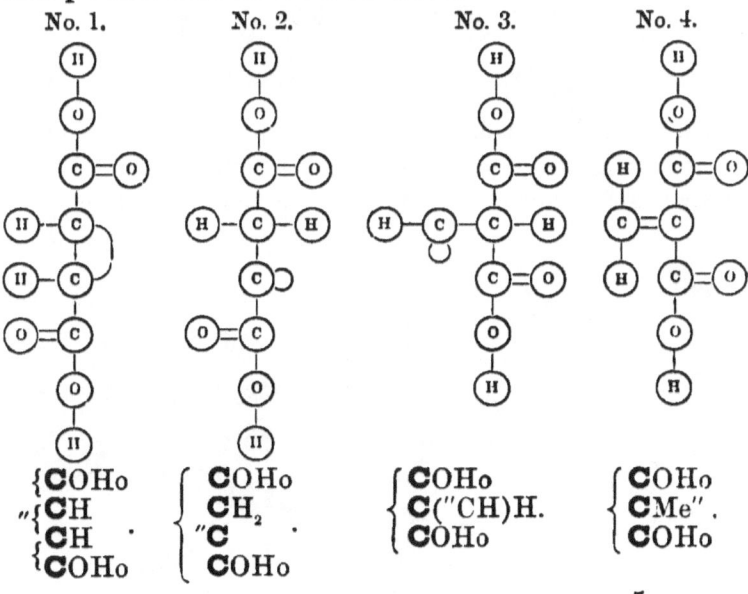

No. 1.　　No. 2.　　No. 3.　　No. 4.

$$\left\{\begin{array}{l}\mathbf{COHo}\\{}_{''}\mathbf{CH}\\\mathbf{CH}\\\mathbf{COHo}\end{array}\right.\quad\left\{\begin{array}{l}\mathbf{COHo}\\\mathbf{CH_2}\\{}_{''}\mathbf{C}\\\mathbf{COHo}\end{array}\right.\cdot\quad\left\{\begin{array}{l}\mathbf{COHo}\\\mathbf{C}(''\mathbf{CH})\mathbf{H.}\\\mathbf{COHo}\end{array}\right.\quad\left\{\begin{array}{l}\mathbf{COHo}\\\mathbf{CMe}''.\\\mathbf{COHo}\end{array}\right.$$

Of these formulæ, Nos. 1 and 2 represent fumaric and maleic acids. Data are still wanting to enable its own particular formula to be assigned to each of these acids; but the first two formulæ must belong to fumaric and maleic acids, because both these acids yield succinic acid under the influence of nascent hydrogen, thus :—

No. 1.

$$"\begin{cases} COHo \\ CH \\ CH \\ COHo \end{cases} + H_2 = \begin{cases} COHo \\ CH_2 \\ CH_2 \\ COHo \end{cases}$$

No. 2.

$$\begin{cases} COHo \\ CH_2 \\ "C \\ COHo \end{cases} + H_2 = \begin{cases} COHo \\ CH_2 \\ CH_2 \\ COHo \end{cases}$$

Succinic acid.

Succinic acid.

That succinic acid contains ethylene $\left("\begin{cases} CH_2 \\ CH_2 \end{cases} \right)$ and not ethylidene $\left(\begin{cases} CH_3 \\ "CH \end{cases} \right)$, is proved by its formation from ethylenic cyanide (p. 289); but formulæ Nos. 3 and 4 give, by the addition of two atoms of hydrogen, the formula of the hitherto unknown isosuccinic acid containing ethylidene, thus :—

No. 3.

$$\begin{cases} COHo \\ C("CH)H \\ COHo \end{cases} + H_2 = \begin{cases} COHo \\ CMeH. \\ COHo \end{cases}$$

No. 4.

$$\begin{cases} COHo \\ CMe" \\ COHo \end{cases} + H_2 = \begin{cases} COHo \\ CMeH. \\ COHo \end{cases}$$

Isosuccinic acid. (Unknown.)

Isosuccinic acid.

Fumaric acid combines directly with bromine, producing dibromosuccinic acid. Maleic acid also combines directly with bromine, producing isodibromosuccinic acid. The following formulæ show the nature of this isomerism :—

Corresponding to No. 1.

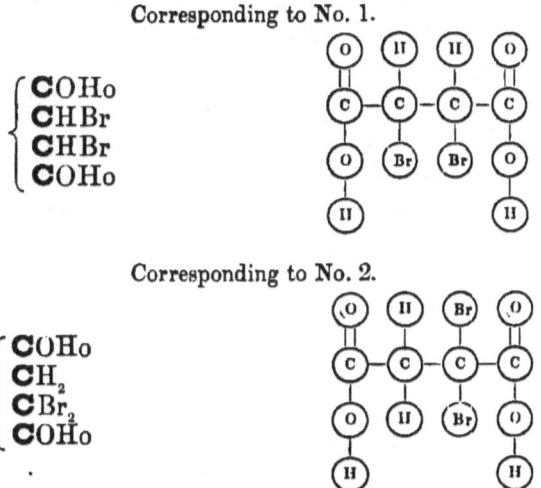

$\begin{cases} COH_0 \\ CHBr \\ CHBr \\ COH_0 \end{cases}$

Corresponding to No. 2.

$\begin{cases} COH_0 \\ CH_2 \\ CBr_2 \\ COH_0 \end{cases}$

When acted upon by nascent hydrogen, both these isomeric bromo-acids produce, as might be expected, the same succinic acid :—

$$\begin{cases} COH_0 \\ CHBr \\ CHBr \\ COH_0 \end{cases} + H_4 = \begin{cases} COH_0 \\ CH_2 \\ CH_2 \\ COH_0 \end{cases} + 2HBr :$$

Succinic   Hydrobromic
acid.   acid.

$$\begin{cases} COH_0 \\ CH_2 \\ CBr_2 \\ COH_0 \end{cases} + H_4 = \begin{cases} COH_0 \\ CH_2 \\ CH_2 \\ COH_0 \end{cases} + 2HBr.$$

Succinic   Hydrobromic
acid.   acid.

Inasmuch as formulæ Nos. 1 and 2 belong to fumaric and maleic acids, it follows that one of the two remaining constitutional formulæ must be that of isomaleic acid. It is impossible at present to determine which of these formulæ is to be assigned to this acid; but in any case there can scarcely be a doubt that isomaleic acid, when treated with nascent hydrogen, will yield isosuccinic acid, as shown above.

Of a similar character is the relation subsisting between the isomeric acids of this series containing five atoms of carbon, viz. itaconic acid, citraconic acid, and mesaconic acid. There are no less than eleven possible formulæ for this five-carbon group of acids; but the three individual formulæ belonging to the three known acids cannot at present be determined. The following four formulæ will serve as specimens of the whole, and as illustrations of the cause of isomerism in these acids:—

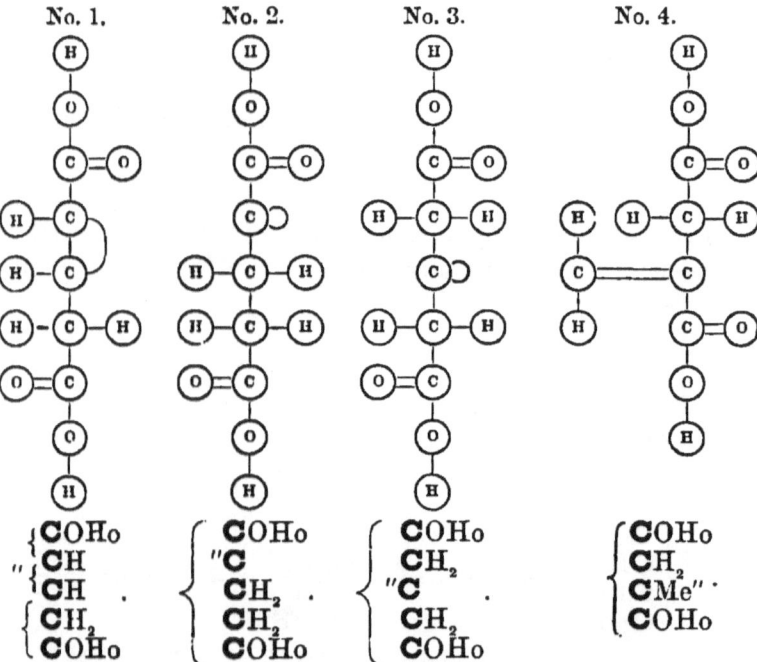

No. 1.

$$\begin{cases} COHo \\ ''\begin{cases} CH \\ CH \end{cases} \\ \begin{cases} CH_2 \\ COHo \end{cases} \end{cases}.$$

No. 2.

$$\begin{cases} COHo \\ ''C \\ CH_2 \\ CH_2 \\ COHo \end{cases}.$$

No. 3.

$$\begin{cases} COHo \\ CH_2 \\ ''C \\ CH_2 \\ COHo \end{cases}.$$

No. 4.

$$\begin{cases} COHo \\ CH_2 \\ CMe'' \\ COHo \end{cases}.$$

Itaconic, citraconic, and mesaconic acids stand in the same relation to pyrotartaric acid as fumaric and maleic acids occupy with regard to succinic acid; for, when submitted to the action of nascent hydrogen, they all yield the same pyrotartaric acid, and it is therefore highly probable that the first three of the above formulæ belong to these acids. The formula of pyrotartaric acid is

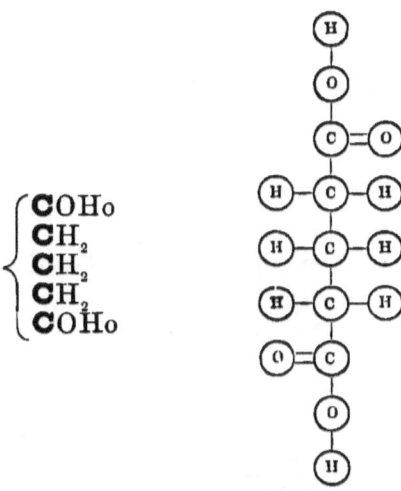

$$\left\{\begin{array}{l} \textbf{CO}Ho \\ \textbf{CH}_2 \\ \textbf{CH}_2 \\ \textbf{CH}_2 \\ \textbf{CO}Ho \end{array}\right.$$

Treated with bromine, the three acids yield three isomeric brominated acids, which are transformed by nascent hydrogen into the same pyrotartaric acid.

CHAPTER XLV.

THE ACIDS.

3. *MALIC OR LACTOID SERIES.*

General formula ...... $\left\{\begin{array}{l} C_nH_{2n-1}Ho(COHo) \\ C_nH_{2n}(COHo) \end{array}\right.$ or $\left\{\begin{array}{l} \textbf{CO}Ho \\ C_nH_{2n-1}Ho. \\ \textbf{CO}Ho \end{array}\right.$

Only two acids belonging to this series are known, viz. tartronic acid and malic acid. Like lactic acid, they both contain an atom of *non-oxatylic* hydroxyl:—

Tartronic acid...... $\begin{cases} \textbf{C}O\text{Ho} \\ \textbf{C}H\text{Ho.} \\ \textbf{C}O\text{Ho} \end{cases}$

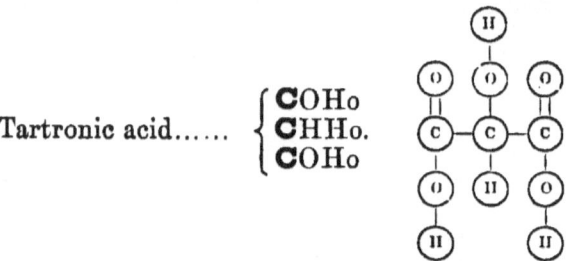

This acid may be regarded as the product of the oxidation of glycerin, although it has not yet been so produced. It is obtained by the gentle oxidation of tartaric acid (p. 352).

Malic acid $\begin{cases} \textbf{C}O\text{Ho} \\ \textbf{C}Me\text{Ho.} \\ \textbf{C}O\text{Ho} \end{cases}$

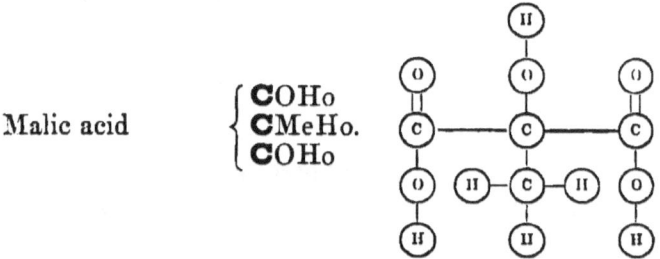

Malic acid may be regarded as the product of the oxidation of the hitherto undiscovered butyl glycerin, $\begin{cases} \textbf{C}H_2\text{Ho} \\ \textbf{C}Me\text{Ho.} \\ \textbf{C}H_2\text{Ho} \end{cases}$

This acid is contained in apples, and in many other fruits.

When gently heated with potassic hydrate, hydrogen is evolved, potassic oxalate and acetate being produced:—

$$\begin{cases} \textbf{C}O\text{Ko} \\ \textbf{C}Me\text{Ho} \\ \textbf{C}O\text{Ko} \end{cases} + \text{KHo} = \begin{cases} \textbf{C}O\text{Ko} \\ \textbf{C}O\text{Ko} \end{cases} + \begin{cases} \textbf{C}H_3 \\ \textbf{C}O\text{Ko} \end{cases} + H_2.$$

| Potassic malate. | Potassic hydrate. | Potassic oxalate. | Potassic acetate. |

4. TARTARIC OR GLYOXYLOID SERIES.

This series contains at present only two members; but these have very numerous isomers, which have been studied, however, only in the case of the first. Like the glyoxylic series, these acids contain two atoms of *non-oxatylic* hydroxyl.

$$\text{Tartaric acid} \ldots\ldots\ldots \left\{ \begin{array}{l} CO Ho \\ CHHo \\ CHHo\text{·} \\ COHo \end{array} \right.$$

$$\text{Homotartaric or glyco-} \atop \text{malic acid} \ldots\ldots\ldots \left\{ \begin{array}{l} COHo \\ CH_2 \\ CHHo. \\ CHHo \\ COHo \end{array} \right.$$

The following are the varieties of tartaric acid:—

1. Dextrotartaric or common tartaric acid.
2. Lævotartaric acid.
3. Racemic acid.
4. Inactive tartaric acid.
5. Metatartaric acid.

The internal arrangement is known only in the case of ordinary and inactive tartaric acids, which are both represented by the above formula.

The difference between dextro- and lævo-tartaric acids is only physical, as far as they are at present investigated.

Racemic acid is a compound of dextro- and lævo-tartaric acids. It may be produced by uniting them, and may again be resolved into them.

Inactive tartaric acid cannot be resolved into dextro- and lævo-tartaric acids.

It may be obtained by the action of water on argentic dibromosuccinate:—

$$\left\{ \begin{array}{l} CO Ago \\ CHBr \\ CHBr \\ CO Ago \end{array} \right. + 2OH_2 = \left\{ \begin{array}{l} COHo \\ CHHo \\ CHHo \\ COHo \end{array} \right. + 2AgBr.$$

Argentic dibromo-   Water.   Inactive tartaric   Argentic
succinate.   acid.   bromide.

The converse of this reaction is the transformation of tartaric acid into succinic acid by means of hydriodic acid (see p. 343).

Metatartaric acid is produced by fusing ordinary tartaric acid.

Reactions.—1. Ordinary tartaric acid, when treated with powerful oxidizing agents, gives formic acid.

2. Under the influence of very gentle oxidizing agents, tartronic acid is formed (p. 350).

$$\left\{\begin{array}{l}\mathbf{CO}\text{Ho}\\ \mathbf{CHH}\text{o}\\ \mathbf{CHH}\text{o}\\ \mathbf{CO}\text{Ho}\end{array}\right. + \mathbf{O_2} = \left\{\begin{array}{l}\mathbf{CO}\text{Ho}\\ \mathbf{CHH}\text{o}\\ \mathbf{CO}\text{Ho}\end{array}\right. + \mathbf{CO_2} + \mathbf{OH_2}.$$

| Tartaric acid. | | Tartronic acid. | Carbonic anhydride. | Water. |

3. Heated with fused potassic hydrate, tartaric acid gives potassic oxalate and acetate, but without evolution of hydrogen:—

$$\left\{\begin{array}{l}\mathbf{CO}\text{Ho}\\ \mathbf{CHH}\text{o}\\ \mathbf{CHH}\text{o}\\ \mathbf{CO}\text{Ho}\end{array}\right. + 3\mathbf{KH}\text{o} = \left\{\begin{array}{l}\mathbf{CO}\text{Ko}\\ \mathbf{CO}\text{Ko}\end{array}\right. + \left\{\begin{array}{l}\mathbf{CH_3}\\ \mathbf{CO}\text{Ko}\end{array}\right. + 3\mathbf{OH_2}.$$

| Tartaric acid. | Potassic hydrate. | Potassic oxalate. | Potassic acetate. | Water. |

CHAPTER XLVI.

THE ACIDS.

TRIBASIC ACIDS.

THE tribasic acids all contain three atoms of oxatyl. They may be divided into the following four series, each series being, however, at present only represented by one acid.

1. Tricarballylic or acetoid series :—

$$\text{Tricarballylic acid} \ldots \left\{\begin{array}{l}\mathbf{CH_2}(\mathbf{CO}\text{Ho})\\ \mathbf{CH}(\mathbf{CO}\text{Ho})\\ \mathbf{CH_2}(\mathbf{CO}\text{Ho})\end{array}\right..$$

2. Aconitic or acryloid series :—

Aconitic acid .... $''\begin{cases} CH(COHo) \\ C(COHo) \\ CH_2(COHo) \end{cases}$.

3. Citric or lactoid series :—

Citric acid .... $\begin{cases} CHHo(COHo) \\ CH(COHo) \\ CH_2(COHo) \end{cases}$.

4. Desoxalic or glyoxyloid series :—

Desoxalic acid ...... $\begin{cases} COHo \\ CHHo \\ CHo(COHo) \\ COHo \end{cases}$ ·

Tricarballylic acid is trihydric. It is obtained from glycerin by replacing the hydroxyl in the latter by cyanogen, and then acting upon the tricyanhydrin, so formed, by potassic hydrate:—

$$\begin{cases} CH_2(CN''') \\ CH(CN''') \\ CH_2(CN''') \end{cases} + 3KHo + 3OH_2 = \begin{cases} CH_2(COKo) \\ CH(COKo) \\ CH_2(COKo) \end{cases} + 3NH_3.$$

Tricyan-   Potassic   Water.   Potassic tricarb-   Ammonia.
hydrin.   hydrate.   allylate.

Aconitic acid (*equisetic acid, citridic acid*) is trihydric. It is found in the roots and leaves of monkshood, and may also be obtained by heating citric acid. Heated to 160° it is converted into itaconic acid :—

$$''\begin{cases} CH(COHo) \\ C(COHo) \\ CH_2(COHo) \end{cases} = ''\begin{cases} COHo \\ CH \\ CH \\ CH_2 \\ COHo \end{cases} + CO_2.$$

Aconitic acid.   Itaconic   Carbonic
     acid.   anhydride.

Citric acid is tetrahydric, and contains, therefore, like lactic acid, one atom of *non-oxatylic* hydroxyl. It is found in the free state in oranges, lemons, citrons, and many other fruits, also in the potato and the onion.   By the graduated application of heat, citric acid yields aconitic, itaconic, and citraconic acids. At the earlier stage of the decomposition it also yields acetone. Heated with fused potassic hydrate it gives potassic oxalate and acetate :—

$$\begin{cases} \textbf{CHHo(COHo)} \\ \textbf{CH(COHo)} \\ \textbf{CH}_2\textbf{(COHo)} \end{cases} + 4\textbf{KHo} = \begin{cases} \textbf{COKo} \\ \textbf{COKo} \end{cases} + 2 \begin{cases} \textbf{CH}_3 \\ \textbf{COKo} \end{cases} + 3\textbf{OH}_2.$$

Citric acid.   Potassic   Potassic   Potassic   Water.
hydrate.   oxalate.   acetate.

Desoxalic acid is pentahydric, and contains, like glyoxylic acid, two atoms of *non-oxatylic* hydroxyl.   It is obtained from the product of the action of sodium upon oxalic ether.

CHAPTER XLVII.

THE ANHYDRIDES.

THE anhydrides are compounds obtained from the acids, by the abstraction of the hydrogen of their hydroxyl, together with sufficient oxygen to form water.

For every two atoms of hydrogen and one of oxygen thus abstracted from hydroxyl, there will obviously remain one atom of oxygen, which, as a dyad element, exactly satisfies the two bonds vacated by the hydroxyl :—

$$2\text{Ho} = \textbf{OH}_2 + \text{O}''.$$

Hydroxyl.   Water.

On this account, two molecules of a monohydric acid are required to form one molecule of anhydride, thus :—

$$2\textbf{CMeOHo} = \begin{cases} \textbf{CMeO} \\ \text{O} \\ \textbf{CMeO} \end{cases} + \textbf{OH}_2.$$

Acetic acid.   Acetic   Water.
anhydride.

The anhydrides of those monobasic and dibasic acids which contain one and two atoms of hydroxyl have alone been investigated.

They may be divided into the following classes :—

1. Anhydrides of the monohydric mono-basic acids ....
$$\begin{cases} C_nH_{2n+1} \\ CO \\ O \\ CO \\ C_nH_{2n+1} \end{cases}$$
and
$$\begin{cases} C_nH_{2n-7} \\ CO \\ O \\ CO \\ C_nH_{2n-7} \end{cases}.$$

2. Anhydrides of the dihydric monobasic acids ....
$$\begin{cases} C(C_nH_{2n+1})_2 \\ \qquad\qquad O. \\ CO \end{cases}$$

3. Anhydrides of the dihydric dibasic acids ....
$$\begin{cases} CO\!\!-\!\!\rceil \\ C_nH_{2n}\ O. \\ CO\!\!-\!\!\rfloor \end{cases}$$

1. ANHYDRIDES OF THE MONOHYDRIC MONOBASIC ACIDS.

These are known only in the acetic and benzoic series.

They bear the same relation to the acids from which they are derived as the ethers to the alcohols.

The residues of different acids can unite to form mixed anhydrides analogous to the mixed ethers. Aceto-benzoic anhydride is a body of this class.

Formation.—By the action of the chloracids, or so-called chlorides, of the monad negative radicals on the potassic salts of the acids :—

$$C(C_nH_{2n+1})OKo + C(C_nH_{2n+1})OCl = \begin{cases} C(C_nH_{2n+1})O \\ O \\ C(C_nH_{2n+1})O \end{cases} + KCl.$$

Potassic salt.   Chloracid.   Anhydride.   Potassic chloride.

Reaction.—In contact with water they are converted into the corresponding acids :—

$$\begin{cases} C(C_nH_{2n+1})O \\ O \\ C(C_nH_{2n+1})O \end{cases} + OH_2 = 2C(C_nH_{2n+1})OHo.$$

Anhydride.   Water.   Acid.

The following is a list of the anhydrides belonging to this class :—

			Fusing-point.	Boiling-point.
Acetic anhydride	$\begin{cases} \mathbf{C}MeO \\ O \\ \mathbf{C}MeO \end{cases}$ or $\begin{cases} \mathbf{C}(CH_3)O \\ O \\ \mathbf{C}(CH_3)O \end{cases}$		——	138°.
Propionic anhydride ...	$\begin{cases} \mathbf{C}EtO \\ O \\ \mathbf{C}EtO \end{cases}$ or $\begin{cases} \mathbf{C}(C_2H_5)O \\ O \\ \mathbf{C}(C_2H_5)O \end{cases}$		——	165°.
Butyric anhydride	$\begin{cases} \mathbf{C}PrO \\ O \\ \mathbf{C}PrO \end{cases}$ or $\begin{cases} \mathbf{C}(C_3H_7)O \\ O \\ \mathbf{C}(C_3H_7)O \end{cases}$		——	about 190°:
Valeric anhydride	$\begin{cases} \mathbf{C}BuO \\ O \\ \mathbf{C}BuO \end{cases}$ or $\begin{cases} \mathbf{C}(C_4H_9)O \\ O \\ \mathbf{C}(C_4H_9)O \end{cases}$		——	about 215°.
Caproic anhydride......	$\begin{cases} \mathbf{C}AyO \\ O \\ \mathbf{C}AyO \end{cases}$ or $\begin{cases} \mathbf{C}(C_5H_{11})O \\ O \\ \mathbf{C}(C_5H_{11})O \end{cases}$		——	——
Œnanthylic anhydride.	$\begin{cases} \mathbf{C}CpO \\ O \\ \mathbf{C}CpO \end{cases}$ or $\begin{cases} \mathbf{C}(C_6H_{13})O \\ O \\ \mathbf{C}(C_6H_{13})O \end{cases}$		42°.	310°.
Benzoic anhydride......	$\begin{cases} \mathbf{C}(C_6H_5)O \\ O \\ \mathbf{C}(C_6H_5)O \end{cases}$	....	——	——
Acetobenzoic anhydride	$\begin{cases} \mathbf{C}MeO \\ O \\ \mathbf{C}(C_6H_5)O \end{cases}$	....	——	120°·
Caprylic anhydride ...	$\begin{cases} \mathbf{C}(C_7H_{15})O \\ O \\ \mathbf{C}(C_7H_{15})O \end{cases}$	....	below 0°.	about 290°.
Pelargonic anhydride..	$\begin{cases} \mathbf{C}(C_8H_{17})O \\ O \\ \mathbf{C}(C_8H_{17})O \end{cases}$	....	+5°.	——
Palmitic anhydride ...	$\begin{cases} \mathbf{C}(C_{15}H_{31})O \\ O \\ \mathbf{C}(C_{15}H_{31})O \end{cases}$	....	53°·8.	——

2. *ANHYDRIDES OF THE DIHYDRIC MONOBASIC ACIDS.*

Formation.—By applying heat to a dihydric monobasic acid, thus :—

$$\left\{ \begin{array}{l} \mathbf{CMeHHo} \\ \mathbf{COHo} \end{array} \right. = \left\{ \begin{array}{l} \mathbf{CMeH} \\ \mathbf{CO} \end{array} \right\} O + \mathbf{OH_2}.$$

Lactic acid.   Lactide.   Water.
(Lactic anhydride.)

Reaction.—Boiled with water, and especially with alkalies, they reproduce the acids from which they were derived :—

$$\left\{ \begin{array}{l} \mathbf{CMeH} \\ \mathbf{CO} \end{array} \right\} O + \mathbf{OH_2} = \left\{ \begin{array}{l} \mathbf{CMeHHo} \\ \mathbf{COHo} \end{array} \right. .$$

Lactide.   Water.   Lactic acid.

3. *ANHYDRIDES OF THE DIHYDRIC DIBASIC ACIDS.*

Formation.—By the action of heat, or of substances having an attraction for water, upon the dihydric dibasic acids :—

$$\left\{ \begin{array}{l} \mathbf{COHo} \\ \mathbf{C_2H_4} \\ \mathbf{COHo} \end{array} \right. = \left\{ \begin{array}{l} \mathbf{CO} \\ \mathbf{C_2H_4} \\ \mathbf{CO} \end{array} \right\} O + \mathbf{OH_2}.$$

Succinic acid.   Succinic anhydride.   Water.

Reaction.—Like the anhydrides of the first and second classes, they unite with water, reproducing the acids from which they were derived.

CHAPTER XLVIII.

THE KETONES.

THE ketones may be regarded as derived from the fatty acids, by the substitution of the hydroxyl of the latter by a monad alcohol radical; they thus resemble the aldehydes in constitution :—

$$\left\{ \begin{array}{l} CH_3 \\ COHo \end{array} \right. \qquad \left\{ \begin{array}{l} CH_3 \\ COH \end{array} \right. \qquad \left\{ \begin{array}{l} CH_3 \\ COMe \end{array} \right.$$

Acetic   Acetic   Acetone.
acid.   aldehyde.

The ketones are also correctly viewed as compounds of carbonic oxide with monad alcohol radicals, thus :—

$$COMe_2.$$
Acetone.

By the action of nascent hydrogen upon the ketones, they are converted into secondary alcohols, whilst the aldehydes, under the same treatment, yield primary alcohols :—

$$\left\{ \begin{array}{l} CH_3 \\ COMe \end{array} \right. + H_2 = \left\{ \begin{array}{l} CH_3 \\ CMeHHo \end{array} \right.$$

Acetone.   Isopropylic
alcohol.

Ketones, unlike aldehydes, do not oxidize spontaneously; neither do they reduce ammoniacal solution of argentic oxide.

Like aldehydes, many of them combine with hydric potassic or hydric sodic sulphite.

Formation.—1. By the action of the zinc compounds of the positive monad radicals upon chloracids :—

$$2 \left\{ \begin{array}{l} C_nH_{2n+1} \\ COCl \end{array} \right. + Zn(C_nH_{2n+1})_2 = 2 \left\{ \begin{array}{l} C_nH_{2n+1} \\ CO(C_nH_{2n+1}) \end{array} \right. + ZnCl_2$$

Chloracid.   Zinc compound.   Ketone.   Zincic
chloride.

2. By the action of sodic ethide and its homologues on carbonic oxide :—

$$CO + 2Na(C_nH_{2n+1}) = \left\{ \begin{array}{l} C_nH_{2n+1} \\ CO(C_nH_{2n+1}) \end{array} \right. + Na_2.$$

Carbonic   Sodium compound.   Ketone.
oxide.

3. By the distillation of the salts of the fatty acids :—

$$2 \left\{ \begin{array}{l} C_nH_{2n+1} \\ COKo \end{array} \right. = \left\{ \begin{array}{l} C_nH_{2n+1} \\ CO(C_nH_{2n+1}) \end{array} \right. + COKo_2.$$

Potassic salt   Ketone.   Potassic
of fatty acid.   carbonate.

4. By distilling together salts of two different fatty acids, ketones containing two different basylous radicals are obtained:—

$$\left\{ \begin{array}{l} \mathbf{CEtH_2} \\ \mathbf{COKo} \end{array} \right. + \left\{ \begin{array}{l} \mathbf{CH_3} \\ \mathbf{COKo} \end{array} \right. = \left\{ \begin{array}{l} \mathbf{CEtH_2} \\ \mathbf{COMe} \end{array} \right. + \mathbf{COKo_2}.$$

Potassic   Potassic   Propylmethyl   Potassic
butyrate.   acetate.   ketone.   carbonate.

5. Ketones are also produced by the consecutive action of sodium and the iodides of the C_nH_{2n+1} radicals on acetic ether, the product so obtained being subsequently boiled with alcoholic solution of potassic hydrate:—

$$4\left\{ \begin{array}{l} \mathbf{CH_3} \\ \mathbf{COEto} \end{array} \right. + \mathbf{Na_2} = 2\left\{ \begin{array}{l} \mathbf{CH_3} \\ \mathbf{CO} \\ \mathbf{CHNa} \\ \mathbf{COEto} \end{array} \right. + \mathbf{2EtHo} + \mathbf{H_2};$$

Acetic ether.   Ethylic sodacetone   Alcohol.
carbonate.

$$\left\{ \begin{array}{l} \mathbf{CH_3} \\ \mathbf{CO} \\ \mathbf{CHNa} \\ \mathbf{COEto} \end{array} \right. + \mathbf{EtI} = \left\{ \begin{array}{l} \mathbf{CH_3} \\ \mathbf{CO} \\ \mathbf{CHEt} \\ \mathbf{COEto} \end{array} \right. + \mathbf{NaI};$$

Ethylic sodacetone   Ethylic   Ethylic ethacetone   Sodic
carbonate.   iodide.   carbonate.   iodide.

$$\left\{ \begin{array}{l} \mathbf{CH_3} \\ \mathbf{CO} \\ \mathbf{CHEt} \\ \mathbf{COEto} \end{array} \right. + \mathbf{2KHo} = \left\{ \begin{array}{l} \mathbf{CH_3} \\ \mathbf{CO} \\ \mathbf{CEtH_2} \end{array} \right. + \mathbf{EtHo} + \mathbf{COKo_2}.$$

Ethylic ethacetone   Potassic   Ethylated   Alcohol.   Potassic
carbonate.   hydrate.   acetone.   carbonate.

$$\text{Ethylated acetone} \ldots\ldots \left\{ \begin{array}{l} \mathbf{CH_3} \\ \mathbf{CO} \\ \mathbf{CEtH_2} \end{array} \right. = \left\{ \begin{array}{l} \mathbf{CEtH_2} \\ \mathbf{COMe} \end{array} \right. .$$

In this compound one atom of hydrogen in acetone has been displaced by ethyl.

A second atom may be displaced in the following manner:—

$$2\left\{ \begin{array}{l} \mathbf{CH_3} \\ \mathbf{COEto} \end{array} \right. + \mathbf{Na_2} = \left\{ \begin{array}{l} \mathbf{CH_3} \\ \mathbf{CO} \\ \mathbf{CNa_2} \\ \mathbf{COEto} \end{array} \right. + \mathbf{EtHo} + \mathbf{H_2}:$$

Acetic   Ethylic disodace- Alcohol
ethe   tone carbonate.

$$\left\{\begin{array}{l}\mathbf{CH_3}\\\mathbf{CO}\\\mathbf{CNa_2}\\\mathbf{COEto}\end{array}\right. + 2\mathbf{EtI} = \left\{\begin{array}{l}\mathbf{CH_3}\\\mathbf{CO}\\\mathbf{CEt_2}\\\mathbf{COEto}\end{array}\right. + 2\mathbf{NaI}:$$

Ethylic disodacetone carbonate.   Ethylic iodide.   Ethylic diethacetone carbonate.   Sodic iodide.

$$\left\{\begin{array}{l}\mathbf{CH_3}\\\mathbf{CO}\\\mathbf{CEt_2}\\\mathbf{COEto}\end{array}\right. + 2\mathbf{KHo} = \left\{\begin{array}{l}\mathbf{CH_3}\\\mathbf{CO}\\\mathbf{CEt_2H}\end{array}\right. + \mathbf{EtHo} + \mathbf{C}\mathbf{OKo_2}.$$

Ethylic diethacetone carbonate.   Potassic hydrate.   Diethylated acetone.   Alcohol.   Potassic carbonate.

Diethylated acetone ......$\left\{\begin{array}{l}\mathbf{CH_3}\\\mathbf{CO}\\\mathbf{CEt_2H}\end{array}\right. = \left\{\begin{array}{l}\mathbf{CEt_2H}\\\mathbf{COMe}\end{array}\right.$

The following is a list of the names, constitutional formulæ, and boiling-points of those ketones which are best known:—

Name	Formula	Boiling-point.
Acetone	$\left\{\begin{array}{l}\mathbf{CH_3}\\\mathbf{COMe}\end{array}\right.$	56°.
Methylated actone (*Ethyl acetyl, methyl acetone.*)	$\left\{\begin{array}{l}\mathbf{CMeH_2}\\\mathbf{COMe}\end{array}\right.$	81°.
Isomeric. Dimethylated acetone. (*Ethyl acetone*)	$\left\{\begin{array}{l}\mathbf{CMe_2H}\\\mathbf{COMe}\end{array}\right.$	93°·5.
Ethylated acetone	$\left\{\begin{array}{l}\mathbf{CEtH_2}\\\mathbf{COMe}\end{array}\right.$	101°.
Propione. (*Ethyl propionyl.*)	$\left\{\begin{array}{l}\mathbf{CMeH_2}\\\mathbf{COEt}\end{array}\right.$	101°.
Isomeric. Methyl valeral	$\left\{\begin{array}{l}\mathbf{CPrH_2}\\\mathbf{COMe}\end{array}\right.$	120°.
Ethyl butyral	$\left\{\begin{array}{l}\mathbf{CH_3}\\\mathbf{COBu}\end{array}\right.$	128°.
Isomeric. Diethylated acetone.	$\left\{\begin{array}{l}\mathbf{CEt_2H}\\\mathbf{COMe}\end{array}\right.$	138°.
Butyrone	$\left\{\begin{array}{l}\mathbf{CEtH_2}\\\mathbf{COPr}\end{array}\right.$	144°.

Little is known of the ketones of the C_nH_{2n-7} series. Two of them have been obtained:—

Benzophenone $\left(\begin{matrix} C_6H_5 \\ CO(C_6H_5) \end{matrix} \right)$, the ketone of benzoic acid, is obtained by heating potassic benzoate.

Methyl benzone or *methyl benzoyl* $\left(\left\{ \begin{matrix} CH_3 \\ CO(C_6H_5) \end{matrix} \right) \right.$ is prepared by distilling together calcic acetate and benzoate.

CHAPTER XLIX.

ETHEREAL SALTS.

THESE compounds correspond to the metallic oxysalts of the acids.

The acids from which they are derived may be either mineral or organic; but the base must always be organic. The haloid ethereal salts are excluded from this family; they have been already described as haloid ethers.

The ethereal salts are produced by reactions analogous to those employed for the preparation of metallic salts:—

$$\left\{ \begin{matrix} CH_3 \\ COHo \end{matrix} \right. + KHo = \left\{ \begin{matrix} CH_3 \\ COKo \end{matrix} \right. + OH_2.$$

| Acetic acid. | Potassic hydrate. | Potassic acetate. | Water. |

$$\left\{ \begin{matrix} CH_3 \\ COHo \end{matrix} \right. + EtHo = \left\{ \begin{matrix} CH_3 \\ COEto \end{matrix} \right. + OH_2.$$

| Acetic acid. | Ethylic hydrate. | Ethylic acetate. | Water. |

But as the hydrates of the organic radicals do not act upon acids so energetically as potassic hydrate, it is often advisable to employ the acid in the form of a potassic salt, and the radical as a sulphoacid, thus, with acids of the acetic series:—

$$SO_2Ho(C_nH_{2n+1}O). + \left\{ \begin{matrix} C_nH_{2n+1} \\ COKo \end{matrix} \right. = \left\{ \begin{matrix} C_nH_{2n+1} \\ CO(C_nH_{2n+1}O) \end{matrix} \right. + SO_2HoKo.$$

| Sulphoacid. | Potassic salt. | Ethereal salt. | Hydric potassic sulphate. |

R

Monobasic acids form only one ethereal salt with each monacid alcohol; and this salt is always neutral.

With diacid alcohols they each form two ethereal salts, and with triacid alcohols three ethereal salts. These are also neutral. Thus with acetic acid we have :—

Acetic salt of a monacid alcohol :—

$$\left\{ \begin{array}{l} \mathbf{CH_3} \\ \mathbf{CO\dot{E}to} \end{array} \right.$$

Ethylic acetate.

Acetic salts of a diacid alcohol :—

$$\left\{ \begin{array}{l} \mathbf{CH_2Ho} \\ \mathbf{CH_2\text{-}O\text{-}CMeO} \end{array} \right. \qquad \left\{ \begin{array}{l} \mathbf{CH_2\text{-}O\text{-}CMeO} \\ \mathbf{CH_2\text{-}O\text{-}CMeO} \end{array} \right.$$

Monacetic glycol.   Diacetic glycol.

Acetic salts of a triacid alcohol :—

$$\left\{ \begin{array}{l} \mathbf{CH_2Ho} \\ \mathbf{CHHo} \\ \mathbf{CH_2\text{-}O\text{-}CMeO} \end{array} \right. \quad \left\{ \begin{array}{l} \mathbf{CH_2\text{-}O\text{-}CMeO} \\ \mathbf{CHHo} \\ \mathbf{CH_2\text{-}O\text{-}CMeO} \end{array} \right. \quad \left\{ \begin{array}{l} \mathbf{CH_2\text{-}O\text{-}CMeO} \\ \mathbf{CH\text{-}O\text{-}CMeO} \\ \mathbf{CH_2\text{-}O\text{-}CMeO} \end{array} \right.$$

Monacetin.   Diacetin.   Triacetin.

Dibasic acids form, with monacid alcohols, two series of ethereal salts :—

1. Acid ethereal salts, as :—

Succinethylic acid .... $\left\{ \begin{array}{l} \mathbf{COEto} \\ \mathbf{C_2H_4} \\ \mathbf{CO\dot{H}o} \end{array} \right.$

2. Neutral ethereal salts, as :—

Ethylic succinate..................... $\left\{ \begin{array}{l} \mathbf{COEto} \\ \mathbf{C_2H_4} \\ \mathbf{CO\dot{E}to} \end{array} \right.$

In the same manner, tribasic acids form with monacid alcohols three series of ethereal salts, the first two of which are acid, and the third neutral.

Prolonged contact with water generally decomposes the ethereal salts, liberating the radicals of the bases in the form of alcohols :—

$$\mathbf{SO_2Meo_2} \ + \ \mathbf{2OH_2} \ = \ \mathbf{SO_2Ho_2} \ + \ \mathbf{2MeHo}.$$

Methylic   Water.   Sulphuric   Methylic
sulphomethylate.   acid.   alcohol.

Ebullition with potassic hydrate, especially when the latter is dissolved in alcohol, effects this transformation very speedily:—

$$\left\{ \begin{matrix} \mathbf{CH_3} \\ \mathbf{COEto} \end{matrix} \right. \quad + \quad \mathbf{KHo} \quad = \quad \left\{ \begin{matrix} \mathbf{CH_3} \\ \mathbf{COKo} \end{matrix} \right. \quad + \quad \mathbf{EtHo}.$$

Ethylic acetate. Potassic hydrate. Potassic acetate. Ethylic alcohol.

CHAPTER L.

ORGANIC COMPOUNDS CONTAINING TRIAD AND PENTAD NITROGEN OR THEIR ANALOGUES.

THIS numerous family may be divided into two great classes:—

1. Compounds of triad nitrogen, phosphorus, arsenic, and antimony.

2. Compounds of pentad nitrogen, phosphorus, arsenic, and antimony.

I. COMPOUNDS OF TRIAD NITROGEN AND OF ITS ANALOGUES.

This class may be again subdivided as follows:—

Positive.	Neutral.	Negative.
1. Amines.	1. Amides.	1. Imides and
2. Phosphines.	2. Alkalamides.	nitrides.
3. Arsines.	3. Trichlorinated and	
	tribrominated amines.	
4. Stibines.	4. Haloid compounds	
5. Oxybases.	of oxybases.	

Of these the Amines and Amides are the most important.

POSITIVE OR BASYLOUS SECTION.

1. *THE AMINES.*

The Amines are commonly termed organic bases or artificial alkaloids; they are divided into—

> A. Monamines.
> B. Diamines.
> C. Triamines.
> D. Tetramines.

The last two have been but little investigated.

A. *MONAMINES.*

There are three kinds of monamines :—

> a. Primary monamines.
> β. Secondary monamines.
> γ. Tertiary monamines.

a. *Primary Monamines.*

General formulæ.

Methyl or C_nH_{2n+1} series .... $\mathbf{N}(C_nH_{2n+1})H_2$.
Vinyl or C_nH_{2n-1} series .... $\mathbf{N}(C_nH_{2n-1})H_2$.
Phenyl or C_nH_{2n-7} series .... $\mathbf{N}(C_nH_{2n-7})H_2$.

Formation.—1. By the reduction of the nitro-substitution compounds of the hydrides of the radicals by sulphuretted hydrogen, ammonic sulphide, zinc and sulphuric acid, or iron and acetic acid :—

$$\mathbf{N}(C_6H_5)O_2 \ + \ 3\mathbf{S}H_2 \ = \ \mathbf{N}(C_6H_5)H_2 \ + \ 2\mathbf{O}H_2 \ + \ S_3.$$

NitrobenzoL   Sulphuretted   Aniline.   Water.
hydrogen.

2. By treating cyanic ethers with boiling solution of potassic

hydrate. The reaction is perfectly analogous to the decomposition of cyanic acid with potassic hydrate:—

$$\mathbf{CN'''Ho} \quad + \quad 2KHo \quad = \quad \mathbf{COKo_2} \quad + \quad \mathbf{NH_3}.$$

| Cyanic acid. | Potassic hydrate. | Potassic carbonate. | Ammonia. |

$$\mathbf{CN'''Eto} \quad + \quad 2KHo \quad = \quad \mathbf{COKo_2} \quad + \quad \mathbf{NEtH_2}.$$

| Ethylic cyanate. | Potassic hydrate. | Potassic carbonate. | Ethylamine. |

3. By the action of the haloid ethers of the monad positive radicals upon ammonia, and subsequent action of potassic hydrate upon the product so formed:—

$$\mathbf{NH_3} \quad + \quad EtI \quad = \quad \mathbf{NEtH_3I}.$$

| Ammonia. | Ethylic iodide. | Ethylammonic iodide. |

$$\mathbf{NEtH_3I} \quad + \quad KHo \quad = \quad \mathbf{NEtH_2} \quad + \quad KI \quad + \quad \mathbf{OH_2}.$$

| Ethylammonic iodide. | Potassic hydrate. | Ethylamine. | Potassic iodide. | Water. |

The following are a few of the primary monamines:—

Methylamine............	$\mathbf{NMeH_2}$ or $\mathbf{N(CH_3)H_2}$.
Ethylamine	$\mathbf{NEtH_2}$ or $\mathbf{N(C_2H_5)H_2}$.
Amylamine	$\mathbf{NAyH_2}$ or $\mathbf{N(C_5H_{11})H_2}$.
Allylamine..............	$\mathbf{NAllH_2}$ or $\mathbf{N(C_3H_5)H_2}$.
Phenylamine (*Aniline*)	$\mathbf{NPhH_2}$ or $\mathbf{N(C_6H_5)H_2}$.

Reaction.—Treated with nitrous acid, they evolve nitrogen and yield the corresponding alcohols:—

$$\mathbf{NPhH_2} \quad + \quad \mathbf{NOHo} \quad = \quad PhHo \quad + \quad N_2 \quad + \quad \mathbf{OH_2}.$$

| Phenylamine. | Nitrous acid. | Phenylic alcohol. | | Water. |

β. Secondary Monamines.

General formulæ.

Methyl or C_nH_{2n+1} series	$\mathbf{N(C_nH_{2n+1})_2H}$.
Vinyl or C_nH_{2n-1} series............	$\mathbf{N(C_nH_{2n-1})_2H}$.
Phenyl or C_nH_{2n-7} series	$\mathbf{N(C_nH_{2n-7})_2H}$.

The secondary monamines are derived from ammonia by the displacement of two atoms of hydrogen by monad positive radicals. They are sometimes called *Imidogen bases*.

Formation.—By the action of the haloid compounds of the

monad positive radicals on the primary monamines, and subsequent treatment with potassic hydrate :—

$$\mathbf{N}EtH_2 \ + \ EtI \ = \ \mathbf{N}Et_2H_2I.$$

Ethylamine.   Ethylic   Diethylammonic
                       iodide.   iodide.

$$\mathbf{N}Et_2H_2I \ + \ KHo \ = \ \mathbf{N}Et_2H \ + \ KI \ + \ \mathbf{O}H_2.$$

Diethylammonic   Potassic   Diethylamine.   Potassic   Water.
iodide.   hydrate.   iodide.

By using the iodide of a radical different from that already contained in the primary monamine, secondary monamines may be formed containing two different radicals, thus :—

$$\mathbf{N}PhH_2 \ + \ EtI \ = \ \mathbf{N}EtPhH_2I.$$

Phenylamine.   Ethylic   Ethylphenylammonic
(Aniline.)   iodide.   iodide.

$$\mathbf{N}EtPhH_2I \ + \ KHo \ = \ \mathbf{N}EtPhH \ + \ KI \ + \ \mathbf{O}H_2.$$

Ethylphenylam-   Potassic   Ethylphenylamine.   Potassic   Water.
monic iodide.   hydrate.   (Ethylaniline.)   iodide.

The following secondary monamines are known :—

Dimethylamine............ $\mathbf{N}Me_2H$ or $\mathbf{N}(CH_3)_2H.$
Diethylamine .... $\mathbf{N}Et_2H$ or $\mathbf{N}(C_2H_5)_2H.$
Methylethylamine ...... $\mathbf{N}MeEtH$ or $\mathbf{N}(CH_3)(C_2H_5)H.$
Ethylamylamine .... $\mathbf{N}EtAyH$ or $\mathbf{N}(C_2H_5)(C_5H_{11})H.$
Ethylphenylamine ...... $\mathbf{N}EtPhH$ or $\mathbf{N}(C_2H_5)(C_6H_5)H.$
Piperidine .... $\mathbf{N}(C_5H_{10})''H.$
Conine .... $\mathbf{N}(C_8H_{14})''H.$

γ. *Tertiary monamines.*

Formation.—By acting upon the secondary monamines with the iodides of the monad positive radicals, and subsequent treatment with potassic hydrate :—

$$\mathbf{N}Et_2H \ + \ EtI \ = \ \mathbf{N}Et_3HI.$$

Diethylamine.   Ethylic   Triethylammonic
                     iodide.   iodide.

$$\mathbf{N}Et_3HI \ + \ KHo \ = \ \mathbf{N}Et_3 \ + \ KI \ + \ \mathbf{O}H_2.$$

Triethylammonic   Potassic   Triethylamine.   Potassic   Water.
iodide.   hydrate.   iodide

By varying the radicals, tertiary monamines with several different radicals may be formed. The following are a few of the known tertiary monamines :—

Trimethylamine	$\mathbf{N}Me_3$ or	$\mathbf{N}(CH_3)_3$.
Triethylamine	$\mathbf{N}Et_3$ or	$\mathbf{N}(C_2H_5)_3$.
Triamylamine	$\mathbf{N}Ay_3$ or	$\mathbf{N}(C_5H_{11})_3$.
Methyl-ethyl-phenylamine	$\mathbf{N}MeEtPh$ or	$\mathbf{N}(CH_3)(C_2H_5)(C_6H_5)$.
Pyridine		$\mathbf{N}(C_5H_5)'''$.
Picoline		$\mathbf{N}(C_6H_7)'''$.
Lutidine		$\mathbf{N}(C_7H_9)'''$.
Collidine		$\mathbf{N}(C_8H_{11})'''$.
Parvoline		$\mathbf{N}(C_9H_{13})'''$.

The constitution of the triad radicals contained in the last five bases is not known.

Tertiary monamines, when acted upon by the iodides of monad positive radicals, yield iodides which are not decomposed by potassic hydrate. In this manner tertiary monamines may be distinguished from primary and secondary monamines. The three may be distinguished from each other by the alternate action of ethylic iodide and potassic hydrate : thus, as we have just seen, tertiary monamines are recognized by producing immediately iodides which are not decomposed by potassic hydrate ; a secondary monamine, however, produces an iodide decomposable by potassic hydrate ; but the base thus liberated is tertiary, and will therefore be transformed immediately into the stable iodide by a second application of ethylic iodide. A primary monamine requires three applications of ethylic iodide and potassic hydrate to produce the same result.

B. *DIAMINES.*

Formation.—The diamines are formed by coupling together two atoms of nitrogen in two molecules of ammonia, or of a primary or secondary monamine, by a dyad radical, which at the same time takes the place of two atoms of hydrogen ; thus :—

$$\left\{\begin{array}{l} \mathbf{NH_2} \\ Et'' \\ \mathbf{NH_2} \end{array}\right. \qquad \left\{\begin{array}{l} \mathbf{NH} \\ Et''_2 \\ \mathbf{NH} \end{array}\right. \qquad \left\{\begin{array}{l} \mathbf{N} \\ Et''_3 \\ \mathbf{N} \end{array}\right. \text{ or } \left\{\begin{array}{l} \mathbf{NEt''} \\ Et'' \\ \mathbf{NEt''} \end{array}\right.$$

Primary   Secondary   Tertiary diamine.
diamine.   diamine.

This reaction is effected by treating ammonia or a primary or secondary monamine with the haloid salt (preferably a bromide) of the dyad radical, thus :—

$$2\mathbf{NH_3} \quad + \quad Et''Br_2 \quad = \quad \left\{\begin{array}{l} \mathbf{NH_3Br} \\ Et'' \\ \mathbf{NH_3Br} \end{array}\right.$$

Ammonia.   Ethylenic   Ethylene-diammonic
       dibromide.   dibromide.

When the salts of ethylene diammonium are decomposed by potassic hydrate, an oxide of the basylous radical is produced, thus :—

$$\left\{\begin{array}{l} \mathbf{NH_3Br} \\ Et'' \\ \mathbf{NH_3Br} \end{array}\right. + \quad 2KHo \quad = \quad \left\{\begin{array}{l} \mathbf{NH_3} \\ Et'' \\ \mathbf{NH_3} \end{array}\right\} O \; + \; \mathbf{OH_2} \; + \; 2KBr.$$

Ethylene-diam-   Potassic   Ethylene-diam-   Water.   Potassic
monic dibromide.   hydrate.   monic oxide.   bromide.

In this respect the diamines differ from the monamines.

Urea and its derivatives belong to the class of diamines.

These compounds are produced by boiling a solution of ammonic cyanate or ethylammonic cyanate, or a homologue of the latter. In these compounds, the two atoms of nitrogen are held together by the dyad radical carbonyl, $\mathbf{CO}$:—

$$\mathbf{CN'''(N^vH_4O)} \quad = \quad \left\{\begin{array}{l} \mathbf{NH_2} \\ \mathbf{CO} \\ \mathbf{NH_2} \end{array}\right.$$

Ammonic cyanate.   Urea.

$$\mathbf{CN'''(N^vEtH_3O)} \quad = \quad \left\{\begin{array}{l} \mathbf{NHEt} \\ \mathbf{CO} \\ \mathbf{NH_2} \end{array}\right.$$

Ethyl-ammonic   Ethylurea.
cyanate.

Ureas in which ethyl and other monad basylous radicals are substituted for hydrogen may also be obtained by the action of ammonia or a monamine on the cyanic ethers, thus :—

$$\mathbf{CN'''Eto} \ + \ \mathbf{NH_3} \ = \ \begin{cases} \mathbf{NHEt} \\ \mathbf{CO} \\ \mathbf{NH_2} \end{cases}.$$

Ethylic   Ammonia.   Ethylurea.
cyanate.

$$\mathbf{CN'''Eto} \ + \ \mathbf{NH_2Et} \ = \ \begin{cases} \mathbf{NHEt} \\ \mathbf{CO} \\ \mathbf{NHEt} \end{cases}.$$

Ethylic   Ethylamine.   Diethylurea.
cyanate.

Reaction.—Urea is decomposed by nitrous anhydride :—

$$\begin{cases} \mathbf{NH_2} \\ \mathbf{CO} \\ \mathbf{NH_2} \end{cases} + \ \mathbf{N_2O_3} \ = \ \mathbf{CO_2} \ + \ 2\mathbf{N_2} \ + \ 2\mathbf{OH_2}.$$

Urea.   Nitrous   Carbonic   Water.
anhydride.   anhydride.

The following is a list of the best-known diamines :—

Ethylene diamine .... $\begin{cases} \mathbf{NH_2} \\ \mathbf{Et''} \\ \mathbf{NH_2} \end{cases}$.

Ethylene diethyl diamine .... $\begin{cases} \mathbf{NHEt} \\ \mathbf{Et''} \\ \mathbf{NHEt} \end{cases}$.

Urea .... $\begin{cases} \mathbf{NH_2} \\ \mathbf{CO} \\ \mathbf{NH_2} \end{cases}$.

Ethyl urea .... $\begin{cases} \mathbf{NHEt} \\ \mathbf{CO} \\ \mathbf{NH_2} \end{cases}$.

Sulphophenylurea .... $\begin{cases} \mathbf{NHPh} \\ \mathbf{CS''} \\ \mathbf{NH_2} \end{cases}$.

THE NATURAL ALKALOIDS.

Of the constitution of these organic bases very little is known. The following is a list of the chief of them, with the sources whence they are derived :—

Alkaloids from Opium.

Morphine......................	$C_{17}H_{19}NO_3$.
Codeine	$C_{18}H_{21}NO_3$, OH_2.
Thebaine	$C_{10}H_{21}NO_3$.
Papaverine	$C_{20}H_{21}NO_4$.
Narcotine......................	$C_{22}H_{23}NO_7$.
Narceine	$C_{23}H_{29}NO_9$.

From Cinchona Bark.

Quinine	$C_{20}H_{24}N_2O_2$.
Cinchonine	$C_{20}H_{24}N_2O$.
Aricine	$C_{23}H_{26}N_2O_4$.

From Tobacco.

Nicotine	$C_{10}H_{14}N_2$.

From Nux vomica.

Strychnine	$C_{21}H_{22}N_2O_2$.
Brucine	$C_{23}H_{26}N_2O_4$.

2, 3, 4. THE PHOSPHINES, ARSINES, AND STIBINES.

These bases cannot be obtained, like the amines, by the displacement of hydrogen in phosphuretted, arseniuretted, and antimoniuretted hydrogen. The tertiary compounds only are known; and they are produced by reactions of which the following may be regarded as types :—

$$\mathbf{As}Na_3 \quad + \quad 3EtI \quad = \quad \mathbf{As}Et_3 \quad + \quad 3NaI.$$

| Sodic arsenide. | Ethylic iodide. | Triethyl-arsine. | Sodic iodide. |

$$3\mathbf{Zn}Et_2 \quad + \quad 2PCl_3 \quad = \quad 2PEt_3 \quad + \quad 3\mathbf{Zn}Cl_2.$$

| Zincic ethide. | Phosphorous trichloride. | Triethyl-phosphine. | Zincic chloride. |

Unlike the amines, these bodies have a powerful affinity for oxygen, in consequence of which they are generally spontaneously inflammable.

5. OXYBASES.

These compounds are only known in the arsenic series.

Arsenious oxybases.

Only one of these, cacodylic oxide, has been carefully investigated.

By the distillation of potassic acetate with arsenious anhydride, a compound known as cacodyl, $'\mathbf{As}''_2Me_4$, is produced. This substance may also be prepared by the action of methylic iodide upon an alloy of sodium and arsenic containing $'\mathbf{As}''_2Na_4$:—

$$'\mathbf{As}''_2Na_4 \quad + \quad 4MeI \quad = \quad '\mathbf{As}''_2Me_4 \quad + \quad 4NaI.$$

| Sodic arsenide. | Methylic iodide. | Cacodyl. | Sodic iodide. |

By allowing cacodyl to absorb oxygen slowly, an oily liquid containing cacodylic oxide ($\mathbf{As}_2Me_4O$) is formed.

This oxybase does not appear to unite with oxygen acids, but it is attacked by hydrochloric acid, forming cacodylic chloride :—

$$\left\{ \begin{array}{l} \mathbf{As}Me_2 \\ O \\ \mathbf{As}Me_2 \end{array} \right. \quad + \quad 2HCl \quad = \quad 2\mathbf{As}Me_2Cl \quad + \quad \mathbf{O}H_2.$$

| Cacodylic oxide. | Hydrochloric acid. | Cacodylic chloride. | Water. |

Cacodylic oxide, when exposed to moist air, absorbs water and oxygen, forming cacodylic acid :—

$$\mathbf{As_2Me_4O} \quad + \quad O_2 \quad + \quad \mathbf{OH_2} \quad = \quad 2\mathbf{AsMe_2OHo}.$$

Cacodylic   Water.   Cacodylic
oxide.   acid.

CHAPTER LI.

ORGANIC COMPOUNDS OF TRIAD NITROGEN AND OF ITS ANALOGUES (continued).

NEUTRAL SECTION.

1. THE AMIDES.

THESE compounds are formed by the substitution of amidogen (NH_2) for the oxatylic hydroxyl of organic acids. They are most conveniently written on the diadelphic type, but may also be formulated upon the ammonia type.

If the acid contain only one atom of oxatyl, a monamide is the result; if two atoms of oxatyl are present in the acid, a diamide is generally formed, &c. Secondary and tertiary compounds can also be produced, as in the case of the amines; but they belong to the negative section of this family.

A. MONAMIDES.

I. Primary Monamides.

Acetamide :—

$$\left\{ \begin{array}{l} \mathbf{CMeO} \\ \mathbf{NH_2} \end{array} \right. \text{ or } \mathbf{NH_2(CMeO)}, \text{ or } \left\{ \begin{array}{l} \mathbf{CH_3} \\ \mathbf{COAd} \end{array} \right.$$

Chloracetamide :—

$$\left\{ \begin{array}{l} \mathbf{C(CH_2Cl)O} \\ \mathbf{NH_2} \end{array} \right. \text{ or } \mathbf{NH_2[C(CH_2Cl)O]}, \text{ or } \left\{ \begin{array}{l} \mathbf{CH_2Cl} \\ \mathbf{COAd} \end{array} \right.$$

Benzamide :—

$$\left\{ \begin{array}{l} \mathbf{C(C_6H_5)O} \\ \mathbf{NH_2} \end{array} \right. \text{ or } \mathbf{NH_2[C(C_6H_5)O]}, \text{ or } \left\{ \begin{array}{l} \mathbf{C(C_5H_3)H_2} \\ \mathbf{COAd} \end{array} \right.$$

Formation.—1. By the distillation of the ammonic salts of the monobasic acids :—

$$\left\{ \begin{array}{l} CH_3 \\ CO(N^vH_4O) \end{array} \right. = \left\{ \begin{array}{l} CH_3 \\ CO(N'''H_2) \end{array} \right. + OH_2.$$

Ammonic acetate.   Acetamide.   Water.

2. By the action of ammonia upon the chloracids :—

$$\left\{ \begin{array}{l} CH_3 \\ COCl \end{array} \right. + NH_3 = \left\{ \begin{array}{l} CH_3 \\ CO(N'''H_2) \end{array} \right. + HCl.$$

Acetylic chloride.   Ammonia.   Acetamide.   Hydrochloric acid.

3. By the action of ammonia on the ethereal salts of the monobasic acids :—

$$\left\{ \begin{array}{l} CH_3 \\ COEto \end{array} \right. + NH_3 = \left\{ \begin{array}{l} CH_3 \\ CO(N'''H_2) \end{array} \right. + EtHo.$$

Ethylic acetate.   Ammonia.   Acetamide.   Alcohol.

Reactions.—1. Boiled with aqueous solutions of acids, the primary monamides yield ammonic salts and acids :—

$$\left\{ \begin{array}{l} CH_3 \\ CO(N'''H_2) \end{array} \right. + HCl + OH_2 = NH_4Cl + \left\{ \begin{array}{l} CH_3 \\ COHo. \end{array} \right.$$

Acetamide.   Hydrochloric acid.   Water.   Ammonic chloride.   Acetic acid.

2. Boiled with potassic hydrate, ammonia is evolved, and a potassic salt, corresponding to the amide, is formed.

$$\left\{ \begin{array}{l} CH_3 \\ CO(N'''H_2) \end{array} \right. + KHo = NH_3 + \left\{ \begin{array}{l} CH_3 \\ COKo. \end{array} \right.$$

Acetamide.   Potassic hydrate.   Ammonia.   Potassic acetate.

II. *Secondary Monamides.*

Diacetimide ... $N(CMeO)_2H$ or $\left\{ \begin{array}{l} CH_3 \\ CO \\ NH \\ CO \\ CH_3 \end{array} \right.$

Succinimide ...... $NH \left[\left\{ \begin{array}{l} CO \\ C_2H_4 \\ CO \end{array} \right\} \right]'' $ or $\left\{ \begin{array}{l} CO- \\ Et''\ (N'''H)''. \\ CO- \end{array} \right.$

These bodies possess a negative character, and are treated of under the negative section of this class as *imides* (p. 376).

Tertiary monamides are little known. They are the *nitrides* (see p. 376).

B. *DIAMIDES.*

The diamides may be regarded as derived from two molecules of ammonia, by the substitution of a dyad negative radical for two atoms of hydrogen ; or they may be considered as formed by the substitution of amidogen for the hydroxyl contained in the two atoms of oxatyl in dibasic acids :—

Primary Diamides.

Oxamide   ... $\left\{ \begin{array}{l} \mathbf{NH_2} \\ \mathbf{(C_2O_2)''} \\ \mathbf{NH_2} \end{array} \right.$ or $\left\{ \begin{array}{l} \mathbf{COAd} \\ \mathbf{COAd} \end{array} \right.$

Succinamide... $\mathbf{N_2H_4(C_4H_4O_2)''}$ or $\mathbf{N_2H_4} \left[\left\{ \begin{array}{l} \mathbf{CO} \\ \mathbf{Et''} \\ \mathbf{CO} \end{array} \right\}'' \right.$, or $\left\{ \begin{array}{l} \mathbf{COAd} \\ \mathbf{Et''} \\ \mathbf{COAd} \end{array} \right.$.

Formation.—1. By the action of heat upon the neutral ammonic salts of dibasic acids :—

$$\left\{ \begin{array}{l} \mathbf{CO(N^vH_4O)} \\ \mathbf{CO(N^vH_4O)} \end{array} \right. = \left\{ \begin{array}{l} \mathbf{CO(N'''H_2)} \\ \mathbf{CO(N'''H_2)} \end{array} \right. + \mathbf{2OH_2.}$$

    Ammonic oxalate.   Oxamide.   Water.

2. By the action of ammonia on the ethereal salts of dibasic acids :—

$$\left\{ \begin{array}{l} \mathbf{COEto} \\ \mathbf{COEto} \end{array} \right. + \mathbf{2NH_3} = \left\{ \begin{array}{l} \mathbf{CO(N'''H_2)} \\ \mathbf{CO(N'''H_2)} \end{array} \right. + \mathbf{2EtHo.}$$

  Ethylic   Ammonia.   Oxamide.   Alcohol.
  oxalate.

3. By the action of ammonia on the chloro-dibasic acids :—

$$\mathbf{4NH_3} + \left\{ \begin{array}{l} \mathbf{COCl} \\ \mathbf{Et''} \\ \mathbf{COCl} \end{array} \right. = \left\{ \begin{array}{l} \mathbf{CO(N'''H_2)} \\ \mathbf{Et''} \\ \mathbf{CO(N'''H_2)} \end{array} \right. + \mathbf{2NH_4Cl.}$$

  Ammonia.   Succinylic   Succinamide.   Ammonic
                chloride.   chloride.

The *secondary* and *tertiary diamides* are but little known.

C. *TRIAMIDES.*

Primary Triamides.—The primary triamides may be regarded as derived from tribasic acids by the substitution of amidogen for the hydroxyl contained in the three atoms of oxatyl of these acids, or as derived from three molecules of ammonia by the displacement of three atoms of hydrogen by the residue of a tribasic acid. A good example of a triamide is

Citramide...... $\begin{cases} \mathbf{C}\text{HHo(COAd)} \\ \mathbf{C}\text{H(COAd)} \\ \mathbf{C}\text{H}_2\text{(COAd)} \end{cases}$ or $\mathbf{N}_3\text{H}_8(\text{C}_6\text{H}_5\text{O}_4)'''.$

Citramide is formed by the action of ammonia on ethylic citrate.

Secondary and *tertiary triamides* have not yet been formed.

2. *THE ALKALAMIDES.*

These compounds are intermediate between the amines and the amides. They are derived from ammonia by the substitution of part of the hydrogen by positive, and part by negative radicals; and inasmuch as two atoms at least of hydrogen must be so substituted, no primary alkalamide can exist.

Secondary and tertiary monalkalamides, dialkalamides, and trialkalamides are known.

Ethyl acetamide .... $\mathbf{N}\text{HEt(CMeO)}.$
Ethyl diacetamide..................... $\mathbf{N}\text{Et(CMeO)}_2.$
Diethyl oxamide .... $\mathbf{N}_2\text{H}_2\text{Et}_2(\text{C}_2\text{O}_2)''.$
Diphenyl-carbonyl-oxalyl diamide . $\mathbf{N}_2(\text{C}_6\text{H}_5)_2(\text{CO})''(\text{C}_2\text{O}_2)''.$
Citryl-triphenyl-triamide .... $\mathbf{N}_3\text{H}_3(\text{C}_6\text{H}_5)_3(\text{C}_6\text{H}_5\text{O}_4)'''.$

The alkalamides incline towards basicity in their character, their degree of alkalinity being about equal to that of urea.

3. *·THE TRICHLORINATED AND TRIBROMI-NATED AMINES.*

If the hydrogen in an amine be gradually substituted by

chlorine or bromine, the basic power of the amine gradually diminishes, and finally a neutral compound is obtained.

This reaction has been studied in the case of aniline, which loses basic energy by the successive displacement of two atoms of hydrogen, and finally becomes neutral by the substitution of three atoms of chlorine or bromine for three of hydrogen :—

$NH_2(C_6H_5)$.　　$NH_2(C_6H_4Cl)$.　　$NH_2(C_6H_3Cl_2)$.　　$NH_2(C_6H_2Cl_3)$.
Aniline.　　　　Chloraniline.　　　　Dichloraniline.　　　　Trichloraniline.

4. *THE HALOID COMPOUNDS OF OXYBASES.*

These bodies are only known in the arsenic series; they are formed by the action of chlorine, bromine, or iodine upon cacodyl and its homologues, or of hydrochloric acid, hydrobromic acid, or hydriodic acid upon the oxybases.

General formula......　$As(C_nH_{2n+1})_2Cl$.

NEGATIVE SECTION.
THE IMIDES AND NITRIDES.

General formula...　$\begin{cases} \text{of imides...} & NH(C_nH_{2n-1}O)_2, \\ \text{of nitrides.} & N(C_nH_{2n-1}O)_3. \end{cases}$

Formation.—By the action of chloracids (the so-called chlorides of negative radicals) upon amides :—

$NH_2(CMeO)$　+　$CMeOCl$　=　$NH(CMeO)_2$　+　HCl.
Acetamide.　　　　Acetylic　　　　Diacetimide.　　　Hydrochloric
　　　　　　　chloride.　　　　　　　　　　　　　acid.

A repetition of this reaction gives acetylic nitride.

An imide may also be formed by the substitution of a dyad negative radical for two atoms of hydrogen in ammonia, thus :—

Succinimide......... $NH(C_4H_4O_2)''$, or $NH\left[\begin{cases} CO \\ Et'' \\ CO \end{cases}\right]''$.

These bodies have hitherto received but little attention.

CHAPTER LII.

II. *COMPOUNDS OF PENTAD NITROGEN AND OF ITS ANALOGUES.*

This class of compounds contains the following series :—

Positive.	Neutral.	Negative.
1. Caustic Nitrogen bases.	1. Salts of Amines.	1. Organic arsenic acids, oxychlorides, and chlorides.
2. „ Phosphorus bases.	2. „ Phosphines.	
3. „ Arsenic bases.	3. „ Arsines.	
4. „ Antimony bases.	4. „ Stibines.	
5. Oxyarsenic bases.	5. „ Oxyarsenic bases.	2. Organic antimonic acids.
6. Oxyantimonic bases.	6. „ Oxyantimonic bases.	

POSITIVE or BASYLOUS COMPOUNDS.

1. *Caustic Nitrogen Bases.*—

General formula...... $N(C_nH_{2n+1})_4Ho.$

In each radical n must be a positive integer. The radicals need not be all of the same atomic weight.

Formation.—By the action of argentic hydrate upon the iodides of the compound ammoniums :—

$$NEt_4I \ + \ AgHo \ = \ NEt_4Ho \ + \ AgI.$$

Tetrethylammonic iodide.   Argentic hydrate.   Tetrethylammonic hydrate.   Argentic iodide.

2. *Caustic Phosphorus Bases.*

3. *Caustic Arsenic Bases.*

4. *Caustic Antimony Bases.*—

By displacing the **N** in the above general formula and in the equation by **P, As,** and **Sb,** the constitution and formation of these three series of compounds will be expressed.

5. *Oxyarsenic Bases.*—These bodies, which are diacid bases, are obtained by the slow oxidation of the tertiary monarsines :—

$$As(C_nH_{2n+1})_3 \ + \ O \ = \ As(C_nH_{2n+1})_3O.$$

Tertiary monarsine.   Oxyarsenic base.

6. *Oxyantimonic Bases.*—These are formed in a manner exactly analogous to that in which the oxyarsenic bases are produced.

NEUTRAL COMPOUNDS.

1. *Salts of Amines.*

General formulæ :—

$$\mathbf{N}(C_nH_{2n+1})(C_mH_{2m+1})_3Cl.$$
$$\mathbf{N}_2(C_nH_{2n})''(C_mH_{2m})''_2H_2Cl_2.$$
$$\mathbf{N}_3(C_nH_{2n-1})'''(C_mH_{2m-1})'''_2H_3Cl_3.$$

In the first formula m may $=0$; in the second, C_mH_{2m} may be displaced by H_2; and in the third, C_mH_{2m-1} may be substituted by H_3.

Formation.—Like the analogous compounds of ammonia, the salts of the amines are formed by the direct union of acids with the amines without elimination of water, thus :—

$$\mathbf{N}EtH_2 \quad + \quad HCl \quad = \quad \mathbf{N}EtH_3Cl.$$

Ethylamine.   Hydrochloric   Ethylammonic
           acid.   chloride.

The haloid salts of the amines may also be produced by the union of the haloid ethers of the monad positive radicals with the amines (for reaction see p. 365).

Character.—The salts of the diamines and triamines are often found to contain only one molecule of acid, instead of two or three as shown in the above general formulæ, which indicate the composition of the normal salts. The nitrogen atoms are in such cases united together by one of the bonds of each, besides being linked by the polyad radicals, thus :—

$$\mathbf{'N}^{iv}_2(C_nH_{2n})''(C_mH_{2m})''_2H(NO_3); \quad {}^{iv}(\mathbf{N}_3)^{xi}(C_nH_{2n-1})'''(C_mH_{2m-1})'''_2H$$

The difference between these two classes of salts will be rendered more evident by a comparison of the following graphic and symbolic formulæ :—

Normal Salts.

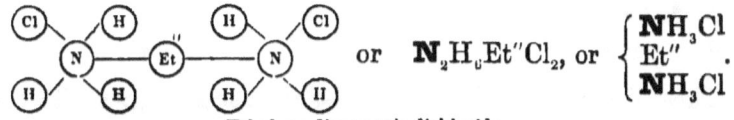

　　or　$\mathbf{N}_2\mathbf{H}_6\mathbf{Et}''\mathbf{Cl}_2$, or　$\begin{cases}\mathbf{NH}_3\mathbf{Cl}\\ \mathbf{Et}''\\ \mathbf{NH}_3\mathbf{Cl}\end{cases}$.

Ethylene-diammonic dichloride.

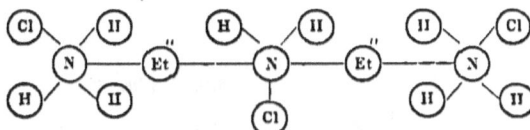

Diethylene-triammonic trichloride.

or　$\mathbf{N}_3\mathbf{H}_8\mathbf{Et}''_2\mathbf{Cl}_3$　or　$\begin{cases}\mathbf{NH}_3\mathbf{Cl}\\ \mathbf{Et}''\\ \mathbf{NH}_2\mathbf{Cl}.\\ \mathbf{Et}''\\ \mathbf{NH}_3\mathbf{Cl}\end{cases}$

Monacid Salts.

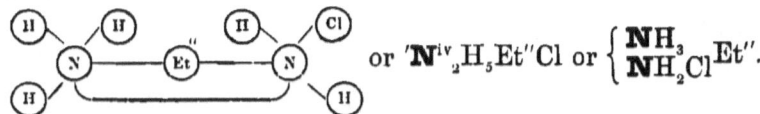

　or　$'\mathbf{N}^{iv}_2\mathbf{H}_5\mathbf{Et}''\mathbf{Cl}$ or $\begin{cases}\mathbf{NH}_3\\ \mathbf{NH}_2\mathbf{Cl}\end{cases}\mathbf{Et}''$.

Ethylene-diammonic monochloride.

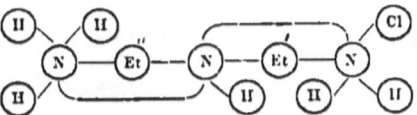

Diethylene-triammonic monochloride.

or　$^{iv}(\mathbf{N}_3)^{xi}\mathbf{H}_6\mathbf{Et}''_2\mathbf{Cl}$　or　$\begin{cases}\mathbf{NH}_3\mathbf{Et}''\\ \mathbf{NH}\\ \mathbf{NH}_2\mathbf{Cl}\end{cases}\mathbf{Et}'''$.

Diacid Salt.

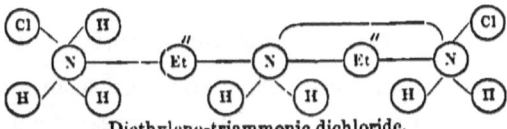

Diethylene-triammonic dichloride.

or　$''(\mathbf{N}_3)^{xiii}\mathbf{H}_7\mathbf{Et}''_2\mathbf{Cl}_2$　or　$\begin{cases}\mathbf{NH}_3\mathbf{Cl}\,\mathbf{Et}''\\ \mathbf{NH}_2\\ \mathbf{NH}_2\mathbf{Cl}\end{cases}\mathbf{Et}'''$.

2. *Salts of Phosphines.*
3. *Salts of Arsines.*
4. *Salts of Stibines.*—

These three series of salts all present close analogies with the salts of the amines both in constitution and in the mode of their formation.

5. *Salts of Oxyarsenic Bases.*

$$\mathbf{As}(C_nH_{2n+1})_3Cl_2.$$

Formation.—By the action of acids on the oxyarsenic bases:—

$$\mathbf{As}Me_3O \quad + \quad 2HCl \quad = \quad \mathbf{As}Me_3Cl_2 \quad + \quad \mathbf{O}H_2.$$

Arsenic   Hydrochloric   Arsenic trimetho-   Water.
trimethoxide.   acid.   dichloride.

6. *Salts of Oxyantimonic Bases.*

These resemble the previous salts in formation and constitution.

NEGATIVE or CHLOROUS COMPOUNDS.

1. *Organic Arsenic Acids, Oxychlorides, and Chlorides.*

The following are the principal bodies of this class :—

Monomethylarsenic acid............	$\mathbf{As}MeOHo_2.$
Arsenic oxydichlormethide	$\mathbf{As}MeOCl_2.$
Arsenic tetrachlormethide	$\mathbf{As}MeCl_4.$
Cacodylic acid	$\mathbf{As}Me_2OHo.$
Cacodylic trichloride	$\mathbf{As}Me_2Cl_3.$

2. *Organic Antimonic Acids.*

No exploration of this series has yet been made. The members of it will doubtless be found to present close analogies with the corresponding series of arsenic compounds.

CHAPTER LIII.

ORGANOMETALLIC BOBIES.

THIS term is applied to a family of compounds in which an organic radical is united directly with a metal; and it serves to distinguish them from other organic compounds containing metals, in which the metal and organic radical are indirectly united or linked to each other.

Thus zincic ethide is an organometallic body, while zincic ethylate and zincic succinate are organic bodies containing metals:—

Zincic ethide ... $\mathbf{Zn}Et_2$.

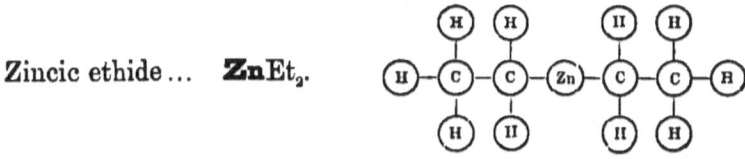

Zincic ethylate... $\mathbf{Zn}Eto_2$.

Zincic succinate ... $\begin{cases} \mathbf{CO} \\ \mathbf{C_2H_4} \ \ \mathbf{Zno''}. \\ \mathbf{CO} \end{cases}$

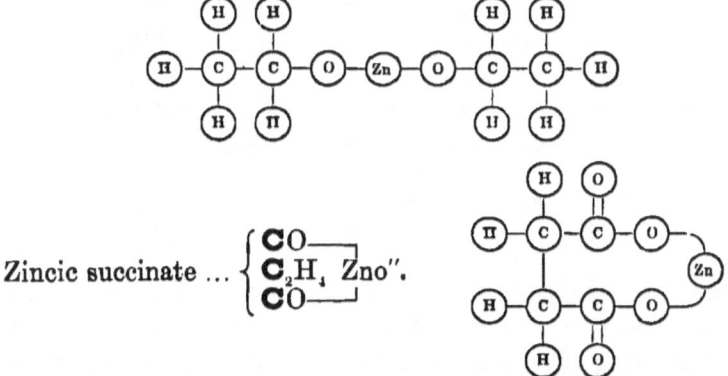

Many organic compounds containing metals are the derivatives of organometallic bodies; thus zincic ethide by oxidation yields zincic ethylate—

$$\mathbf{Zn}Et_2 \quad + \quad O_2 \quad = \quad \mathbf{Zn}Eto_2;$$

Zincic
ethide.   Zincic
ethylate.

and by further oxidation zincic ethylate can be converted into zincic acetate—

$$\begin{matrix} \mathbf{CMeH_2} \\ \mathbf{CMeH_2} \end{matrix}\mathbf{Zno''} + 2O_2 = \begin{matrix} \mathbf{CMeO} \\ \mathbf{CMeO} \end{matrix}\mathbf{Zno''} + 2\mathbf{OH_2}.$$

Zincic ethylate.   Zincic acetate.   Water.

Another instance of the derivation of organic bodies containing metals from organometallic bodies is seen in the formation of potassic propionate by the action of potassic ethide upon carbonic anhydride :—

$$\mathbf{CMeH_2K} \ + \ \mathbf{CO_2} \ = \ \left\{ \begin{matrix} \mathbf{CMeH_2} \\ \mathbf{COKo} \end{matrix} \right. .$$

Potassic   Carbonic   Potassic
ethide.   anhydride.   propionate.

Formation of organometallic bodies.

Organometallic bodies are produced in a large number of reactions, which may, however, be classed under the following four heads :—

I. By the union of monad positive radicals *in statu nascenti* with a metal, or by the coalescence of a metal with the iodide of a monad positive radical.

Thus, when zinc and ethylic iodide are heated together to 100° in closed vessels, zincic ethide is formed :—

$$2\mathbf{EtI} \ + \ \mathbf{Zn_2} \ = \ \mathbf{ZnEt_2} \ + \ \mathbf{ZnI_2}.$$

Ethylic   Zincic   Zincic
iodide.   ethide.   iodide.

Sometimes light may be employed instead of heat to effect this change, as in the case of the organo-tin compounds. In the formation of organo-mercury compounds by this method, light is indispensable to the reaction :—

$$\mathbf{EtI} \ + \ \mathbf{Hg} \ = \ \mathbf{HgEtI}.$$

Ethylic   Mercuric
iodide.   ethiodide.

II. By the action of the respective metals alloyed with potassium or sodium upon the iodides of the monad positive radicals.

By this process there is less tendency to form compounds containing both positive radicals and negative elements. Potassium or sodium compounds are never produced in this reaction, because they cannot exist in the presence of ethylic iodide or its homologues. This process is well adapted for the formation of arsenic, antimony, tin, mercury, lead, bismuth, and tellurium compounds:—

$$4EtI \quad + \quad SnNa_4 \quad = \quad SnEt_4 \quad + \quad 4NaI.$$

| Ethylic iodide. | Tin sodium alloy. | Stannic ethide. | Sodic iodide. |

$$2EtI \quad + \quad HgNa_2 \quad = \quad HgEt_2 \quad + \quad 2NaI.$$

| Ethylic iodide. | Sodium amalgam. | Mercuric ethide. | Sodic iodide. |

III. By the action of the zinc compounds of the monad positive radicals upon the haloid compounds, either of the metals themselves, or of their organo- derivatives.

For the production of organometallic bodies containing less positive metals than zinc, this method is generally the most convenient, and is of most universal application. Compounds containing mercury, tin, lead, antimony, and arsenic have been thus produced; but the process has failed when applied to the haloid compounds of copper, silver, platinum, and iron; for, although these bodies are violently acted upon, the organic radicals do not unite with the metal:—

$$SnCl_4 \quad + \quad ZnEt_2 \quad = \quad SnEt_2Cl_2 \quad + \quad ZnCl_2.$$

| Stannic chloride. | Zincic ethide. | Stannic dichlor- ethide. | Zincic chloride. |

$$SnCl_4 \quad + \quad 2ZnEt_2 \quad = \quad SnEt_4 \quad + \quad 2ZnCl_2.$$

| Stannic chloride. | Zincic ethide. | Stannic ethide. | Zincic chloride. |

$$2HgEtI \quad + \quad ZnEt_2 \quad = \quad 2HgEt_2 \quad + \quad ZnI_2.$$

| Mercuric ethiodide. | Zincic ethide. | Mercuric ethide. | Zincic iodide. |

IV. By the displacement of a metal in an organometallic compound by another and more positive metal.

This method has been successfully employed for the formation of the organo-compounds of potassium, sodium, lithium, aluminium, and zinc. In the first three cases the reaction takes place at ordinary temperatures, some of the original compound entering into the composition of the resulting organometallic body :—

$$3\mathbf{Zn}Et_2 \ + \ Na_2 \ = \ 2ZnNaEt_3 \ + \ Zn.$$

Zincic ethide.   Sodic zincic ethide.

$$3\mathbf{Hg}Et_2 \ + \ Al_2 \ = \ '\mathbf{Al}'''_2Et_6 \ + \ 3Hg.$$

Mercuric ethide.   Aluminic ethide.

$$\mathbf{Hg}Ay_2 \ + \ Zn \ = \ \mathbf{Zn}Ay_2 \ + \ Hg.$$

Mercuric amylide.   Zincic amylide.

Reactions of organometallic bodies.

1. The most interesting reaction of the *organo- compounds of the monad metals* is, their transformation into salts of normal fatty acids by the action of carbonic anhydride (see p. 301).

2. The *organo- compounds of potassium and sodium* decompose the iodides of the monad positive radicals in the cold, forming hydrides and dyad radicals :—

$$\mathbf{C_2H_5}Na \ + \ \mathbf{C_2H_5}I \ = \ NaI \ + \ \mathbf{C_2H_5}H \ + \ \mathbf{C_2H_4}.$$

Sodic ethide.   Ethylic iodide.   Sodic iodide.   Ethylic hydride.   Ethylene.

3. The *organo- compounds of zinc* are decomposed by water, with formation of the hydrides of the radicals :—

$$\mathbf{Zn}Et_2 \ + \ 2OH_2 \ = \ \mathbf{Zn}Ho_2 \ + \ 2EtH.$$

Zincic ethide.   Water.   Zincic hydrate.   Ethylic hydride.

4. By the slow action of dry oxygen, they pass through two stages of oxidation :—

$$\mathbf{Zn}Et_2 \ + \ O \ = \ \mathbf{Zn}EtEto ;$$

Zincic ethide.   Zincic ethide ethylate.

$$\mathbf{Zn}EtEto \ + \ O \ = \ \mathbf{Zn}Eto_2.$$

Zincic ethide ethylate.   Zincic ethylate.

5. Monad negative elements, such as iodine, remove successively the two atoms of ethyl :—

$$\mathbf{ZnEt_2} \quad + \quad \mathbf{I_2} \quad = \quad \mathbf{ZnEtI} \quad + \quad \mathbf{EtI} :$$

| Zincic ethide. | | | Zincic ethiodide. | Ethylic iodide. |

$$\mathbf{ZnEtI} \quad + \quad \mathbf{I_2} \quad = \quad \mathbf{ZnI_2} \quad + \quad \mathbf{EtI}.$$

| Zincic ethiodide. | | | Zincic iodide. | Ethylic iodide. |

6. The organo-zinc compounds are extremely useful for the
displacement of chlorine or its analogues by ethyl or its homologues :—

$$2\mathbf{PCl_3} \quad + \quad 3\mathbf{ZnEt_2} \quad = \quad 2\mathbf{PEt_3} \quad + \quad 3\mathbf{ZnCl_2}.$$

| Phosphorous trichloride. | Zincic ethide. | Triethyl-phosphine. | Zincic chloride. |

$$\mathbf{SiCl_4} \quad + \quad 2\mathbf{ZnEt_2} \quad = \quad \mathbf{SiEt_4} \quad + \quad 2\mathbf{ZnCl}.$$

| Silicic chloride. | Zincic ethide. | Silicic ethide. | Zincic chloride. |

$$2\left\{\begin{matrix}\mathbf{C_2H_4Cl}\\ \mathbf{O}\\ \mathbf{C_2H_4Cl}\end{matrix}\right. \quad + \quad \mathbf{ZnEt_2} \quad = \quad 2\left\{\begin{matrix}\mathbf{C_2H_4Et}\\ \mathbf{O}\\ \mathbf{C_2H_4Cl}\end{matrix}\right. \quad + \quad \mathbf{ZnCl_2} :$$

| Chlorether. | Zincic ethide. | Ethylo-chlorether. | Zincic chloride. |

$$2\left\{\begin{matrix}\mathbf{C_2H_4Et}\\ \mathbf{O}\\ \mathbf{C_2H_4Cl}\end{matrix}\right. \quad + \quad \mathbf{ZnEt_2} \quad = \quad 2\left\{\begin{matrix}\mathbf{C_2H_4Et}\\ \mathbf{O}\\ \mathbf{C_2H_4Et}\end{matrix}\right. \quad + \quad \mathbf{ZnCl_2}.$$

| Ethylo-chlorether. | Zincic ethide. | Diethylated ethylic ether. | Zincic chloride. |

Diethylated ethylic ether is isomeric with butylic ether, and
contains the radical methylo-ethylated methyl (see p. 205). By
oxidation it would doubtless give methylated acetone (p. 360).

7. Oxygen may also be displaced in a similar manner.
Thus :—

$$2'\mathbf{N''_2O_2} \quad + \quad \mathbf{ZnEt_2} \quad = \quad \left.\begin{matrix}\mathbf{N_2OEt\text{-}O}\\ \mathbf{N_2OEt\text{-}O}\end{matrix}\right\}\mathbf{Zn''} \quad \text{or} \quad \begin{matrix}\mathbf{N_2EtO}\\ \mathbf{N_2EtO}\end{matrix}\mathbf{Zno''}.$$

| Nitric oxide. | Zincic ethide. | Zincic dinitro-ethylate. |

This compound is analogous to zincic propionate, the latter
containing two atoms of tetrad carbon in the place of the two
tetrad pairs of nitrogen atoms :—

Zincic propionate ... $$\left.\begin{matrix}\mathbf{COEt\text{-}O}\\ \mathbf{COEt\text{-}O}\end{matrix}\right\}\mathbf{Zn''} \quad \text{or} \quad \begin{matrix}\mathbf{CEtO}\\ \mathbf{CEtO}\end{matrix}\mathbf{Zno''}.$$

8. An analogous reaction is observed with sulphurous anhydride :—

$$2SO_2 \quad + \quad ZnMe_2 \quad = \quad \begin{matrix} SOMe\text{-}O \\ SOMe\text{-}O \end{matrix} \bigg] Zn'' \quad \text{or} \quad \begin{matrix} SMeO \\ SMeO \end{matrix} Zno''.$$

Sulphurous   Zincic   Zincic methyldithionate.
anhydride.   methide.

9. When ethylic borate is acted upon by zincic methide, the ethoxyl becomes replaced by methyl :—

$$BEto_3 \quad + \quad 3ZnMe_2 \quad = \quad BMe_3 \quad + \quad 3ZnMeEto.$$

Ethylic   Zincic   Boric   Zincic methide
borate.   methide.   methide.   ethylate.

10. When ethylic oxalate is heated with zincic ethide, and water afterwards added, diethoxalic ether is formed :—

$$\begin{cases} COEto \\ COEto \end{cases} + ZnEt_2 + 2OH_2 = \begin{cases} CEt_2Ho \\ COEto \end{cases} + ZnHo_2 + EtHo.$$

Ethylic   Zincic   Water.   Diethoxalic   Zincic   Alcohol.
oxalate.   ethide.   ether.   hydrate.

11. By the action of ammonia, or of certain amines and amides, zincic ethide exchanges its zinc for hydrogen :—

$$ZnEt_2 \quad + \quad 2NH_3 \quad = \quad ZnAd_2 \quad + \quad 2EtH.$$

Zincic   Ammonia.   Zincic amide.   Ethylic
ethide.   hydride.

12. The organo-zinc compounds, by losing one of their organic radicals, become monad radicals, as shown by the following formulæ :—

Methylozincic dinitrome- } $'N''_2OMe\text{-}O\text{-}(ZnMe)$.
thylate ....

Ethylozincic dinitroethy- { $'N''_2OEt\text{-}O\text{-}(ZnEt)$.
late ....

Ethylic ethylo-zincic di- { $\begin{matrix} CEt_2\text{-}O\text{-}(ZnEt) \\ COEto \end{matrix}$.
ethoxalate ....

13. *Mercuric ethide*, when treated with bromine, loses one-half of its ethyl, which is displaced by the negative element —

$$HgEt_2 \quad + \quad Br \quad = \quad HgEtBr \quad + \quad EtBr.$$

Mercuric   Mercuric   Ethylic
ethide.   ethobromide.   bromide.

14. *Mercuric methide*, when submitted to the action of mercuric iodide, yields mercuric methiodide :—

$$HgMe_2 \ + \ HgI_2 \ = \ 2HgMeI.$$

Mercuric methide.   Mercuric iodide.   Mercuric methiodide.

The hydrates corresponding to the mercuric ethobromide and methiodide have been produced. They are powerful caustic bases, of the formulæ

$$HgEtHo \ and \ HgMeHo.$$

Mercuric ethohydrate.   Mercuric methohydrate.

15. The *organo-stannous* compounds unite directly with negative elements, passing into stannic bodies :—

$$SnEt_2 \ + \ I_2 \ = \ SnEt_2I_2.$$

Stannous ethide.   Stannic iododiethide.

16. *Hypostannic organo-compounds* undergo a similar transformation :—

$$'Sn'''_2Et_6 \ + \ I_2 \ = \ 2SnEt_3I :$$

Hypostannic ethide.   Stannic iodotriethide.

$$SnEt_3I \ + \ I_2 \ = \ SnEt_2I_2 \ + \ EtI.$$

Stannic iodotriethide.   Stannic iododiethide.   Ethylic iodide.

17. *Hypostannic ethodiniodide* is formed by the action of iodine upon stannic ethodimethide :—

$$2SnEt_2Me_2 \ + \ I_6 \ = \ 'Sn'''_2Et_4I_2 \ + \ 4MeI.$$

Stannic ethodimethide.   Hypostannic ethodiniodide.   Methylic iodide.

18. *Stannic ethide*, when treated with hydrochloric acid, yields stannic chlorotriethide and ethylic hydride :—

$$SnEt_4 \ + \ HCl \ = \ SnEt_3Cl \ + \ EtH.$$

Stannic ethide.   Hydrochloric acid.   Stannic chlorotriethide.   Ethylic hydride.

The oxide and hydrate corresponding to the stannic chlorotriethide are known ; their formulæ are :—

$$Oxide \ \dots \begin{cases} SnEt_3 \\ O \\ SnEt_3 \end{cases} ; \quad Hydrate\dots \ SnEt_3Ho.$$

s 2

These compounds, and the salts which they form, correspond in composition, constitution, and, to a certain extent, in properties, with the compounds of methyl:—

	Alcohol.	Haloid ether.	Ether.
Methylic	$CH_3Ho.$	$CH_3Cl.$	$\begin{cases} CH_3 \\ O \\ CH_3 \end{cases}$.
Stanntriethylic ...	$SnEt_3Ho.$	$SnEt_3Cl.$	$\begin{cases} SnEt_3 \\ O \\ SnEt_3 \end{cases}$.

19. *Stannic chlorodiethide* is readily reduced to stannous ethide by the action of zinc:—

$$SnEt_2Cl_2 \ + \ Zn \ = \ SnEt_2 \ + \ ZnCl_2.$$

| Stannic chlorodiethide. | | Stannous ethide. | Zincic chloride. |

20. *Perplumbic ethide* resembles stannic ethide in its reactions; thus with hydrochloric acid it yields perplumbic chlorotriethide and ethylic hydride:—

$$PbEt_4 \ + \ HCl \ = \ PbEt_3Cl \ + \ EtH.$$

| Perplumbic ethide. | Hydrochloric acid. | Perplumbic chlorotriethide. | Ethylic hydride. |

21. *Perplumbic triethohydrate* ($PbEt_3Ho$) is a powerful base, forming salts with acids.

22. The *organo-tellurium compounds* form oxides and salts. The following are the formulæ of tellurium ethide and some of its compounds:—

Tellurium ethide .... $TeEt_2.$
Tellurous diethoxide .... $TeEt_2O.$
Tellurous diethiodide .... $TeEt_2I_2.$
Tellurous diethosulphate......... $TeEt_2(S^{vi}O_4)''.$

Constitution of Organometallic Bodies.

The organometallic compounds are constituted on the types of the metals they contain. It was, in fact, the study of these bodies which first led to the doctrine of the atomicity of elements. They afford striking examples of monad, dyad, triad, tetrad, pentad, and hexad types.

The organic derivatives of the monad metals are formed on the type of potassic chloride (KCl) :—

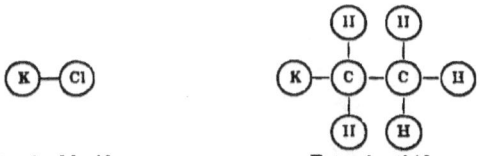

Potassic chloride.   Potassic ethide.

The organo-zinc, cadmium, magnesium, and mercury compounds are formed upon the type of zincic chloride ($ZnCl_2$):—

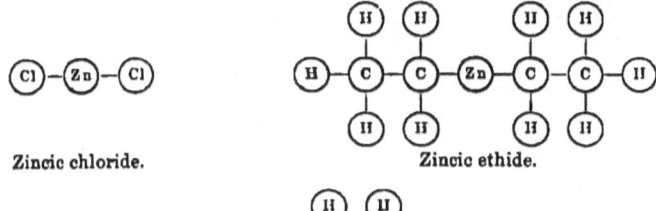

Zincic chloride.   Zincic ethide.

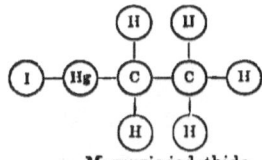

Mercuric iodethide.

The organo-aluminic compounds are formed upon the type of aluminic chloride ($'Al'''_2Cl_6$):—

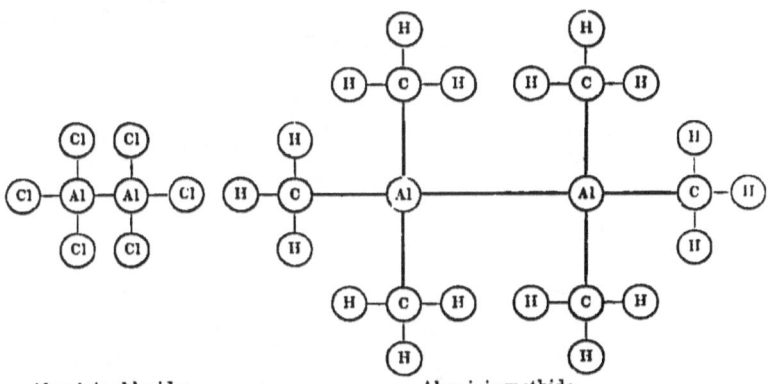

Aluminic chloride.   Aluminic methide.

The organo-tin compounds are formed upon the three types $"Sn"Cl_2$, $'Sn'''_2Cl_8$, and $SnCl_4$,—the first resembling the zincic chloride type, and the second the aluminic chloride type (see p. 389) :—

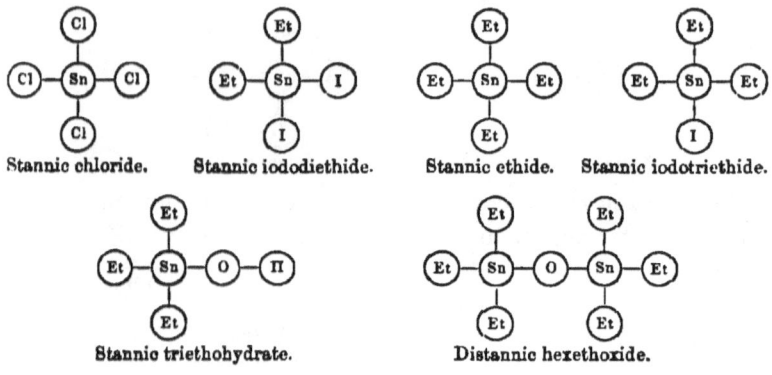

Stannic chloride.   Stannic iododiethide.   Stannic ethide.   Stannic iodotriethide.

Stannic triethohydrate.   Distannic hexethoxide.

The inorganic types of the organo-tellurium series are **Te**Cl$_2$ and **TeO$_2$** :—

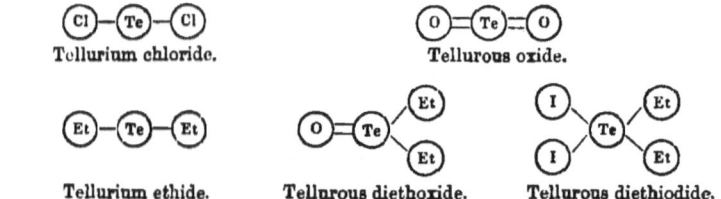

Tellurium chloride.   Tellurous oxide.

Tellurium ethide.   Tellurous diethoxide.   Tellurous diethiodide.

The organo-arsenic, antimony, and bismuth compounds are derived from the types $'$**As**$''_2$**S**$''_2$, **As**Cl$_3$, **As**OHo$_3$, **Sb**Cl$_3$, **Sb**Cl$_5$, **Bi**Cl$_3$, and **BiO$_2$**Ho (see pp. 370, 377, and 380) :—

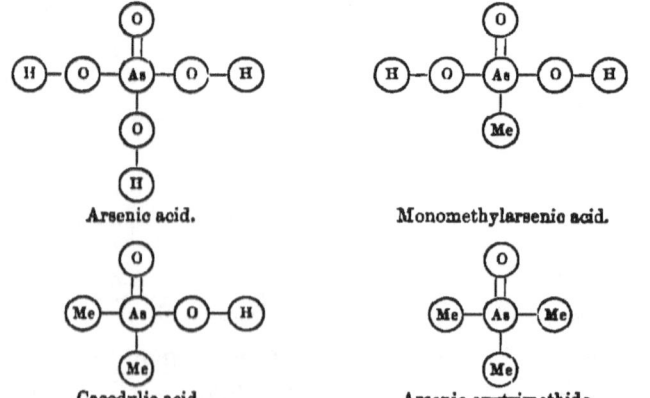

Arsenic acid.   Monomethylarsenic acid.

Cacodylic acid.   Arsenic oxytrimethide.

The effect of the substitution of positive for negative radicals in compounds is well exhibited in the case of arsenic acid, $AsOHo_3$, as illustrated in the above graphic representations. By the substitution of one atom of methyl for hydroxyl, a well-defined acid (less negative, however, than arsenic acid) is produced, monomethyl arsenic acid, $AsOMeHo_2$. By the replacement of a second atom of hydroxyl by methyl, a very feeble acid, cacodylic acid, $AsOMe_2Ho$, is obtained. By the replacement of the third atom of hydroxyl by methyl, the acid properties are completely destroyed, a feeble base, the arsenic oxytrimethide, being formed, $AsOMe_3$. Finally, by the substitution of methyl and hydroxyl for the remaining atom of oxygen there is produced a powerful base, tetramethylarsenic hydrate, $AsMe_4Ho$.

The following is a list of the principal organometallic bodies at present known:—

I. Organo- compounds containing monad metals:—

Potassio-zincic methide .... $KMe,ZnMe_2$.
Potassio-zincic ethide.............. $KEt,ZnEt_2$.
Sodio-zincic ethide .... $NaEt,ZnEt_2$.
Lithio-zincic ethide .... $LiEt,ZnEt_2$.
Lithio-mercuric ethide .... $LiEt,HgEt_2$.

II. Organo- compounds containing dyad metals:—

Magnesic ethide .... $MgEt_2$.
Zincic methide .... $ZnMe_2$.
Zincic ethide .... $ZnEt_2$.
Zincic amylide...................... $ZnAy_2$.
Mercuric methide .... $HgMe_2$.
Mercuric ethide .... $HgEt_2$.
Mercuric methiodide .... $HgMeI$.
Mercuric ethonitrate .... $HgEt(N'O_3)$.
Stannous ethide .... $''Sn''Et_2$.
Tellurium methide .... $TeMe_2$.

III. Organo- compounds containing triad metals :—

These compounds belong to the 11th family of organic bodies, and have been treated of at p. 370.

IV. Organo- compounds of tetrad metals :—

Stannic methide .... $SnMe_4$.

Stannic iodotrimethide .... $SnMe_3I$.

Stannic iododimethide .... $SnMe_2I_2$.

Hypostannic ethide .... $'Sn'''_2Et_6$.

Stannic ethylodimethide .... $SnEt_2Me_2$.

Hypostannic ethodiiodide .... $'Sn'''_2Et_4I_2$.

Perplumbic ethide .... $PbEt_4$.

Perplumbic chlorotriethide ...... $PbEt_3Cl$.

V. Organo- compounds of pentad metals :—

These bodies belong to the 11th family of organic compounds, and have been already treated of at p. 377 and 380.

INDEX.

s 5

THE END.

Printed by TAYLOR and FRANCIS, Red Lion Court, Fleet Street.

Students' Class-Books.

AN ELEMENTARY TEXT-BOOK OF THE MICROSCOPE,
including a Description of the Methods of Preparing and Mounting
Objects.

BY J. W. GRIFFITH, M.D., F.L.S., &c.

Post 8vo, with 12 Coloured Plates, 7s. 6d.

"A capital book."—*Medical Times*, April 16, 1864.

MANUAL OF CHEMICAL QUALITATIVE ANALYSIS.

By A. B. NORTHCOTE, F.C.S., and ARTHUR H. CHURCH, F.C.S.

Post 8vo, 10s. 6d.

HANDBOOK OF CHEMICAL MANIPULATION.

By C. GREVILLE WILLIAMS, F.R.S. 15s.

ELEMENTARY COURSE OF GEOLOGY, MINERALOGY,
AND PHYSICAL GEOGRAPHY.

By Professor ANSTED, M.A., F.R.S., &c.

Second Edition, 12s.

ELEMENTARY COURSE OF BOTANY:
Structural, Physiological, and Systematic.

By Professor HENFREY, F.R.S.

12s. 6d.

MANUAL OF BRITISH BOTANY.

By Professor BABINGTON, M.A., F.R.S., &c.

Fifth Edition, 10s. 6d.

GENERAL OUTLINE OF THE ORGANIZATION OF THE
ANIMAL KINGDOM.

By Professor T. RYMER JONES, F.R.S.

8vo, Third Edition, £1 11s. 6d.

LECTURE NOTES FOR CHEMICAL STUDENTS:
Embracing both Mineral and Organic Chemistry.

By EDWARD FRANKLAND, F.R.S., &c.

Post 8vo, 12s.

JOHN VAN VOORST, 1 Paternoster Row.